UNDERGROUND INFRASTRUCTURE OF URBAN AREAS 3

Underground Infrastructure of Urban Areas 3

Editors

Cezary Madryas, Andrzej Kolonko,
Beata Nienartowicz & Arkadiusz Szot
Institute of Civil Engineering, Wrocław University of Technology, Wrocław, Poland

CRC Press is an imprint of the
Taylor & Francis Group, an **informa** business
A BALKEMA BOOK

CRC Press/Balkema is an imprint of the Taylor & Francis Group, an informa business

Typeset by V Publishing Solutions Pvt Ltd., Chennai, India
Printed and bound in Great Britain by CPI Group (UK) Ltd, Croydon, CR0 4YY

Published by: CRC Press/Balkema
P.O. Box 11320, 2301 EH Leiden, The Netherlands
e-mail: Pub.NL@taylorandfrancis.com
www.crcpress.com – www.taylorandfrancis.com

ISBN: 978-1-138-02652-0 (Hbk + CD-ROM)
ISBN: 978-1-315-72295-5 (eBook PDF)

Underground Infrastructure of Urban Areas 3 – Madryas et al. (Eds)
© 2015 Taylor & Francis Group, London, ISBN 978-1-138-02652-0

Table of contents

Underground Infrastructure of Urban Areas 3 – Madryas et al. (Eds)
© 2015 Taylor & Francis Group, London, ISBN 978-1-138-02652-0

Preface

The development of modern urban areas increasingly requires the use of underground space. Research in the underground infrastructure of cities should be primarily devoted to studies aimed at developing integration models of underground transportation systems with network systems for media transportation in the newly developed underground space. This requires primarily the identification and description of the technical condition of the underground infrastructure in cities. Failure to undertake or delay in research work—both theoretical and in laboratories and on existing structures—will prevent improvement of the standard of living and sustainable city development which is closely related to improved quality of freight and passenger transportation.

In order to meet contemporary expectations, it is necessary to build transportation tunnels, underground parking lots, underground rail in cities which results in a need to construct, reconstruct and modernise many elements of network infrastructure. As the international experiences shows, the scale of the issue is gigantic and adjustment of cities to the expectations of their inhabitants with respect to living standards requires enormous intensification of efforts not only in terms of science (theoretical) but also practical, concerning modernised operation programs of underground infrastructure facilities and modern designs with respect to modernisation and expansion thereof. The situation is similar in the large cities in Poland where the researches in this direction are intensified. The results of these studies are already visible. Examples are:

- the construction of a road tunnel in Gdansk,
- the completion of the tunnels of the second line of the subway tunnels in Warsaw,
- the construction of utility tunnel under the Vistula river in Warsaw,
- analysis towards the construction of underground railways in other cities, such as Kraków and Wrocław,
- renovation and construction of many of sewerage collectors using trenchless technology.

Special attention should be focused on the strategy to adjust the sewage network to municipal needs which is due to the fact that combined sewers in the structure—contrary to other pipelines—very often have very large diameters, often similar to the diameter of underground transportation tunnels. Further, it is a system that occupies the largest part of underground structures in cities (without a developed underground rail system) if we include inspection chambers, pumping stations, holding tanks and other facilities. Therefore, in order to adjust cities to intensified underground developments it is necessary first to undertake research work to improve the processes of operation, modernisation and potential reconstruction of sewage systems. Without such efforts, sustainable development of urban agglomerations will not be possible.

The monography is a compilation of work by many authors who presented the results of their analytical work of many years concerning designing, construction and operation of underground facilities. Many authors also identified the development directions required for an optimum use of underground space in the studied cities. Thanking the authors for their intellectual contribution to the monography, I trust that it will be of interest to people interested in underground infrastructure in urban agglomerations and will be helpful to their work.

Cezary Madryas
Main Editor

Underground Infrastructure of Urban Areas 3 – Madryas et al. (Eds)
© 2015 Taylor & Francis Group, London, ISBN 978-1-138-02652-0

Sponsors

PLATINUM SPONSOR

HERRENKNECHT AG

GOLD SPONSOR

HOBAS System Polska
Sp. z o.o

SILVER SPONSORS

AMITECH Poland
Sp. z o.o.

BLEJKAN
Sp. z o.o.

Centrum Technologiczne BETOTECH
Sp. z o.o.

MC-Bauchemie
Sp. z o.o.

TINES S.A.

ViaCon Polska
Sp. z o.o.

Uponor Infra
Sp. z o.o.

OTHER SPONSORS

BETONSTAL
Sp. z o.o.

Dolnośląska Okręgowa Izba
Inżynierów Budownictwa

HABA-Beton
Johann Bartlechner
Sp. z o.o.

METRO WARSZAWSKIE
Sp. z o.o.

ZPB Kaczmarek Polska
Sp. z o.o. S.K.A

Underground Infrastructure of Urban Areas 3 – Madryas et al. (Eds)
© 2015 Taylor & Francis Group, London, ISBN 978-1-138-02652-0

Trenchless installation technologies of sea outfalls, intakes and landfalls

D. Petrow-Ganew & L. Zur Linde
Herrenknecht AG, Schwanau, Germany

T. Abel
Wroclaw University of Technology, Wroclaw, Poland

ABSTRACT: This paper is a technical description of the technologies for trenchless installation of sea outfalls, seawater intakes and lanfalls. Trenchless installation technologies have less impact on the environment and the existing infrastructure than other methods applying open cut trenching. Additionally, the pipeline is better protected against damage and therefore has a higher life expectancy. For the installation of sea outfalls and seawater intakes, Horizontal Directional Drilling (HDD), Pipe Jacking (Microtunnelling) and Segment Lining methods may be applied depending on the site and ground conditions, pipe diameter and length to be installed as well as on constructability and economic factors. Herrenknecht manufactures and supplies all types of tunnelling machines and ancillary equipment to facilitate different installation concepts for landfall seawater intakes outfalls. For these concepts, the following trenchless installation techniques can be applied:

- Pipe Jacking, Microtunnelling
- Segment Lining
- Direct Pipe/Easy Pipe
- Horizontal Directional Drilling (HDD).

For the installation technologies mentioned above project examples are given and the mode of installation described by the help of figures. The advantages and disadvantages are given for each of the installation methods.

1 INTRODUCTION

The situation in coastal areas is becoming critical. More and more people are moving to the world´s urban centers. Big cities are turning into megacities with more than 5 million inhabitants. Most of these megacities grow up in coastal areas. Cities like Jakarta, Lagos, Manila, Mumbai, Bangkok, New York, Osaka-Kobe, Rio de Janeiro and Shanghai have one thing in common: they are all located on narrow stretches of coast line. More than 40 percent of the world´s population already live within 100 kilometres of the sea, another 40 percent live in cities along the rivers. Thus, the majority of the world population lives close to waters. The challenge to match in these places is to provide people with sufficient potable water and to handle sewage disposal to assure health and quality of life.

In the construction of utility infrastructures in coastal areas or in river regions, tunneled Outfalls, Intakes and Landfalls are en effective and sustainable method. With the help of sea outfalls (including river outfalls in the following) wastewater can be transported away from the coastline or river bank and discharged at locations where diffusion, dispersion and decomposition are enhanced. The municipal wastewater may be fully treated, pre-treated or untreated.

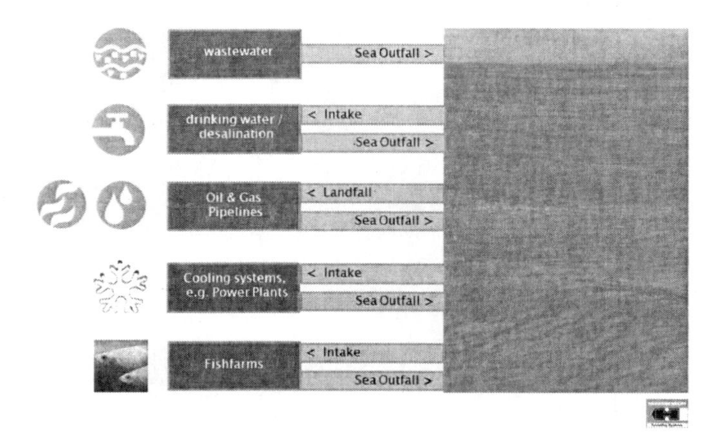

Figure 1. Application fields of sea outfalls, intakes and landfalls.

Water intakes can be constructed, for example, to supply water to desalination or power plants. If no beach or sandy floor exists near the plant location, or if the site conditions are inadequate for infiltration, a tunnelled offshore intake system is the ideal choice. The world-wide growing demand for oil and gas makes the construction of pipelines on and offshore necessary. Please find hereafter an overview of application fields for tunnelled pipeline systems from the mainland into the Sea and vice versa.

2 ADVANTAGES OF TRENCHLESS INSTALLATION TECHNIQUES

2.1 *Minimized impact on environment*

Sea outfalls or seawater intakes may generally be installed by trenchless or open-cut trenching methods. In comparison to open-cut methods of pipe installation, trenchless installation techniques reduce impact on environment to a minimum. The installation takes completely place underground, without harming the marine wildlife. Sea water quality remains untouched, emissions and vibrations caused by the pipeline installation are considerably reduced.

2.2 *Minimized impact on existing infrastructure*

Trenchless marine pipeline installation can also be applied in densely populated areas, where only little space is available for jobsite installation. Due to the displacement of the "jobsite" underground, life in coastal areas can continue, tourism and shipping traffic is not affected. Existing pipe networks can be conserved.

2.3 *Benefits for construction time and pipeline life cycle*

The installation works of a pipeline underground do hardly depend on external conditions like the weather, storms, ebb and flood or sediment transport. This makes the trenchless technology a safe and reliable construction method. In the long run, the installation underground even extends the pipeline life cycle, the pipeline remains protected underground against damage by ships or sabotage, with a lower risk of settlement and a higher seismic safety.

Figure 2. Function principle of pipe jacking marine outfall technology.

2.4 *Influence of currents on pipelines laid on the sea bed*

Pipelines laid by open-cut techniques on the sea bottom are directly influenced by hydrodynamic forces resulting from currents. Typical currents caused by orbital movements of wave particles and near-coast currents can damage pipes especially during heavy sea or storms.

Hydrodynamic forces cause erosion, transport and accretion of seabed material. Offshore buildings in general influence the currents and may lead to heavy erosion or accretion near the buildings. Pipelines laid on the sea bottom may increase the current velocity and thus turbulences. A two-phase flow of sand and water under the pipeline forms. Scours develop, which depend on the following parameters:

- Vertical current velocity profile
- Turbulence
- Wave reflexion
- Bed material
- Bed roughness.

If the pipeline is laid directly on the sea bottom, scouring can lead to a large free span of the pipeline between two supports. If the pipeline is laid on concrete supports, the scouring may cause the supports and the pipeline to sink. The bending radius of the pipeline between two supports may be exceeded and pipeline may break.

3 TECHNOLOGIES FOR TRENCHLESS INSTALLATION OF OFFSHORE LINES

3.1 *Pipe jacking and microtunnelling*

3.1.1 *Installation of pipeline*

A launch shaft is excavated at the land side of the pipeline. The dimensions and the design of this shaft may vary according to the specific requirements of the site with economics being a key factor. A thrust wall is constructed to provide a reaction against which to jack. The initial alignment of the pipe jack is obtained by accurately positioning guide rails within the thrust pit on which the pipes are laid.

Powerful hydraulic jacks are used to push the jacking pipes through the ground. At the same time, excavation at the tunnel face is taking place within a steerable shield. The Herrenknecht remote controlled microtunnelling machines are operated from a control panel in a container which is located on surface next to the launch shaft. This is an advantage regarding safety regulations, because no staff has to work in the tunnel during construction. The position of the remote-controlled machine is supervised by a guidance system.

Today, the developed tunnelling technique enables the realization of long distance advances, also in difficult ground conditions. The following possibilities reduce friction and jacking forces.

Automatic lubrication system.

During the pipe jacking process the whole pipeline is pushed through the ground. Rising friction forces between the surrounding ground and the pipe string lead to increasing jacking forces. However the maximum jacking force is strictly limited by the maximum permissible pipe load. To reduce friction, the pipeline should be lubricated continuously. Bentonite suspensions act as lubricants during the pipe jacking process. They are mixed in bentonite plants at the job site and are pumped into the tunnel via hoses or pipes. Through injection nozzles within the jacking pipes, the lubricant is squeezed into the annular gap between the pipeline and the surrounding ground. Thus, the jacking forces can be reduced considerably.

Reduced jacking forces optimize the performance of the pipe jacking process in terms of lower pipe loads and thus longer jacking distances.

A new generation of our bentonite lubrication system enables the automatic distribution of bentonite suspensions along the alignment, while controlling and recording the injected bentonite volumes along each pipe at the same time. This permits to select and adjust the desired bentonite volume of each meter along the tunnel route, also considering changes in geology. Monitoring, control and recording of relevant data, such as bentonite volumes, pressures and friction forces, is done in the control container where the machine operator supervises the pipe jacking process.

Use of interjacking stations.

Another, mostly additional possibility to keep jacking forces down in difficult ground or long drive lengths, is the use of intermediate jacking stations. They are installed to reduce and distribute the jacking forces, thereby reducing the forces applied by the main hydraulic jacks in the start pit and secondly on the thrust wall. The interjack stations are integrated in several distances in the pipeline and serve to separately advance the pipeline in sections.

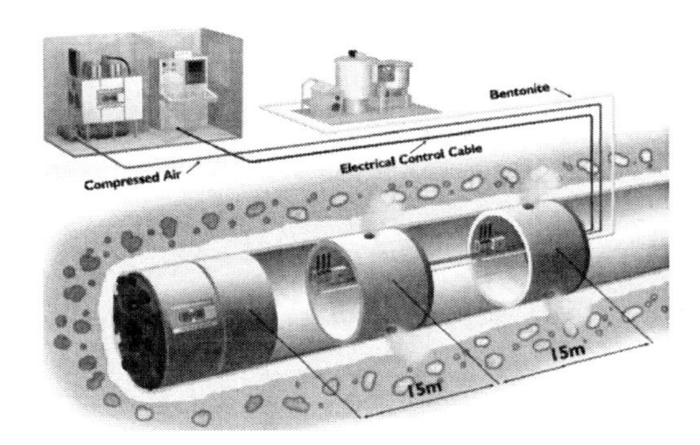

Figure 3. Installation of the automatic lubrication system in the tunnel.

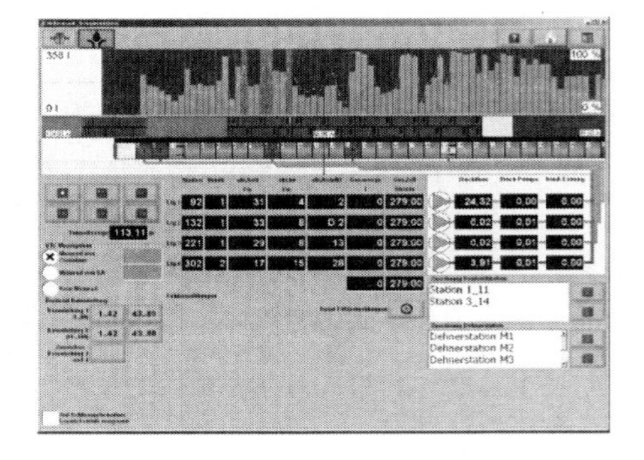

Figure 4. Monitoring display in control container.

Access to cutting wheel | Airlock system

A deciding factor for attainable drive length is the lifetime of the cutting tools. Due to strongly worn cutting tools the tunnelling performance will be decreased, and when cutting tools are worn-out, they have to be changed. Herrenknecht has designed tunnelling machines, which give the possibility to enter the excavation chamber to inspect the cutting head and to change worn cutting tools. If necessary, an airlock can be used to be independent from possible ground water in front of the tunnelling machine. Consequently, longer drives can be realized.

3.1.2 *Machine recovery*

Tunnelling machines to be used for Sea Outfalls are equipped with an additional recovery module, consisting of a steel can with bulkhead to close the machine and hydraulic cylinders to separate tunnel and machine. The supply of hydraulic oil for these cylinders is done by divers and connected from the outside skin of the recovery module. After complete installation of the tunnel, the seaside end of the pipeline is mostly closed with a bulkhead equipped with a valve.

In most cases the tunneling equipment has to be recovered and lifted up to the surface. Therefore, the jacking machine is equipped with lifting eyes on its upper side. There are two ways to lift the tunnelling machine:

- A barge with a crane is moored at the position from which the jacking machine shall be recovered. To be more independent from the sea, a jack-up platform with crane can be installed which is able to lift higher weights than a floating barge. The crane is connected to the lifting eyes of the jacking machine by means of a spreader beam. The connection has to be carried out with the help of divers. The jacking machine is lifted to surface by the crane.

Figure 5. Interjacking installed in pipe.

Figure 6. Interjacking station in operation.

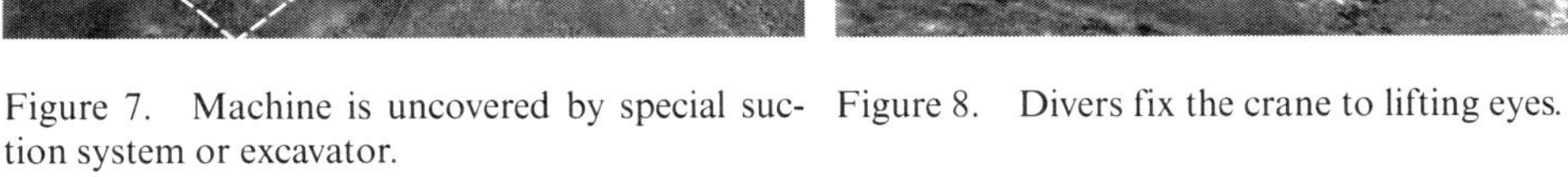

Figure 7. Machine is uncovered by special suction system or excavator.

Figure 8. Divers fix the crane to lifting eyes.

Figure 9. Lifting tunnelling machine to a barge in the Newa River, Russia.

Figure 10. Recovery of machine by a jack-up platform.

Figure 11. Recovery of tunnelling machine via airbags, lifting by crane in harbor (Sorek, Israel).

- Another possibility to lift the machine from seabed to water surface is the application of airbags. These are fixed by divers to the lifting eyes of the machine. A compressor installed on a ship or barge on the surface inflates the number of airbags needed to lift the weight of the machine. Water level fluctuations caused by ebb and flood may be considered to reduce the lifting height. The barge or a ship transports the jacking machine to the next harbour, where it can be taken out of the water by a high-capacity crane.

3.1.3 *Reference project: Outfall Golden Sands, Bulgaria*

Contractor	Ponsstroyengineering EOOD, Struma Imoti
Use of sealine	Waste Water Treatment Plant outlet
No. of sealines	2 x 200 m
Machine type	AVN700XC
Method of installation	Pipe Jacking
Pipes	GRP Hobas Pipe OD860 mm
Ground water level	up to 10-18m
Geology	Sand, big rocks clay shells
Ground water level	up to 12-18m
Best performance	35,75 m per day

Figure 12. AVN700 equipment in the start shaft.

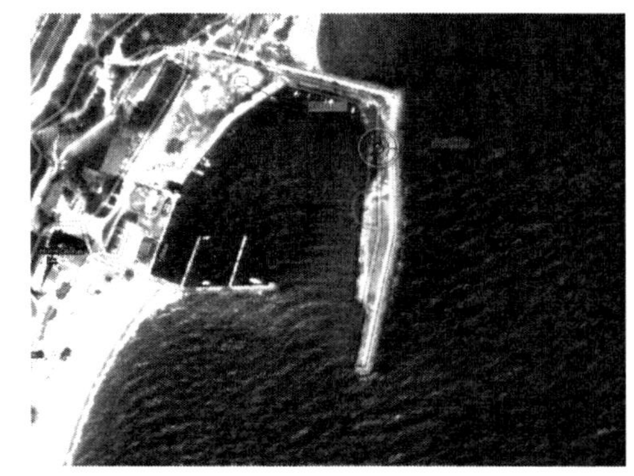

Figure 13. Deep 18 m start shaft on the breakwater.

Golden Sands is one of the most popular Bulgarian Black Sea summer resort. During summer time up to 60.000 tourists are visiting this place. For the last decade hundreds of modern hotels have been build, however the underground infrastructure was not able to keep up the enormous city grow. Thanks to European Union Golden Sands were granted 48 Mil € funding for the modernization of its wastewater treatment plant. Part of this project was see outfall. Tender for the outfall part was gained by local Bulgarian consortium, which decided to rent the whole AVN700XC microtunnelling equipment from Herrenknecht AG. Design utilize 2 drives of 200 m from start shaft located of the stormwater breaker. Depth of the shaft was 18 m. Due to perfect cooperation between all involving companies 200 m outfall drive was perform in only 7 days.

3.2 *Segment Lining*

Segment lining is a tunnel construction method mainly used for marine tunnels with larger diameters (>ID 2000 mm) and/or longer distances. The tunnel lining is assembled in the tailskin of the tunnelling machine and consists of segments, which are assembled to a complete segment ring. The reinforced concrete segments are transported to the machine on a rail system. The excavated material is transported to the surface via a rail-bound or slurry system.

3.2.1 *Function principle of Segment Lining*

Within this support method, hydraulic cylinders are situated in the tunnelling machine. These cylinders represent the advance unit. After having completed an advance cycle, e.g. achievement of the length of segment, a number of jacking cylinders is retracted in order to create a sufficient space for the first segment. The other cylinders maintain their contact to the ring already built in order to prevent the shield from moving backwards due to the earth pressure. A so called erector picks the segments up and puts them to the correct position. After completion of one segment ring, all jacking cylinders are extended for a further advance cycle. By this method of tunnel lining the already built tunnel frame remains on the spot in the ground and serves as abutment for the advance. Grout is injected into the overcut which is created by the cutting wheel during excavation.

3.2.2 *Segment Lining for tunneled outfalls*

For the construction of tunneled outfalls, the segmental lining tunnelling machine is equipped with a recovery module integrated in the machine structure, similar to pipe jacking. Segment Lining uses the jacking cylinders to separate machine from the tunnel after completion of the last segment ring. Segment Lining can be applied for inner diameters of

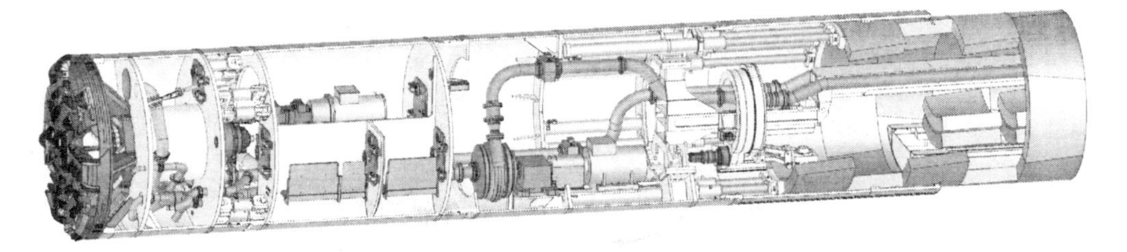

Figure 14. Section through AVND Segment Lining tunnelling machine.

Figure 15. View into a segmentally lined outfall tunnel, ID 2800.

minimum 2000 mm. The advantage of this lining technology is the longer drive distance. No interjacking stations are required, because the tunnel is erected behind the machine. Recovery of the machine takes place similar to the pipe jacking method and can be considered not being more sophisticated.

3.2.3 *Reference project: Gold Coast Desalination Plant, Australia*

Contractor	John Holland Engineering Pty. Ltd. Group
Use of sealine	Desalination Plant
No. of sealines	1. Brine outlet - 2000m 2. Feed line - 2200m
Machine type	AVN 2800 (OD 3440)
Method of installation	Segment Lining
Geology	Quartz, metabasalt (up to 100MPa)
Ground water level max	70m

Since 2001, Australia has been suffering from an unprecedented period of drought. This results in serious consequences to the densely populated coastal areas. One solution to such water shortages is to install desalination plants to convert brackish sea water to potable water.

The new "Gold Coast Desalination Plant" south of Tugun is the first large-scale desalination facility on Australia´s eastern seaboard. Since November 2008, it provides 125 million liters of fresh drinking water a day to the residents of South East Queensland, Australia´s fastest growing region.The Gold Coast Desalination Plant produces fresh drinking water drawn from the Pacific Ocean. Due to environmental and safety considerations, the tunnelled construction was the best choice with minimal impact on environment and the local community. Due to the depth of up to 70 m below rock, clays and sands, surface noise and

Figure 16. View on jobsite, besides Gold Coast Airport, close to the beach.

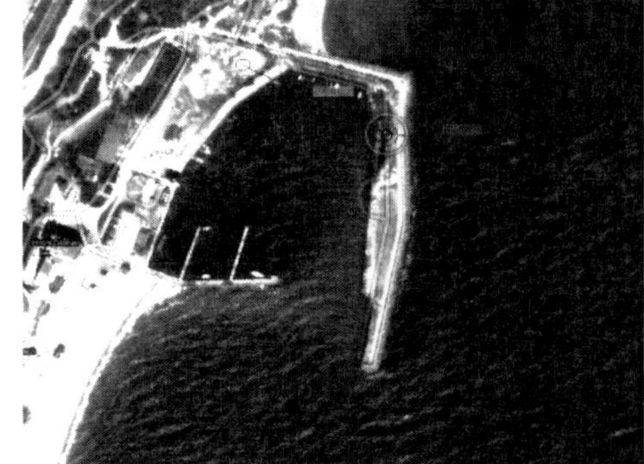

Figure 17. View into 70 m deep start shaft.

vibration was considered as unlikely. Thus, tunnelling was undertaken in accordance with strict environmental standards.

The intake tunnel is 2.2 km long—750 m from the plant to Tugun´s beach and then 1.4 km offshore out to the sea. The brine outlet tunnel is 2 km long—780 m from the plant to the beach and 1.2 km offshore. This tunnel is connected to a 200 m long diffuser on the seafloor to quickly disperse the brine water. Each tunnel included three curves with minimum radii of 400 m through hard rock (up to 100 MPa) and groundwater pressures of up to 7 bar. Two Herrenknecht AVN2800 AH tunnelling machines tunneled the 3.4 m outer diameter tunnels. The gap of 20 cm between the concrete segments and the tunnel was filled with grout. Finally 21,600 concrete segments were used to line the two tunnels. The complete underground mission was accomplished after only eight months of construction.

3.3 *Direct Pipe®-method*

The Direct Pipe®-method for trenchless installation of prefabricated steel pipelines combines microtunnelling technology of excavation with a thrust unit, the so-called Pipe Thruster. Direct Pipe® incorporates the advantages of microtunnelling which enable application in difficult ground conditions, while reducing risks to a minimum. This opens up new application potentials. Originally developed as an auxiliary tool for the pullback of the pipe in the HDD-method, it was presented for the first time at the Hannover Fair in spring 2006. The Pipe Thruster embraces the prefabricated and laid out pipeline and pushes it into the ground in strokes of five meters each. The requisite bore hole is excavated by a slurry microtunnelling machine (AVN) which is mounted at the front of the pipeline.

3.3.1 *Technology principle*

Direct Pipe®-method allows the excavation of the borehole and the simultaneous trenchless installation of a prefab and tested pipeline in one single continuous step. Similar to pipe jacking, the soil is excavated with a Herrenknecht microtunnelling machine. The position along the specified tunnel route is monitored by state of the art techniques of controlled pipe jacking. The excavated material is removed through the slurry circuit placed in the prefab pipeline. The forces which are necessary for pushing the pipeline ahead are exerted by a novel push mechanism known as the Pipe Thruster, an innovation of Herrenknecht AG. The Pipe Thruster operates like the standard acking frame used for jacking concrete pipes and transmits the push forces to the cutterhead via the pipeline.

Just like the microtunnelling method, prior to launching the machine is positioned at the requisite access angle on a launch rail in front of the launch seal. The pipeline is welded to the conical rear section of the machine and mounted on rollers behind the launch pit. The

Figure 18. Direct Pipe® view on lauch shaft with Pipe Thruster and end shaft with AVN machine.

clamping unit of the Pipe Thruster embraces the pipeline and thrusts it into the ground along with the machine. The current maximum pipeline diameter which can be clamped is 60" (OD = 1524 mm). The forces to be anchored depend on the pipeline access angle and the maximum thrust force to be applied. Due to the pullback option of the whole pipeline in case of non-destructible obstacles, the geological risk is minimized.

3.3.2 *Sea bed recovery*

Depending on the project details the sea bed recovery of the Direct Pipe machine ensures the absolute minimum amount of offshore work is required. By deploying a sub sea recovery module at the back of the Direct Pipe machine it can be safely pushed off the end of the pipe enabling it to be lifted to a waiting barge, or floated to the surface and towed to port to be recovered. This procedure, developed by Herrenknecht, removes the need for an offshore shaft to recover the machine.

3.3.3 *Advantages of Direct Pipe®*

Technical advantages of Direct Pipe®

- Single-phase mode of operation.
- Permanent borehole support.
- Borehole collapse breaking the surface is thus not possible.
- The cutting wheel can be equipped for any geological condition.
- Excavated material is transported safely out of the borehole by conveyor lines, sedimentation in the borehole is not possible.
- Curved routes can be tracked, precise to the centimeter, thereby avoiding excessive bending stress on the pipeline and excessive laying forces.
- Operations at the target pit only to dismantle the cutting head.

Economic advantages Direct Pipe®

- Limited work area required at the launch pit and target pit.
- A minimum volume of slurry is required.
- Due to a very small overcut a minimum borehole diameter can be realized, thus minimizing the amount of excavated material.
- Boring time can be reduced, no coupling times are necessary and the pipeline can be laid continuously in one single step.
- No permanent protective piping/casings are necessary.

3.3.4 *Reference project: Brine outlet Rysumer Nacken, Emden, Germany*

Contractor	Meyer & John
Use of River outlet	Sewage line
No. of pipelines	1 x Ø 48" – 283m length
Machine type	AVN 1000 XC + HK 750 PT
Method of installation	Direct Pipe
Geology	sand, silt, clay, stones, wood
Ground water level max.	15m

Table 1. Suitability of trenchless installation technologies for different site conditions.

		Pipe Jacking	Segment Lining	HDD	Direct Pipe
Geology	Cohesive	+	+	+	+
	Clay, silt, Sand	+	+	+	+
	Gravel	+	+	-	+
	Boulders	+	+	-	o
	Mixed	+	+	o	+
	Rock	+	+	+	o
Pipe material	PEHD	-	-	+	-
	Steel	o	o	+	+
	GRP	+	-	+	-
	Concrete	+	+	-	-
Pipe diameter	≤ DN 1000	o	-	+	o
	≥ DN 1200	+	-	o	+
	≥ DN 2000	+	+	-	-
	> DN 3500	o	+	-	-
Laying length	≤ 100 m	+	-	0	+
	100 – 500 m	+	o	+	+
	500 – 1200 m	+	+	+	+
	>1200 m	o	+	+	0

+: particularly suitable.
O: conditionally suitable.
-: not suitable.

Europe, and Germany in particular, are among the largest natural gas importers in the world. Natural gas storage facilities are necessary to counter supply security concerns and deal with large fluctuations in demand. Deep below the surface of the earth, gas can be stored safely in salt formations, whose walls provide particularly tight seals. Underground salt mines are flushed out with water to produce huge cavities which are then used as storage caverns. In order to drain the resulting brine, a 42-kilometer-long pipeline had be laid to the Ems River near the city of Emden.

The final 283 meters of the outflow pipe at the Rysumer Nacken artificial dune field were built using the new single-step Direct Pipe® method. The AVN1000XB Herrenknecht Direct Pipe® machine worked its way through mixed ground of sand, silt, clay, gravel, wood and stones from the launch pit on the shore towards the target pit in the Ems, with a maximum Pipe Thruster power of 750 tonnes. For this single-step pipe laying method, the Herrenknecht TBM, with a drilling diameter of 1,325 millimeters, proceeds with a 48-inch, PE-coated pipeline in tow. Because of the tight space constraints at the construction site, the pipeline could not be laid in one length. For this reason, individual, 36-meter pipe lengths were prepared and joined onto the pipe string. The TBM reached its target with supreme precision; its deviation from the ideal line was a minimal 3 centimeters. After reaching the target pit the TBM was detached from the installed pipeline and recovered. At this stage, the sea outfall was complete. The team proceeded very quickly with performances of up to nine meters per hour and reached the target structure in the Ems River only 7 tunnelling days later.

4 CONCLUSION

As described above tunneled Sea Outfalls, Intakes and Landfalls offer many advantages over the conventional open-trench installations. The application of one of the described installation technologies depends on technical, temporal and commercial factors. Geological ground conditions, specified time and costs targets are the decisive criteria for the selection of the most suitable technology. In the following table, the suitability of the four technologies for different site conditions is compared. Herrenknecht AG supplies the whole scope of solutions to any specific project requirement.

Underground Infrastructure of Urban Areas 3 – Madryas et al. (Eds)
© 2015 Taylor & Francis Group, London, ISBN 978-1-138-02652-0

Special cements and concrete durability

A. Garbacik & W. Drożdż
The Institute of Ceramics and Building Materials, Division in Cracow, Poland

Z. Giergiczny
Technological Center Betotech in Dąbrowa Górnicza, Silesian University of Technology, Gliwice, Poland

ABSTRACT: The paper presents the requirements set for special cements in the current standards. This type of cement, in addition to the basic features, meets the additional essential requirements of the building practice: low LH and very low heat of hydration VLH, resistance to sulphate attack SR and HSR, and a low content of active alkali NA (alkaline corrosion resistant). The application of special cements in the performance of concrete structures avoids the disadvantages of the concrete, and thus contributes to the durability of the concrete structure (long life cycle).

1 INTRODUCTION

Current standard cement PN-EN 197-1:2012 "Cement—Part 1: Composition, specifications and conformity criteria for common cements" allows to produce a wide range of cements with different properties, and consequently, the possibilities of their use in construction industry. This is particularly true of cements containing in its composition mineral additives (CEM II, CEM III, CEM IV, CEM V) especially, as this type of cement has a number of special characteristics desired in the performance of many structures. Special properties of the cement include: low (LH) or very low heat of hydration (VLH), resistance to sulphate attack (SR, HSR) and resistance to alkali corrosion NA.

The use of special cements, which in addition to the basic standard properties satisfy the additional requirements of the characteristic features, enable the design and execution of concrete structures with high durability for the destructive impacts (mainly various types of chemical attack). The use of this type of cement is helpful when performing massive concrete, concrete exposed to chemically aggressive environments, or when using reactive aggregates in the production of concrete or concrete elements. In such cases, the use of special cements avoids the disadvantages of the concrete, and thus, contribute to the stability of the concrete structure (extended life cycle).

Issues of special cements, in light of the requirements of current standards, are the content of hereby work.

2 CEMENTS WITH LOWER HEAT OF HYDRATION (LH, VLH)

The reaction of the cement components with water is of exothermic nature and the emitted heat raises the temperature of the moulded concrete element or building construction. Cements based on the mineral composition and the strength class perform different heat of hardening (hydration). The rate of heat release during the setting and hardening translates into the speed of increasing strength, particularly in the initial period. The kinetics of the heat release by the selected cements is shown in Figure 1.

The ignorance of the size of hydration heat of the applied cement may cause negative effects. The heat emitted from the process of cement hydration can lead to large differences in the temperature between the concrete surface and the interior, which can cause thermal stress. If the thermal stresses exceed the limit strength of concrete, the process of creating microcracks starts. Reduced durability of concrete is the end result of the process. This negative phenomenon is especially observed in the performance of large massive concretes (hydro-technical constructions) and objects with large surface areas (construction of tanks in wastewater treatment plants). It can be prevented through the use of cement with low heat of hydration LH complying with the requirements of PN-EN 197-1 or cement with a very low heat of hydration VLH complying with the requirements of standard EN 14216:*2005* "Cement—Composition, specifications and conformity criteria for very low heat special cements", furthermore, meeting the requirements of shown in Table 1.

From the group of cements available on the market the cements can consist of the ones with a high content of mineral additives (slag cement CEM III, Pozzolanic cement CEM IV, composite cement CEM V) of 32,5 N strength class. Therefore, designing hydraulic

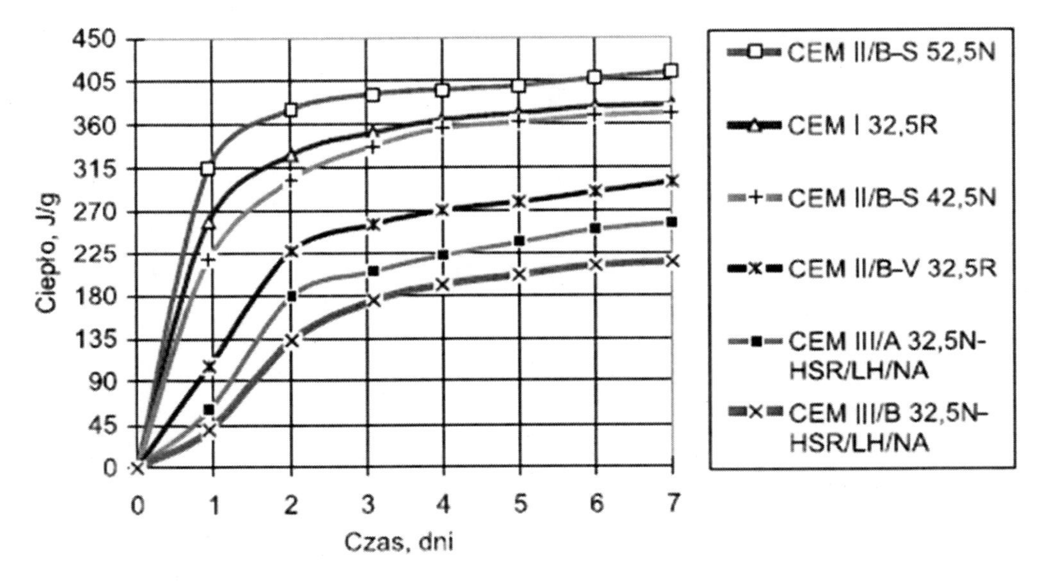

Figure 1. The kinetics of heat release by the selected cements.

Table 1. The requirements for cements with low LH and very low heat of hydration VLH.

Cement group	Cement type	Requirements	Test methods
Cements with low hydration heat LH acc. to PN-EN 197-1	Each cement of common use acc. to PN-EN 197-1	Hydration heat after 41 hours $\leq$ 270 J/g	PN-EN 196-9 Semi adiabatic method
		Hydration heat after 7 days $\leq$ 270 J/g	PN-EN 196-8—Solution method
Cements with very low hydration heat VLH acc. to PN-EN 14216	Slag cement III/B and III/C Pozzolannic cement IV/A and IV/B Composite cement V/A and V/B, 22,5 strength class only.	Hydration heat after 41 hours $\leq$ 220 J/g	PN-EN 196-9 Semi adiabatic method
		Hydration heat after 7 days $\leq$ 220 J/g	PN-EN 196-8—Solution method

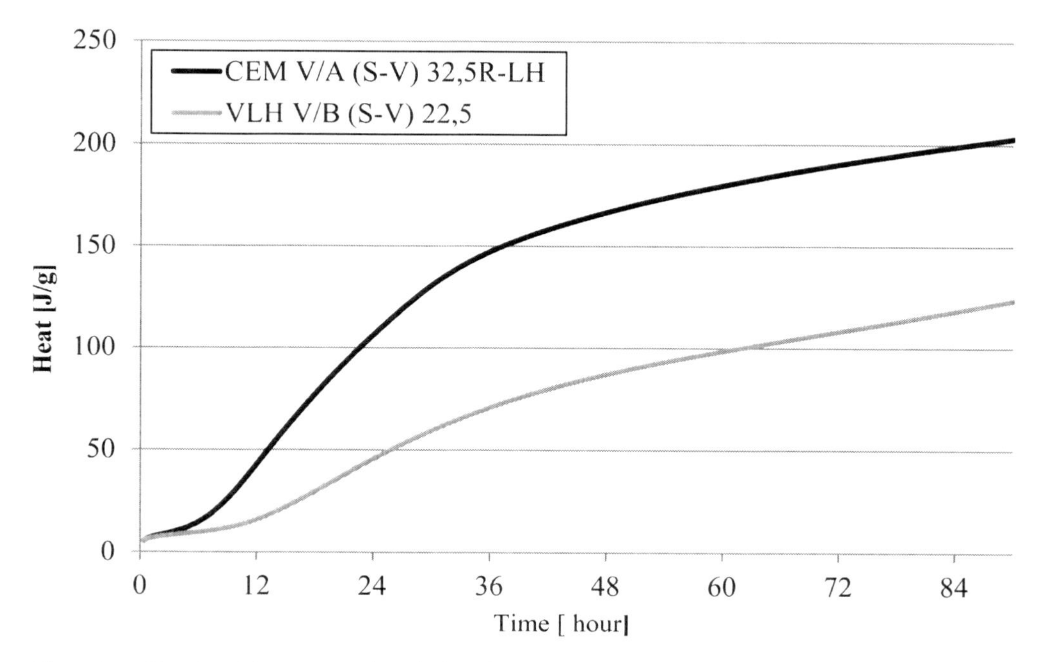

Figure 2. Kinetics of heat generation during the hydration of composite cement CEM V.

structures, large foundation work and objects with large areas it is necessary to recommend the application of these kinds of cement, also, to pay attention to the optimal composition of the cement content in concrete (the higher cement content in the composition of concrete the greater amount of heat generated during hardening).

Figure 2, shows as an example, the difference in the heat generation during the hydration of composite cement CEM V/A and CEM V/B (Dziuk et al. 2012).

Special cements with a very low heat of hydration VLH divide into the following types:

- VLH III—slag cement
- VLH IV—pozzolanic cement
- VLH V—composite cement.

The production of the cements is possible in 22,5 strength class only.

3 CEMENTS RESISTANT TO SULPHATES (SR, HSR)

Concrete is the most exposed to extreme environmental impacts in fields of marine, municipal, environmental (sewage treatment plants, collectors, sewage, landfill, etc.), mining, hydrotechnical and road construction.

There are many factors affecting the corrosion resistance of concrete, however, the microstructure of the hardened cement paste and its phase composition are crucial. Cement paste resistant to chemical corrosion ought to characterize with a low content of calcium hydroxide, and a very low content of hydrated calcium aluminates with a low C/S ratio (Actin 2008, Wu et al. 2002, Dąbrowska et al. 2013).

The structures exposed to groundwater or seawater sulphate corrosion is the most common type of chemical attack. The source of sulphates may also be fertilizers or industrial waste. The lowest resistance of concrete components to sulphate ions is performed by hardened cement paste, particularly portlandite $Ca(OH)_2$ and hydrated calcium aluminates. Therefore, sulphate resistant Portland cements CEM I typically contain less than 3% of tricalcium aluminate C_3A.

It is believed that cements with a high content of mineral additives with pozzolanic and/or hydraulic activity characterize with increased corrosion resistance to chemical attack. This is mainly due to the following (Actin 2008, Giergiczny 2013):

- limitation of the clinker phases are susceptible to corrosion, i.e. tricalcium aluminate in cement composition, which is associated with a lower share of Portland cement clinker in the cement composition on account of cement additive (mainly in case of CEM III ÷ CEM V),
- reduction of $Ca(OH)_2$ in the hardened binder matrix and the increase of C-S-H gel phase share with a low C/S ratio,
- change of the hardened cement paste microstructure as a result of filling in of the pore spaces with in the hardened paste with pozzolanic reaction product and/or hydraulic additive,
- sealing of the structure by the addition of mineral particles of cement composition.

Standardization research carried by the experts at the European Committee for Standardisation CEM over the group of 27 kinds of common cements proved that only a few are classified in the PN-EN 197-1:2012, as a special cements resistant to sulphates SR. The selected group of SR cements are the following: Portland cement CEM I, slag cement CEM III/B, C and pozzolanic cement CEM IV/A, B, complying with the basic requirements of the standard as to guarantee the participation of sulphate resistant constituents (Table 2).

A wide range of studies completed across the country (Chłądzyński et al. 2008), confirmed the increased sulphate corrosion resistance of domestic cements containing in its composition mineral additives. The results of corrosion tests of a few selected cements are given as an example in Figure 3. The test involves observing a change in length (expansion) of cement paste hardened standard samples (the bars with 20 × 20 × 160 mm dimensions), made of the tested cement and maturing for 28 days under standard conditions (water at the temperature of 20°C), immersed in a solution of Na_2SO_4 having a concentration of SO_4^{2-} ions = 16 g/dm^3 for a period of one year.

Cements recognized in the CEN member countries as resistant to sulphates, thus being outside the scope of the European standard EN 197-1, are the subject of the classification and requirements of national standards.

A national standard for special cements PN-B-19707:2013 "Cement—Special Cements. Composition, specifications and conformity criteria", amended in 2013, extends the classification of sulphates resistant special cements HSR by blended Portland cements CEM II/A, B, M (S, V) and composite cements CEM V/A, B (S, V). The requirements set against sulphate resistant cements HSR by the national standard PN-B 19707: 2013 are shown in Table 3.

Table 2. Classification of sulphate resistant cements SR acc. to PN-EN-197-1:2012.

SR cement type	Cement group	Classification criterion
Portland cement CEM I-SR	CEM I-SR 0 CEM I-SR 3 CEM I-SR 5	Content in clinker $C_3A = 0\%$ Content in clinker $C_3A \leq 3,5\%$ Content in clinker $C_3A \leq 5\%$
Slag cement CEM III-SR	CEM III /B-SR CEM III/C-SR	Minimum content of granulated slag only as for common use cements of B, C groups
Pozzolanic cement CEM IV-SR	CEM IV/A-SR CEM IV/B-SR	Content of C_3A in clinker $\leq 9\%$. Siliceous fly ash V and/or natural pozzolana are in the composition of this cement type

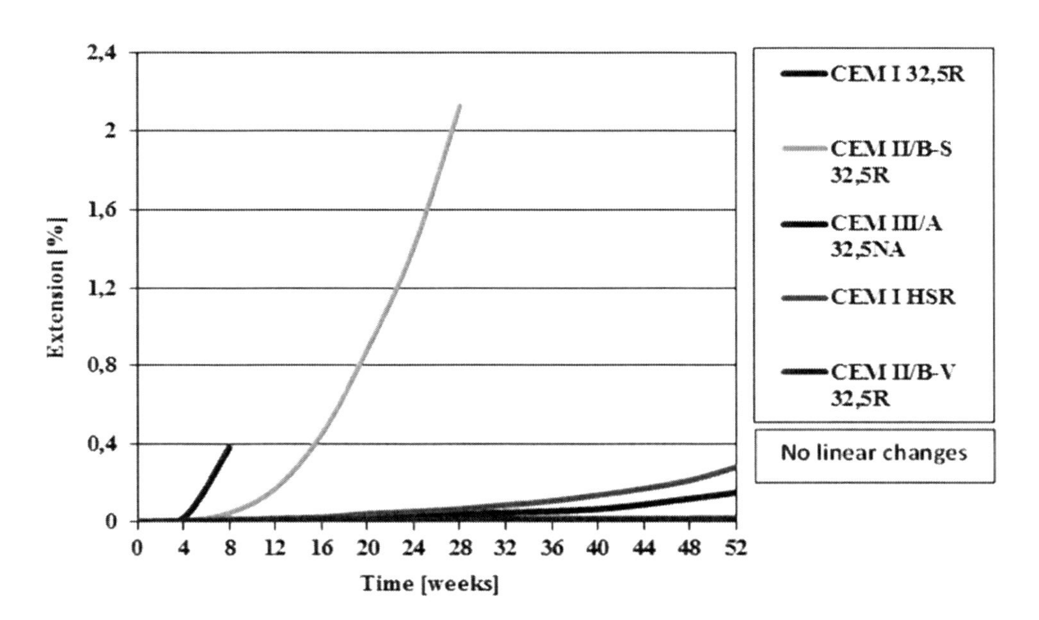

Figure 3. Sulphate resistance of selected cements.

Table 3. Requirements of special sulphate resistant cement HSR acc. to PN-B-19707.

	Requirements		
HSR cement type	Share of non clinker cement component [% by mass]	Resistance to sulphates	Clinker
CEM II/A-V CEM II/A-S CEM II/A-M (S-V) CEM II/B-S	No requirements	Resistance to sulphate attack expressed as the value of expansion in Na_2SO_4 solution after 52 weeks $X_t \leq 0,5\%$ c Test method—Appendix A to standard PN-EN 19707:2013	Content of tricalcium aluminate $C_3A \leq 5\%$
CEM II/B-V	Share of siliceous fly ash $V \geq 25\%$		No requirements
CEM II/B-M (S-V)	Share of siliceous fly ash $V \geq 20\%$		No requirements
CEM III/A	Share of granulated blast furnace slag $S \leq 49\%$		$C_3A \leq 9\%$
CEM III/A	Share of granulated blast furnace slag $S \geq 50\%$		No requirements
CEM V/A (S-V) CEM V/B (S-V)	PN-EN 197-1		No requirements

4 CEMENTS WITH LOW NA ACTIVE ALKALIS CONTENT

For several dozen of years there has been many cases of adverse chemical reactions of the aggregate with the surrounding cement paste, and in particular, its liquid phase.

The most significant of these processes are the sodium and potassium ions present in the concrete pore liquid.

Table 4. Content of alkalis soluble in water in mineral additives.

	Alkali content						Content of soluble alkalis in relations to their total content [%]		
	Total [% by mass]			In aqueous extract [% by mass]					
Fly ash sample	Na_2O	K_2O	Na_2O_{eq}	Na_2O	K_2O	Na_2O_{eq}	Na_2O	K_2O	Na_2O_{eq}
Calcareous fly ash (W)	0,25	0,12	0,33	0,0102	0,0018	0,011	4,08	1,50	3,33
Calcareous fly ash (W)	0,31	0,21	0,45	0,0132	0,0107	0,020	4,26	5,10	4,44
Siliceous fly ash (V)	0,84	3,31	3,02	0,0212	0,0147	0,031	2,52	0,44	1,03
Ground granulated blast furnace slag (S)	0,53	0,39	0,79	0,0068	0,0046	0,010	1,28	1,18	1,27
Portland cement CEM I	0,28	0,58	0,66	0,1752	0,4382	0,464	62,57	75,55	70,30

Table 5. Requirements concerning special low alkaline NA cement, in accordance with PN-B-19707:2013, as well as the alkali content in cement.

	Requirement		
Cement type NA	Share of non-clinker cement component [% by mass]	Total content of alkalis Na_2O_{eq} [% by mass]	Active alkalis $^*Na_2O_{eq}$ [% by mass]
CEM I CEM II/A-LL	No requirements	≤0,60%	0,30–0,47
CEM II/A-V	Share of slag S, fly ash V; $S + V \geq 14\%$	≤1,20%	0,51
CEM II/A-S		≤0,70%	0,48
CEM II/A-M (S-V)		≤1,20%	0,47
CEM II/B-V	Share of siliceous fly ash $V \geq 25\%$	≤1,50%	0,52
CEM II/B-S	PN-EN 197-1	≤0,80%	0,48
CEM II/B-M (S-V)	Share of siliceous fly ash $V \geq 20\%$	≤1,30%	0,51
CEM III/A	Share of granulated blast furnace slag $S \leq 49\%$	≤0,95%	0,28
	Share of granulated blast furnace slag $S \geq 50\%$	≤1,10%	0,34
CEM III/B-C	PN-EN 197-1	≤2,00%	0,18–0,25
CEM IV/A (V)	Share of siliceous fly ash $V \geq 25\%$	≤1,50%	0,48
CEM IV/B (V)	PN-EN 197-1	≤2,00%	0,36
CEM V/A (S-V)	Share of the sum of siliceous fly ash and granulated blast furnace slag $(S + V) \leq 49\%$	≤1,60%	0,28
	Share of the sum of siliceous fly ash and granulated blast furnace slag $(S + V) \geq 50\%$	≤2,00%	0,16
CEM V/B (S-V)	PN-EN 197-1	≤2,00%	0,16–0,21

*Determined according to the method described in standard ASTM C 114-04.

Alkali-aggregate reaction takes place between sodium and potassium ions—contained in the liquid phase present in concrete pores—and a reactive silica present in the aggregates. Na^+ and K^+ ions in the pore fluid can come from cement, mixing water, use of mineral additives and chemical admixtures as well as aggregates. Under appropriate humidity conditions and the concrete temperature a gel of potassium sodium silicate, absorbing water, therefore, increasing the volume and causing destructive stresses (Dunant 2012, Thomas 2011, Owsiak 2002, Owsiak 2008).

The reaction alkali-aggregate is very slow and its negative consequences may become apparent only after several years. The motion of this reaction depends on a number of factors including among others: the amount of active silica and its fineness, concentration of sodium and potassium ions in the liquid phase in the pores of the concrete, etc. Owsiak Z. has described the mechanism of an alkali—silica the reaction in details (Owsiak 2002).

The results of numerous experimental works (have shown, that the use of mineral additives (fly ash, ground granulated blast furnace slag), in cement (concrete) composition can reduce the negative effects caused by alkali-silica reaction, even in the case of using aggregates with high reactivity.

This appears due to the fact that the alkalis contained in slag and fly ash are extremely difficult alkali-soluble (Table 4), which while using these mineral additives leads to the reduction of the of OH^- ions concentration in the concrete pore liquid (Drożdż et al. 2012, Thomas 2011, Owsiak 2008).

Granulated blast furnace slag and fly ash contain most of the alkalis in the glassy phase (insoluble). Only a very small part appears in the form of easily soluble sulphates. Such feature makes siliceous fly ash and granulated blast furnace slag very useful in making alkali corrosion-resistant concrete. It has been documented and confirmed in referrals and technical guidelines in many countries. This fact was also reflected in Poland, by the records of the national standard PN-B-19707:2013 concerning the low alkaline cement NA (Table 5).

It should be noted that the data included in Table 5, i.e. the requirements for maximum alkali content of in cements expressed in Na_2O_{eq} are the result of a high content of sodium and potassium compounds, especially in the siliceous fly ash. However, the very low leachability of alkalis, both fly ash and granulated blast furnace slag, ensures, that the content of active alkalis (available in the concrete pores for the alkali-reactive aggregate reaction) in each standardized low alkali cement NA is below ≤0, 6% (Table 5), which according to the literature on the subject enables the design and manufacture of alkaline corrosion-resistant concrete constructions (Drożdż et al. 2012, Thomas 2011, Owsiak 2008).

5 SUMMARY

Current standards for special cements include: cements with low (LH) and a very low hydration heat (VLH), sulphate resistant cements SR and HSR and cements with low content of active alkali NA. The use of cements with special properties allows the design and construction of concrete structures with high durability, such as concrete subjected to sulphate attack, massive concrete and hydraulic structures, concretes produced with the use of potentially reactive aggregates, etc.

The revised national standard PN-B 19707:2013, takes into account a number of new special sulphate resistant cements HSR and low alkali cements NA of the group of blended Portland cements CEM II/A, B and composite cements CEM V/A, B.

The opportunity of implementing the cements HSR and NA from the groups of Portland cement CEM II and CEM V, on a large-scale, in industrial conditions, will also allow for a reduction of the clinker, and hence, CO_2 emissions in the production of these cements in comparison to energy-intensive and high-emission special Portland cements CEM I.

REFERENCES

Aitcin P.C.: *Binders for durable and sustainable concrete*, wyd. Taylor & Francis Group, 2008.

Chłądzyński S., Garbacik A.: Multicomponent cements in civil engineering Stowarzyszenie Producentów Cementu, Kraków, 2008 (in Polish).

Dąbrowska M., Giergiczny Z.: *Chemical resistance of mortars made with calcareous fly ash*. Roads and Bridges, 12 (2013), pp. 131–146.

Drożdż W., Giergiczny Z.: *Influence of calcareous fly ash in Portland Cement on ASR in concrete*. 18. Ibausil—International Conference on Building Materials, Weimar, 2012, t. 2, s. 319–326.

Dunant C.F., Scrivener K.L.: *Effects of aggregate size on alkali-silica reaction induced expansion*. Cement and Concrete Research, 2012, Vol. 42, pp. 745–751.

Dziuk D., Giergiczny Z., Sokołowski M., Puzak T.: *Concrete resistant to aggressive media Rusing a composite cement CEM V/A*. Underground Infrastructure of Urban Areas 2—Madryas, Nienartowicz &Szot (eds). Taylor&Francis Group, London, 2012, pp. 13–22.

Giergiczny, Z. 2013 *Fly ash in cement and concrete composition*, Gliwice: wyd. Politechniki Śląskiej (in Polish).

Owsiak Z.: *The alkali-silica reaction in concrete*. Ceramics.Polish Ceramic Bulletin, Vol. 72, Kraków, 2002 (in Polish).

Owsiak Z.2008, *The internal sulphate corrosion of concrete*, Kielce: wyd. Politechniki Świętokrzyskiej, Kielce 2008 (in Polish).

Thomas M.D.A.: The effect of supplementary cementing materials on alkali-silica reaction: A review. Cement and Concrete Research, vol. 41, 2011, pp. 209–216.

Wu Z., Naik T.R., *Properties of concrete produced from multicomponent blended cements*, Cement and Concrete Research 32 (2002) 1937–1942.

Underground Infrastructure of Urban Areas 3 – Madryas et al. (Eds)
© 2015 Taylor & Francis Group, London, ISBN 978-1-138-02652-0

Durability of composites made with fly ash—slag cements with a low content of Portland cement clinker

Z. Giergiczny & K. Synowiec
Technological Center Betotech in Dąbrowa Górnicza, Silesian University of Technology, Gliwice, Poland

ABSTRACT: The paper presents results of corrosion resistance tests of mortars made with fly ash—slag cements. Tested mortars were prepared according standard procedure with use of fly ash—slag cement with a low content of Portland cement clinker (20%). Two types of fly ashes were used—siliceous (V) and calcareous (W). Resistance to gaseous media (marking the depth of carbonation) and liquid media as well (investigation of the permeability to chloride ions and sulphates corrosion resistance) were assessed. It was found that the mortars made with fly ash—slag cements are characterized by a higher resistance to aggressive liquid media (water solutions of chlorides and sulphates) but much lower resistance to carbonation in comparison with Portland cement CEM I. Moreover it was found that in longer-term tested mortars are characterized by denser cement matrix, especially when the calcareous fly ash was a component of cement.

1 INTRODUCTION

Durability of cements composites, in addition to compressive strength, is the main criterion of their quality assessment. The measure of durability is resistance to aggressive agents attack. Mortar (concrete) is a capillary-porous composite material and its resistance to aggression depends on accessibility of pores microstructure to harmful media originating from the surrounding environment. Permeability of aggressive liquid and gaseous media might lead to corrosion of the steel reinforcement and cause destruction of reinforced concrete structures. In case of cement matrix particularly aggressive are water solutions of sulphates and chlorides (Chindaprasirt et al. 2008, Glinicki et al. 2010).

Some economic and ecological reasons (reduction of CO_2 emission, consumption of raw materials and energy inputs) lead to changes in the cement industry. Fundamental direction of these changes is use of mineral additives as a main constituents of cement. (Hauer et al. 2010, Schneider et al. 2011, Giergiczny et al. 2011). PN-EN 197–1 "Cement—Part 1: Composition, specifications and conformity criteria for common cements" standard distinguish 27 types of common used cements, and despite of such a wide range of cements, European Committee for Standardization continue works on new possibilities of cement composition with higher content of mineral additives, mainly ground granulated blast furnace slag (S) combined with limestone (L, LL) or siliceous fly ash (V) (Hauer et al. 2010, Schneider et al. 2011). In this respect there is no study of the potential use of calcareous fly ash (W), which in Europe is available in the amount of nearly 60 million tonnes per year (Giergiczny 2013).

One of the most important issue while incorporating new material solutions is durability assessment in terms of resistance to chemically aggressive agents attack (Gliniki et al. 2010). The aim of the research presented in this paper was to evaluate the resistance of mortars made with fly ash—slag cements to chloride ions permeability, sulphates corrosion and carbonation.

2 METHODS OF INVESTIGATION

Approved research program provided determination of the basic properties of the cement, i.e. water demand, initial setting time and soundness according to PN-EN 196-3 "Methods of testing cement. Part 3: Determination of setting times and soundness" as well as compressive strength according to EN 196-1 "Methods of testing cement. Part 1: Determination of strength". Test procedures used to evaluate the corrosion resistance are described in sections 2.1–2.3.

2.1 *Chloride ions permeability*

Chloride ions permeability measurement was carried out according test procedure specified in American standard ASTM C 1202-05 "Electrical Indication of Concrete's Ability to Resist Chloride Ion Penetration (Rapid Chloride Permeability Test—RCPT)". It is based on an evaluation of the resistance of mortar to the chloride ions attack on the basis of the electric charge passing through the sample during the test. Test samples were made of standard mortar (PN-EN 196-1) as cylinder with a diameter of 100 mm and height of 50 mm. Chloride ions permeability test was made after 28 and 90 days of curing in water, each time 2 samples were tested.

In order to ensure tightness, lateral surface of samples was covered with resin and saturated with water in vacuum conditions, then placed between the electrodes (Fig. 1a) behind them there were chamber for solutions of 3% NaCl and 0.3 M NaOH. The electrodes were combined with the device Proove'it (Fig. 1b), which over a period of 6 h maintains a constant voltage of 60 V, a scheme is shown in Figure 1c.

Every 30 minutes during the test value of current intensity was read off. On this basis, according to the formula (1), volume of flow of electric charge Q was set:

$$Q = 900\left(I_0 + 2I_{30} + 2I_{60} + \cdots + 2I_{300} + 2I_{330}\right) \tag{1}$$

where Q—charge that has passed through the sample (C), I_0—initial current intensity (A), I_t—current intensity after t time (in minutes) (A).

Criteria for evaluation of chloride ion permeability acc. American standard ASTM C 1202-05 are shown in Table 1.

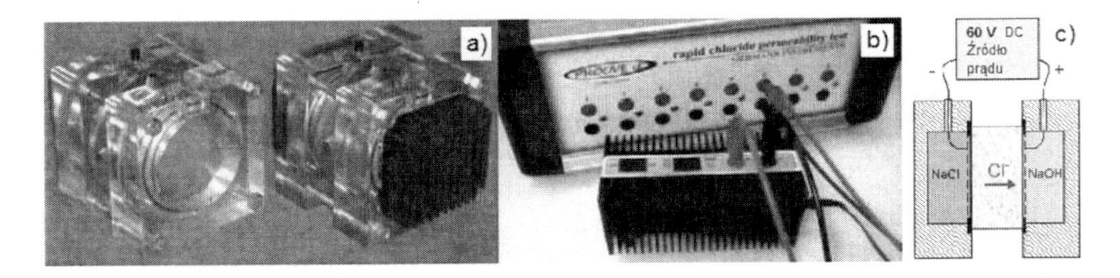

Figure 1. a) Samples placed in between electrodes, b) measuring device Proove'it, c) chloride ion permeability measurement scheme (nr).

Table 1. Criteria for evaluation of chloride ion permeability (ASTM C 1202-05).

Value of the flowing charge (C)	Chloride ion permeability
>4000	High
2000–4000	Medium
1000–2000	Low
100–1000	Very low
<100	Negligible

2.2 *Resistance to sulphates attack*

The test of sulphates aggression resistance was carried out on the base of method acc. Polish standard PN-B-19707 "Cement. Special cement. Composition, specifications and conformity criteria". This method allows to determine the resistance of cement to the sulphates ions attack by measuring the change in length of samples (expansion) and the observation of destructions (cracks, scratches, efflorescence, leaks, discoloration, scaling, etc.) on the surface of samples and compared to samples cured in deionized water.

For each cements tests were carried out on 6 beam samples made of standard mortar with dimension of 20 × 20 × 160 mm (beams with metal ball pins). Samples were prepared acc. standard procedure PN-EN 196-1, although mortar was compacted only by 10 shakes (instead of 60) and placed in form in a single layer (instead of two layers). After demoulded (after 24 or 48 hours) samples were cured in deionized water up to 28 days. Then initial length l_0 of each beam was measured with use of Graff-Kaufmann apparatus. Afterwards 3 beams were placed in deionized water, and the rest in solution of Na_2SO_4 (SO_4^{2-}—ions concentration was 16 ± 0,5 g/dm^3). Every 28 days samples length l_t was measured and changes of external appearance were observed by the time of next 52 weeks. Deionized water for samples curing was not changed by the whole period (only added to keep stable level), while sodium sulphates solution was replaced after every single measurement. Linear changes Δl_t for each beam were calculated in time t. On the base of this average length changes were determined for samples cured in deionized water Δl_t (H_2O) as well as for those cured in sodium sulphates solution Δl_t (Na_2SO_4). Expansion value was calculated acc. formulas (2) and (3).

$$\Delta l_t = \frac{l_t - l_o}{160} \cdot 100\% \tag{2}$$

$$X_t = \Delta l_{t(Na_2SO_4)} - \Delta l_{t(H_2O)} \tag{3}$$

where Δl_t—length change in time t (%), l_o—initial length measured before placing sample in solution (mm), l_t—length in t (mm), 160—nominal length of beam sample (mm), X_t—expansion in time t (%), Δl_t (Na_2SO_4)—average length change in time t for samples cured in Na_2SO_4 solution (%), Δl_t (H_2O)—average length change in time t for samples cured in water (%).

Cement is acknowledged as resistant to sulphates attack if expansion measured in time of 52 weeks is lower than 0,5%. In addition no cracks, fractures, colour changes, efflorescence on the surface of tested samples can be observed.

2.3 *Carbonation depth*

Resistance to carbonation was determined by measurement of carbonation depth acc. procedure described in draft of European standard prEN 12390-12 "Testing hardened concrete—Part 12: Determination of the potential carbonation resistance of concrete. Accelerated carbonation method". Test was carried out on standard mortar samples prepared acc. PN-EN 196-1 (acc. standard draft concrete samples should be used). Rectangular samples with dimension of 40 × 40 × 160 mm were therefore tested.

Designation was made after determined time of curing samples in laboratory conditions in water in temperature of 20 ± 2°C. After reaching certain age mortar beams were broken in order to get two samples. Halves of beams received that way were kept in air-dry condition (temperature 18–25 °C, humidity 50–65%) for 14 days. Then one of them was placed in CO_2 chamber, where CO_2 concentration were 4 ± 0,5%, temperature −20 ± 2°C, and humidity −55 ± 5%. Time of exposure to the CO_2 was 56 days. In the meanwhile second sample (half of beam) was still kept in air-dry conditions, as a reference sample.

After 56 days of exposure to CO_2 each of two samples (halves kept in chamber and reference one) were broken again to receive fresh fracture surface. Then this surface was sprayed with 1% solution of phenolphthalein and left in air-dry conditions. After 1 hour carbonation depth was measured (distance between the edge of sample and border of characteristic pink (purple) colour of phenolphthalein). Measurement was made for each edge in 5 points with use of electronic slide calliper, for greater accuracy (0,1 mm). Then average carbonation depth was calculated for single edge, and then average from all 4 edges, acc. formulas (4) and (5):

$$d_{k1} = \frac{\sum(d_{1-5})}{5} \tag{4}$$

$$d_k = \frac{\sum(d_{k1-4})}{4} \tag{5}$$

where: d_{ki}—average carbonation depth for single edge (mm), $d_{1\ 5}$—single measurement for one edge (mm), d_k—average carbonation depth (mm), $dk_{1\ 4}$—carbonation depth for one edge (mm).

3 COMPONENTS CHARACTERISTICS AND COMPOSITION OF TESTED CEMENTS

Cements for research purposes were prepared by homogenization in established proportion (Table 4) ordinary Portland cement CEM I 52,5R, ground granulated blast furnace slag (S) with calcareous fly ash (W, W+) or siliceous fly ash (V). Calcareous fly ash was used in two forms: raw—supplied upon reception from electrostatic precipitator hopper (W) and mechanically activated (W+) by grinding in laboratory ball mill. Total content of ordinary Portland cement CEM I 52,5 in testes cements was established at the level of 20%. Gypsum as a controller of setting time were added in an amount to obtain a stable content of SO_3, at the level of 2,81%. Physical properties of cements components are presented in Table 2, while chemical composition in Table 3.

4 TEST RESULTS AND DISCUSSION

In the first stage of research program physical and mechanical properties of tested cements and mortars were evaluated. Results are presented properly in Table 5 and in the Figure 2.

Table 2. Physical properties of cement components.

Properties	CEM I 52,5 R	Granulated blast furnace slag (S)	Calcareous fly ash		Siliceous fly ash (V)
			Raw (W)	Ground (W+)	
Density (g/cm^3)	3,07	2,93	2,55	2,71	2,22
Fineness (cm^2/g)	4410	4160	1900	4700	2800

Table 3. Chemical composition of cement components.

Component	Component content (% mass)									
	LOI	SiO_2	Al_2O_3	Fe_2O_3	CaO	MgO	SO_3	Na_2O	K_2O	Cl^-
CEM I 52,5R	2,80	20,05	5,35	2,61	63,42	1,46	2,81	0,18	0,86	0,071
Granulated blast furnace slag (S)	–	37,63	6,84	1,48	45,63	5,33	0,08	0,55	0,56	0,053
Calcareous (W) fly ash	2,67	45,17	20,79	4,58	20,60	1,49	2,96	0,23	0,19	0,001
Siliceous (V) fly ash	1,95	53,25	25,05	6,65	3,86	2,78	0,42	1,11	3,25	0,008

Table 4. Tested cements compositions.

Component	PW1	PW2	PW1+	PW2+	PV1	PV2
CEM I 52,5R	20	20	20	20	20	20
Granulated blast furnace slag (S)	50	30	50	30	50	30
Calcareous (W) fly ash—raw	30	50	–	–	–	–
Calcareous (W+) fly ash—ground	–	–	30	50	–	–
Siliceous (V) fly ash	–	–	–	–	30	50

Table 5. Physical properties of tested cements.

Cement	Water demand (%)	Initial setting time (min)	Soundness (mm)	Fineness (cm^2/g)	Density (g/cm^3)
CEM I	33	190	1	4400	3,07
PW1	32	505	0	3800	2,85
PW2	33	470	2	3500	2,76
PW+1	32	500	1	4200	2,86
PW+2	34	440	0	4200	2,80
PV1	27	470	1	3800	2,68
PV2	26	610	1	3500	2,52

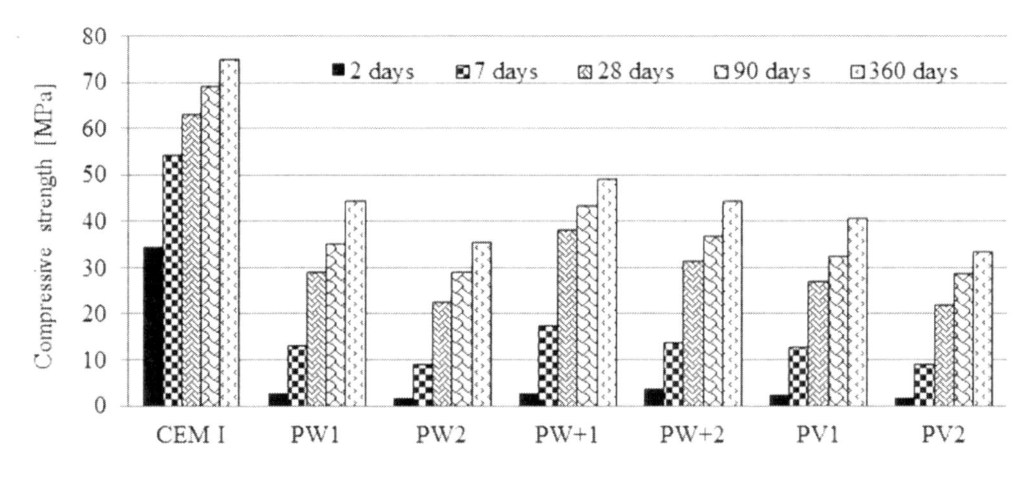

Figure 2. Compressive strength of mortars.

Water demand of cements containing calcareous fly ash (W, W+) is formed at the level determined for ordinary Portland cement CEM I 52,5R (33%). While cements containing siliceous fly ash (V) are characterized by lower water demand (about 4–5%). For all tested cements initial setting time elongation was observed in comparison with ordinary Portland cement CEM I 52,5R. In case of cements containing calcareous fly ash (W, W+) initial setting time was shorter when fly ash content was higher. Reverse relation was observed for cement with siliceous fly ash (PV1, PV2). Requirement for soundness (≤10 mm acc. PN-EN 197-1) was fulfilled by all tested cements.

Compressive strength is the primary parameter which characterize cement (strength class), and in parallel an important criterion to evaluate its quality. In the Figure 2 results of compressive strength tests of cement mortars are presented. Tested cements are characterized by very low initial strength (after 2 and 7 days of curing) and considerable increase of strength in longer terms. That is why a decision on evaluating compressive strength after 90 and 360 days of curing was made up. Falling tendency of strength in function of fly ash content in cement was observed. Compressive strength of mortars with cements containing siliceous fly ash (PV) is formed at the same level as with cements incorporating raw calcareous fly ash (PW). Cements with activated calcareous fly ash (ground PW+) are characterized by the highest compressive strength of all cements in every tested term. It speaks volumes for significant activity of this additive and for visible synergic effect in the system of ground calcareous fly ash—ground granulated blast furnace slag. This effect might be connected with modification of hydration process, and thus with modification of cement matrix microstructure. It should be pointed that due to some difficulties with proper compaction, mortars with cement incorporating 50% of calcareous fly ash were formed at w/c ratio equal 0,6 when raw fly ash was used, and 0,55 when activated one was an ingredient of cement.

Hardened mortars made of tested cements after 28 days of curing were characterized by low (2000 C) and very low (1000 C) chloride ions permeability. However it was observed that while fly ash content in cement is higher resistance to chloride ion penetration is lower. This effect is particularly visible in case of mortars made with cement containing raw calcareous fly ash (PW1, PW2). By comparison, ordinary Portland cement (CEM I52,5R) mortar was characterized by high chloride ions permeability (passing charge value over 4000 C). After 90 days of curing all of tested cements (beyond CEM I 52,5R) met the criterion of very low permeability (passing charge lower than 1000 C). Resistance improvement observed in longer term might result from puzzolanic (and/or hydraulic) activity of fly ashes and granulated blast furnace slag. Those reactions products tighten cement matrix structure, which became less porous, and therefore more hermetically to aggressive media.

In Figure 4 expansion quantity of cement mortars beam samples held in sodium sulphates solution are presented.

On the base of received tests results it was claimed that fly ash—slag cements mortars are characterized by definitely higher sulphates attack resistance than a reference mortar made with ordinary Portland cement CEM I 52,5R, which does not meet the criterion of sulphates resistance HSR acc. Polish standard PN-B 19707 (maximum acceptable expansion level after 52 weeks lower than 0,5%). Moreover, in case of ordinary Portland cement mortar beam samples were intensively fractured (Fig. 5), while none of the rest tested samples has shown any cracks, by the whole testing period (Fig. 6).

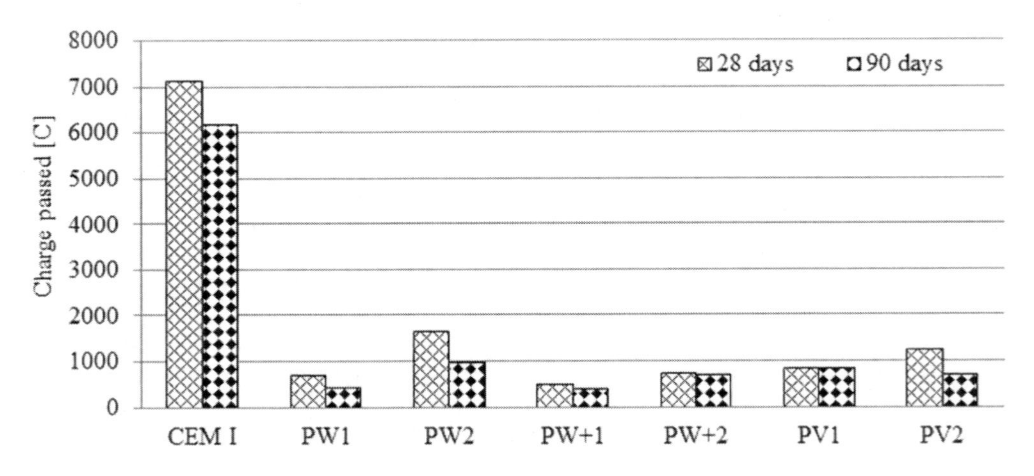

Figure 3. Chloride ions permeability test results.

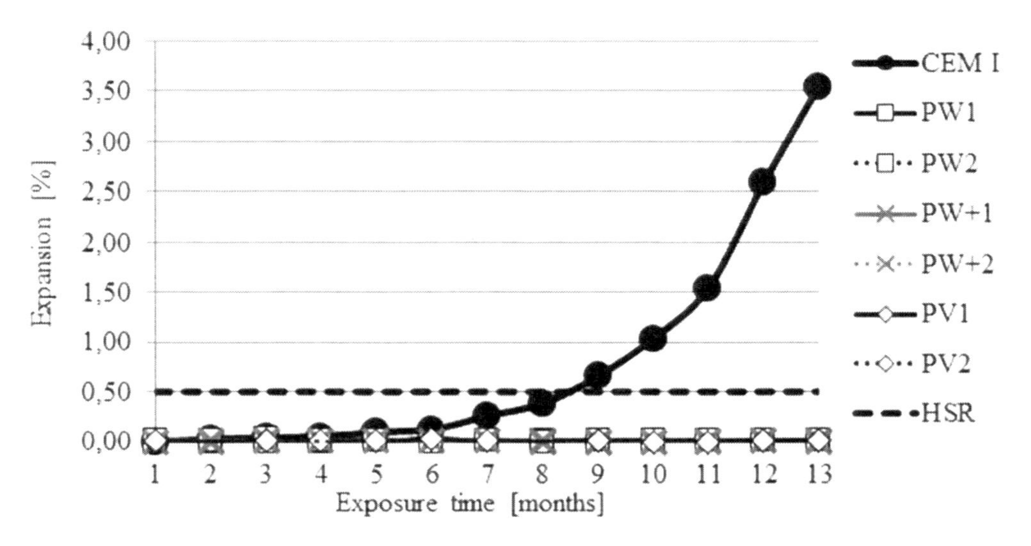

Figure 4. Expansion of mortars caused by sulphates attack.

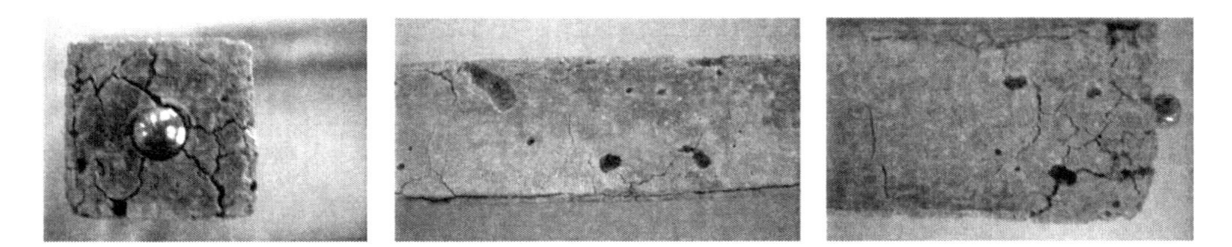

Figure 5. Reference beam samples made of standard mortar (CEM I 52,5R) after 52 weeks of exposure on sulphates attack (large expansion and visible destruction—numerous cracks and fracturing).

Figure 6. Tested beam samples after 52 weeks of exposure on sulphates attack (no signs of destruction).

Carbonation depth depends on the cement composition and the age of hardened mortar. Thus in result it depends on permeability of cement matrix. For all tested cements it was observed, that higher content of fly ash (both calcareous and siliceous) in cement composition results in greater range of carbonation depth. After 28 days of curing the most resistant to carbonation were mortars with cements PW+ (Fig. 7a). Carbonation depth in mortars with cements PW and PV was observed at the very close level. Similar trends were noticed after 90 days of curing (Fig. 7b), although the range of carbonation was observed much smaller. This should be associated with the effect of tightening matrix microstructure with pozzolanic—hydraulic reaction products. Reference mortar made with ordinary Portland cement CEM I 52,5R, tested in parallel, showed no signs of carbonation in any tested terms.

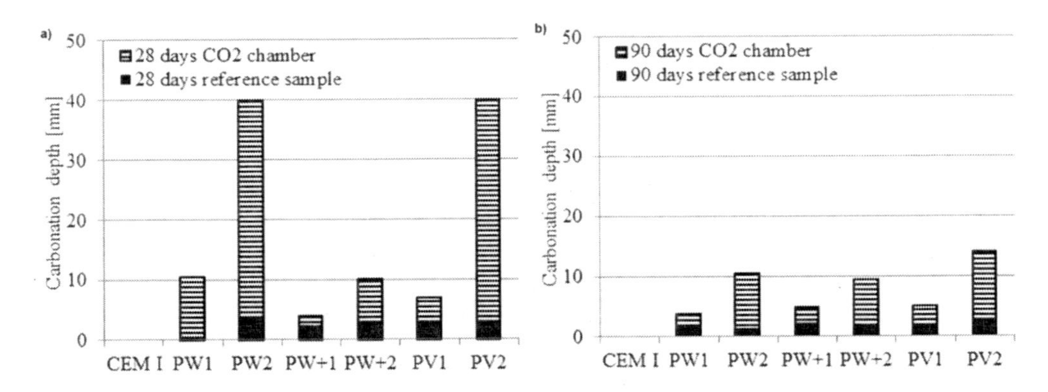

Figure 7. Carbonation depth of mortars a) after 28 days curing, b) after 90 days curing.

5 SUMMARY

Received test results analysis leads to conclusion, that use of multicomponent cements incorporating flay ash (siliceous or calcareous) and ground granulated blast furnace slag results in improvement resistance of cement composites to aggressive liquid media but negatively affects on gases permeability (CO_2).

Following studies results, it was observed that mortars with fly ash—slag cements are characterized by higher resistance to water solutions of chlorides and sulphates attack, although there are more susceptible to carbonation in comparison with ordinary Portland cement CEM I. Moreover it was found, that in longer-terms cement matrix of tested mortars is much denser, especially when the activated (ground) calcareous fly ash was a component of cement. As a result favourable effect on compressive strength as well as durability of cement composites was observed.

REFERENCES

C 1202-05 *Standard Test Method for Electrical Indication of Concrete's Ability to Resist Chloride Ion Penetration.*

Chindaprasirt, P. & Rukzon, S. & Sirivivatnanon, V. 2008. Resistance to chloride penetration of blended Portland cement mortar containing palm oil fuel ash, rice husk ash and fly ash. *Construction and building materials* 22 (2008): 932–938.

Giergiczny, Z. & Garbacik, A. & Drożdż, W. 2011. Synergic effect of non-clinker constituents in Portland composite cements *Cementing a sustainable future. XIII International Congress on the Chemistry of Cement*, Madrid, 2011 p. 49.

Giergiczny, Z. 2013. *Fly ash in cement and concrete composition (Popiół lotny w składzie cementu i betonu)*, Gliwice: wyd. Politechniki Śląskiej (*in Polish*).

Glinicki, M.A. & Jóźwiak-Niedźwiedzka, D. & Gibas, K. 2010. Evaluation of permeability of concrete containing high-calcium fly ash—the research concept. *Energy and environment in technology of building, ceramic, glass and refractory materials. Warsaw—Opole may 2010.*

Hauer, B. & Mueller, Ch. & Severins, K. 2010. New findings concerning the performance of cements containing limestone, granulated blast furnace slag and fly ash as main constituents. *Cement International* Part 1–3/2010: 80–86; Part 2–4/2010: 83–93.

PN-B-19707:2013 *Cement. Special cement. Composition, specifications and conformity criteria.*

PN-EN 196-1:2006 *Methods of testing cement. Part 1: Determination of strength.*

PN-EN 196-3:2006 *Methods of testing cement. Part 3: Determination of setting times and soundness.*

PN-EN 197-1:2002 *Cement—Part 1: Composition, specifications and conformity criteria for common cements.*

pr-EN 12390-12:2010 *Testing hardened concrete. Part 12: Determination of the potential carbonation resistance for concrete. Accelerated carbonation method.*

Schneider, M. & Romer, M.& Tschudin, M. & Bolio, H. 201.1 Sustainable cement production—present and future. *Cement and Concrete Research* 41(2011):642–650.

Underground Infrastructure of Urban Areas 3 – Madryas et al. (Eds)
© 2015 Taylor & Francis Group, London, ISBN 978-1-138-02652-0

Trenchless pipe construction: Benefits by hard and soft evaluation criteria

H. Görg & A. Krüger
Fachgebiet Abwasser- und Abfalltechnik, Universität Siegen, Siegen, Germany

ABSTRACT: Pipeline systems are an essential part of civilian infrastructure. The different methods of pipe rehabilitation contain many basic aspects, special details and unfortunately often some technical limits for a practical use. Although engineers have to know all technical standards, their decisions are often led by efficiency criteria. The traditional open coverage type implies extensive construction sites with use of trucks, excavators and wheel loaders. Furthermore the extent of road rubble, soil and basement excavation affects the process of cost-efficient pipe laying. There are other growing advantages, especially in the eyes of a resident-friendly and sustainable society. With less construction traffic, less machine noise and less emission of hazardous substances the almost invisible working trenchless constructions protect the environment from harmful development. The technical facts and the way, in which some of the soft and hard criteria are arguing for a broad practice, are the main subject of the article.

1 INTRODUCTION

In the last decades the trenchless technologies have gained more and more relevance for the construction of the underground infrastructure. In all the parts of civilian life the trenchless applications have improved their technical performance by many innovations, e.g. in the fields of

- sewers for domestic and storm water,
- pipes for water supply,
- gas pipelines,
- district heating,
- cables for the electric power distribution,
- wire for the telecommunication network,
- ...

In each case the extent of trenchless building is different. There are some distinguishing characters, mostly caused by the transport medium. With reference to the pipe material, the construction basics show some close similarities.

Since a few years the discussion on tight sewages systems and the influence of leaky pipes for the environment has increased. The condition of the sewers intensifies the demand for prompt sanitation, especially in private ground. The exit from nuclear and fossil-fuel energy will require a suitable network for the green energy just as the supply with resources like drinking water should be assured by a working pipeline system in future. Nowadays all procedures of civil engineering have to be regarded under economic aspects.

Globalization encourages the need of a holistic, integrated approach to environmental questions. With regard to the prospective global challenges, which are expressed by the demographical process, climate change and the scarcity of resources, sustainable solutions are requested in all parts of the world, in order to protect further generations from a negative development.

Figure 1. Modern infrastructure in urban terrains (TT, 2013).

2 TRENCHLESS TECHNOLOGIES

2.1 *Gravity pipe systems for sewage*

The sewerage system of the federal republic of Germany consists of more than 500,000 km of public pipes. However the length of private pipelines is twice as much (Berger & Falk 2009). The age, function and quality are heterogeneous. Damages with various features and for various reasons and to a different extent and different consequences induce the operators, to correct the situation sooner or later. The question of tightness is a central point for exfiltration and infiltration through pipeline systems. According to a broad analysis (Berger & Falk 2009) almost 20% of the German public sewerage system is in need of rehabilitation. A much higher damage potential is ascertained in the private service sewers, within an estimated length of three times more than the public sewers. Worst case scenarios (Diederich & Rehling 2005) assume that more than 70% of the German private drainages might be leaky. However the specifications (Depth, base slope, position and accessibility, etc.) of the domestic sewers represent certain constraints with limiting factors.

According to the DWA standards, published by the German Association for Water, Wastewater and Waste DWA (cf. (DWA 2014)), the rehabilitation of sewers is subdivided in:

- repair
- renovation
- renewal (replacement).

The repair is defined as a removal of local pipe damages. There are different methods to repair a broken sewer with manual or with robot systems using a kind of injection, coating or sealing process (epoxy resin, short liner, inside-sleeves). However the choice of the repair technique depends on the accessibility of the damage. Only minor damages can be repaired. On the one hand the repair devices offer an improvement of the sewer status for very attractive terms, but on the other hand they only guarantee a temporary non-durable success. If the maximum load in the meaning of static charge fails, a complete renewal will be unavoidable.

Renovation stands for the improvement of the actual viability of sewers by using the original substance partially or completely. The methods of sewer renovation are normally initiated from a manhole or a launch pit. The covering methods use materials to coat the old pipe, like cement mortar. The lining methods use pipes or hoses to substitute the former pipeline system. Lining methods depend on the locations of the procedures. New pipe systems, which are produced at the factory (short pipe modules, long pipes, spoiled folded pipes on a drum) vary from those which are constructed at the construction site immediately before the placement (spiral-wound pipe lining and the cured in place UV-pipelining). Those tight lining methods without any annular gap between the new and the old pipe (close-fit) enable a much easier reconnection of the laterals. The reduction of the diameter through the liner´s wall thickness might influence the hydraulic capacity of the sewer (WAVIN 2006).

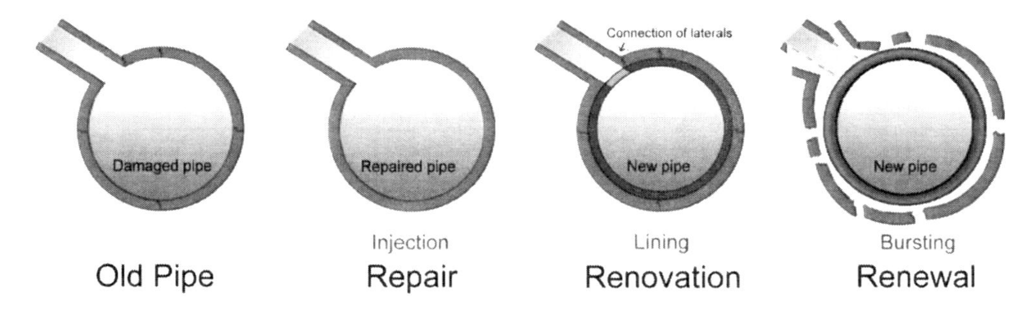

Figure 2. Definitions of sewer rehabilitation (TT 2013; modified).

Renewal (replacement) defines the construction of a new sewage pipeline in an existing or new path, without involving the function of the original sewage pipe. Techniques of renewing are possible in closed coverage and open cut construction types. The open cut replacement can always use a new route. Trenchless reconstructions in new line are possible with uncontrolled and controlled drilling methods like Horizontal Directional Drilling (HDD), micro tunneling, pipe jacking or pipe ramming.

Trenchless technologies using the existing line of the former pipelines use static or dynamic bursting and the pipe-eating. Manholes and most often launch and exit pits are needed to pull or push the new pipe through the host pipe. By pulling the bursting head along the pipe the host material is smashed to many pieces. Shards and fragments are displaced to the betting. Immediately after the process of bursting the process of lining the new pipe is initiated. Only pull-resistant materials such as ductile cast iron or plastics (PE) are qualified as materials for the new pipe (Rameil 2006). The bursting process enables an enlargement of diameter. Thus it is possible to raise the flow capacity. However the mentioned replacement devices destroy the laterals, so that pits (keyholes) in open coverage types are required to restore the junctions. Instead of restoring the laterals along the pipeline route, laterals can also be centralized at the manholes. Similar to burst-lining the technique called pipe eating cracks the old pipe.

2.2 *Pressure pipe systems for water and gas*

Using other technical standards and other technical terms all devices of sewer rehabilitation in principle can be used for pressure pipes, too (DVGW 2006). However there are some typical limitations caused by the worse accessibility to pressure pipes and the medium itself. In contrast to the sewer systems the water and gas net do not have manholes for entering the conduits. The rehabilitation of water pipes must strictly not affect the drinking water quality. Working with gas lines requires a secure site management in order to avoid accidents. The system is run under pressure. Therefore the question of tightness is essential for a risk free operation and there may be a small loss of the transport medium and its pressure.

Due to the small diameters and some clearance concerning the slope the trenchless technologies are qualified very well for pressure pipes, especially in the category of replacement. These circumstances favour especially the trenchless bursting method for pressure pipes. Although not guaranteeing an accurate basic slope, the HDD-Drilling and especially the impact mole for service pipes have been proved and tested many times in the trenchless practice.

The impact mole is often used to construct new service pipes to connect private homes trenchless to the public network. New innovative technologies such as the keyhole-boring use modified variants of the impact mole or the HDD-Drilling to restrict the excavations as much as possible.

2.3 *Electricity grids and telecommunication network*

This branch offers much potential for the future. There is a high demand for a supply of fast internet on the basis of light-wave cables (FFTH-Fibre to home, smart grid). The new German

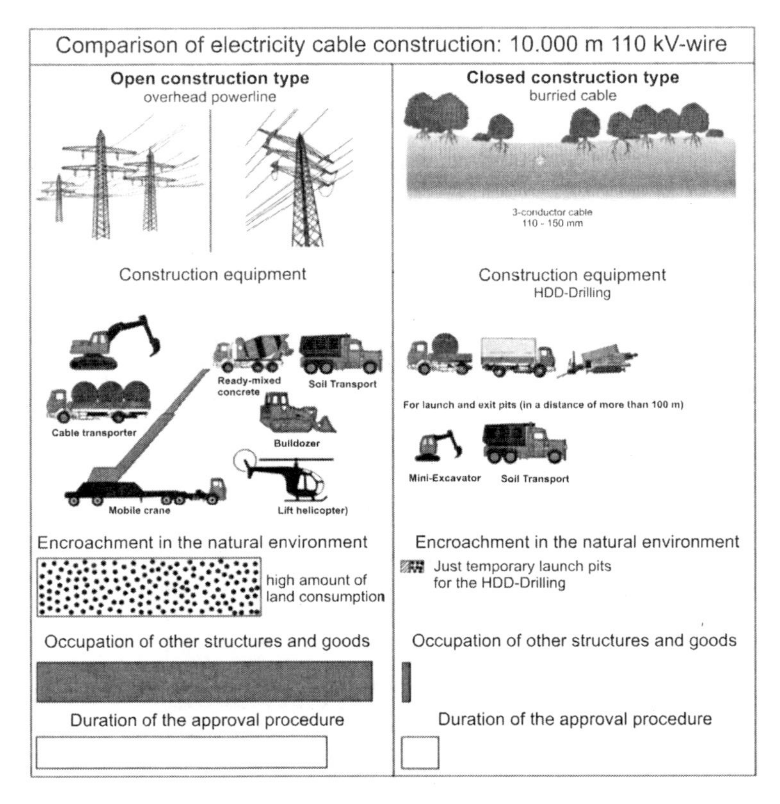

Figure 3. Comparison of electricity cable construction (Bayer 2011; modified).

energy policy promotes renewable energy systems. The design and the construction of onshore and offshore wind farms will entail transport systems for delivering the energy to the consumer.

For the laying of electric cables the non-disruptive method stands in competition with the open building method and with overhead power lines. For telecommunication and electric cables with low or medium voltage underground cables are suitable. Concerning the transport of high-voltage-electricity there seems to be no alternatives to the transport using the overland-aerial cables and utility poles. In connection with the development of the electrical power supply, the adequacy of underground cabling for the transport of power in the range of high voltage is discussed increasingly, if the interests of citizens meet the planning of the electricity supplier. The appearances of the landscape and the issues of flora and fauna and their habitats have to be observed. Figure 3 shows a comparison of possible electricity cable constructions.

However innovations in order to shield the environment from the inductive magnetic fields of electric underground cable are necessary, for example by using new materials for the ducts and cables. This should make the power-transport via underground cable more interesting in the future. The ploughing with the cable plough and also the HDD horizontal directional drilling seem to be practical for the underground construction of cables (Bayer 2011). Existing power lines can be renewed by over drilling in their existing line.

3 THE BENEFITS OF TRENCHLESS TECHNOLOGIES

3.1 *Decision finding*

The local conditions and the constraints on the construction-site will give a clear indication, if a closed or an open coverage type is practicable and efficient. Knowing the basically features of sewer rehabilitation (cf. chapter 2.1) the planners should follow the technical standards. The standards of the DWA contain a decision guideline (Figure 4). However the flowchart does not include the details of each repair-, renovation- or renewal-technique. Under the reflection of typical exclusion criteria (cf. chapters 2.2, 2.3) the question should be asked if a similar strategy is possible/reasonable or not, seems to be expedient for pressure pipes too.

3.2 *Economical efficiency*

Efficiency analyses are very important in all areas of civil engineering. Profitability is described in different ways and the consequences for the participants of construction areas are also various. The awarding authority wants to control the budget with cost saving during the planning phase. The contractor tries to control his budget. Large construction projects need a very detailed planning supported by good cost management. As part of preliminary studies the estimate of costs enables transparent decision making. Preliminary studies often contain dynamic comparative cost methods. This allows the economic evaluation of the costs of different plan variants, in order to consider which variant is most successful. The comparative cost calculation method (e.g. according to the LAWA-Commission "Bund/Länder-Arbeitsgemeinschaft Wasser"—guideline (DWA 2012)) is aiming at the total cost amount of each variant during the given time period. The comparative cost method can consider different scenarios regarding the economic efficiency over the planning horizon.

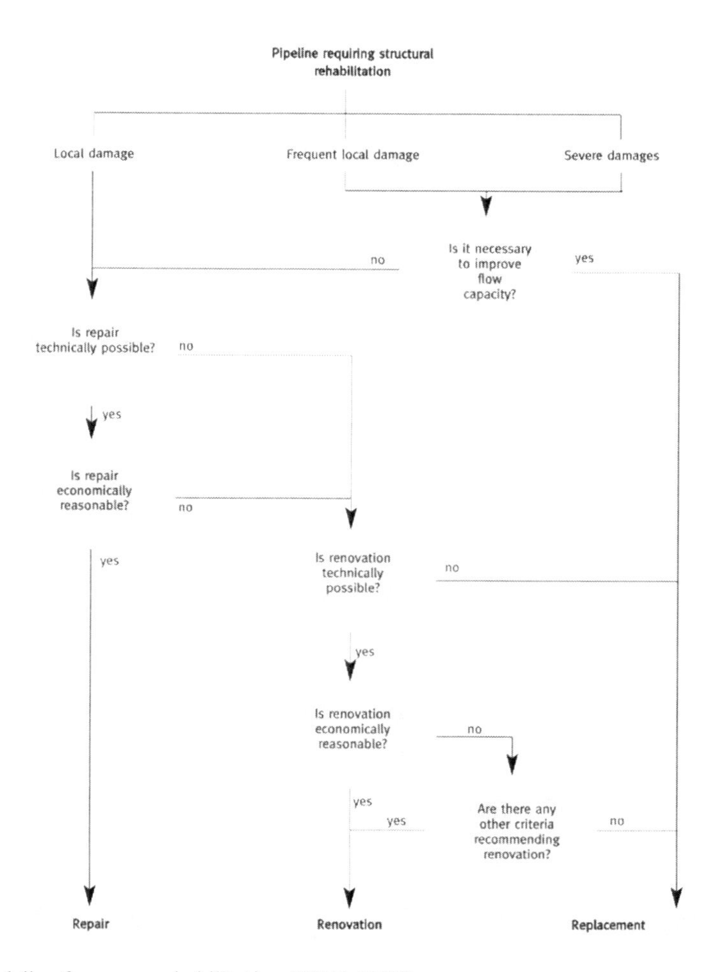

Figure 4. Guideline for sewer rehabilitation (DWA 2013).

The technical and economic life of a product has a high impact for the rating of costs. Neither the economic nor the technical lifetime of repair, renovation and replacement is equal. Therefore the comparative cost method includes the repetition of rehabilitation method as well as a possible combination of rehabilitation devices. Using the financial mathematics all cost flows (first-time investment, reinvestment, running costs) are brought to an equal standard of comparison. Future capital investment for restocking of the asset has to be calculated by discounting the interest. The price inflation has to be taken into account.

The technical suggestions for the life of replacement, renovation and repair devices are defined by the LAWA-Commission (DWA 2012). However the survey of the DWA-Organization under German municipal sewerage operators presents a different picture (Berger & Falk 2009). This survey contains more or less the practical experiences and the technical appraisal of the different technologies.

The attainment of the intended life depends mostly on the construction quality. Figure 5 shows the results of a comparison analysis between the different technologies of sewer rehabilitation.

The break-even point shows the point of time, when the most efficient technology becomes viable. In this case the devices of renovation seems to be most efficient after a use period of 12,5 years. However each job site is a unicum, for that reason general statements are difficult.

Besides the general choice between the repair, renovation and replacement the detailed selection between the different devices must be made. (cf. chapter 2.1) If the decision favours the long lasting replacement technologies the replacement can either be realized by closed or by open coverage type. The cost comparison of the open coverage construction type and the trenchless technologies is affected by the site conditions like (cf. (Görg et al. 2013)).

- the depth of the groundwater table,
- the surface (roads and paths pavement),
- the working panel (places of tree population, water, buildings, boundary, ...),

Table 1. Technical life of sewer rehabilitation (DWA 2012), (Berger & Falk 2009).

	Life according to LAWA-guideline	Life according to DWA-survey
Repair	2–15 years	23 years*
Renovation	25–40 (50) years	47 years*
Renewal	50–80 (100) years	82 years*

* Average mean under the different devices.

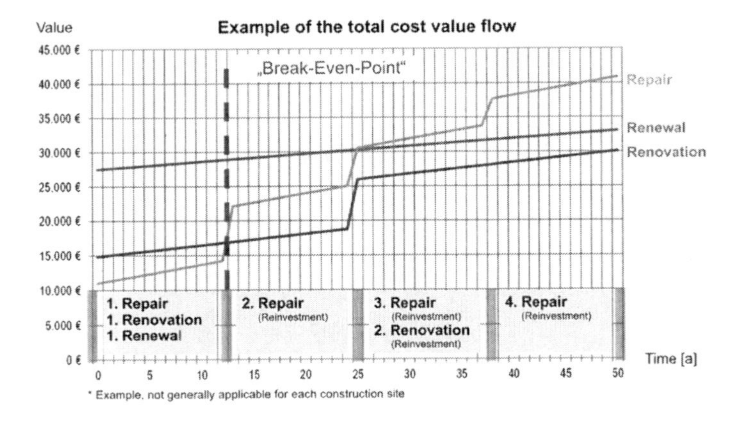

Figure 5. Result of a cost comparison (Görg et al. 2013).

- the soil conditions, the existence of contaminated liabilities,
- the pipeline system (with its depth, position, profile shape, basic material and diameter, manhole distance and installation like laterals, junctions, fittings, ...),
- the existing infrastructure (close-by pipes, connections, receiving waters, ...),
- the building yard equipment (accessibility of the site, traffic regulation and so on ...).

According to many examples (e.g. Reineck 2006) trenchless construction often entails high efficiency.

3.3 *Sustainable efficiency*

New global challenges such as the protection of the climate, the change of demography and the scarcity of natural resources must be taken into account during the decision making process which follows more criteria than a purely economic view of aspects.

Sustainability describes the achievement of goals, which meet the needs of the present without compromising the ability of future generations to meet their own needs (cf. Definition of the Brundtland-Commission, (Brundtland Report 1987)). Economy, ecology and social responsibility should be consistent with each other containing aspects of feasibility, affordability and sustainability. The classic tools for evaluating sustainability like point systems or marks for single criteria are not that useful in the area of pipe construction since the results are less objective and often described by a lack of provable and convincible data.

Over the past few years sustainable benefits are expressed by cost calculations, the so-called indirect costs (also defined as "social costs", "society costs", "opportunity or abatement costs"; cf. (Schmidt 2010), (Hölterhoff 2011) (Scheuble 2012)). Indirect costs quantifying sustainability are becoming increasingly important as aspects of sustainability are gaining importance. (cf. (GSTT 2011)). Indirect costs are often consequential costs, not occurring at the place (job-site), where they are produced and not attributed to the cost producer (GSTT 2011). They only become relevant in an aggregate calculation. If anything, the price is paid by the community without any monetary relevance for the contractor of the site (Hölterhoff 2011). The collective good is a definition in the German water rights, describing the prevention of harmful environmental pollution. So far this definition does not include financial charges for the general public.

Indirect costs often lead to a loss of value or to a decrease of quality. Unfortunately these costs, which are paid by the society (state, city, citizens), and only become obvious at a later time and at a different place. The main problem is the fact that they are difficult to quantify in monetary terms.

But it is possible to monetize the indirect costs, if equipment hours, construction times and the additional fuel consumption due to traffic jams and/or detours are analyzed (GSTT 2011). Thus emission balances can be drawn. However the extent of indirect cost differs from site to site.

There are several examples for indirect costs (cf. (Görg et al. 2013), (Hölterhoff 2011) (RSV 2011)), specified as

- Costs for environmental preservation against emissions as greenhouse gases, fine dust and noise
- Traffic jams because of construction sites on roads
- Residence annoyance for citizens and commerce, also loss of tourism
- Disturbance of an intact road surface with secondary damage on roads and consequential damage on cars
- Ecological costs for subjects of protection (subsidence/damage on buildings, cultural heritage, historical monuments, Curtailing of flora and fauna, water, soil, air)
- Costs of public health service by site accidents or site caused diseases (physical or mental long-term effects); risks of pollutants, noise, ...
- Loss in value on property (e.g. public easement like power poles), including short-term or permanent disturbance of the image of the landscape due to construction sites or buildings.

In the following some of these indicators are considered in terms of their relationship to indirect (social) cost and their favorability for trenchless applications.

In the context of climate change with a global warming, emissions of harmful greenhouse gases are seen increasingly critical. Besides carbon dioxide, fine particles and noise are typical construction site emissions. In the sum of the emissions, the trenchless construction methods perform themselves significantly better than the open cut. The open-cut features an intense material and equipment use, while the closed coverage type in an ideal case entirely renounces excavators, bulldozers, street rollers, transport trucks and road finisher. Open-cut means extra costs for removal, transport and disposal of the cracked road, the excavated soil and the old pipe material. In addition, the transport and installation of new building materials for bedding, backfill and road restoration have to be taken in account. This is all associated with significant emissions. Trenchless technologies shorten the construction period. It is clear that the shorter construction has an impact on the replacement costs of equipment and thus also on the direct construction costs. However less construction time reduces the duration of emissions. Studies assume that the building processes-related carbon dioxide CO_2-emissions from open cut job sites are 3-times higher than those of the trenchless sites (GSTT 2011). Figure 6 shows the results of a scientific study from Austria, which comes to a similar conclusion.

In the construction zone the operator is responsible for secure traffic guidance during the whole construction time. For the purposes of the German Road Traffic Act, the participants of the traffic have to behave in such a way not to harm or put in danger any other person and avoid any perturbation or hindrance above everything inevitable. Because of the cramped construction site this is with the open-cut in urban space not always possible. Traffic lights and one lane construction crossings are often part of urban sites. The side entries and side exits can also restrict the mobility of road users. Construction sites with restricted driving space carry a high risk of danger, especially as the dimensions of cars are becoming bigger. In the construction sites, the speed is usually limited and overtaking is strictly forbidden. Traffic jams with an increased risk of accidents (rear-end collision) by starting and stopping processes or lane change (zipper system) cost time and nerves. They also lead to increased fuel consumption and increased emissions of carbon dioxide. Congestion leads to gridlock and the detriment of all road users. Modern navigation systems or diversion panels can minimize congestion. However, a longer-lasting traffic guidance measure leads to circumstances where the regular traffic uses a large-scale bypass, with greater cost and additional emissions due to higher fuel consumption.

The growing public participation seen at the major German projects Stuttgart 21 (railway station), Berlin's new airport and Hamburg opera have shown that citizens' interests influence more and more construction projects and environmental strategies. Resistance starts first among those, who are directly affected. In many ways residents are affected by construction sites. Emissions of pollutants, dust and noise, restrictions on the accessibility of the plots are good reasons to prefer those methods which perform better in these areas. The citizen has just in its immediate privacy (property) a legitimate interest, to minimize and avoid trouble. The same is true for trade and commerce. Business planning can be a problem if the delivery of merchandise by truck from

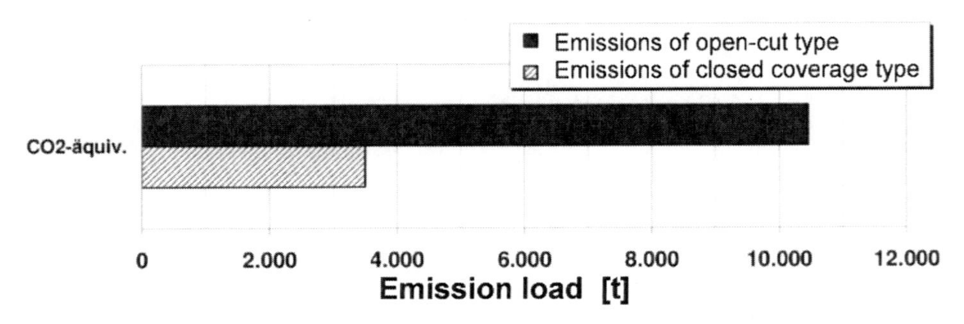

Figure 6. Emissions of greenhouse gases in open and closed coverage type on the example of 500 km of pipes (Schöller & Kitzberger 2011).

the site is obstructed or operations are disrupted. The retail sector relies on frequent clientele (parking) or walk-in customers (sidewalks, shopping areas). Bus stops must be installed at construction sites if necessary. Business location is directly related to the company's success. For that reason, construction sites should not impede the ordinary business activities.

Trenchless techniques can reduce the interference in the functioning road and thus prevent subsequent costs (Krüger et al. 2011). The road status, with potholes and their problems lead to a vehement discussion. In the face of an increasing traffic volume the costs for preserving the streets in good conditions are rising dramatically. In addition to the direct costs to restore the road, damage to roads often become visible at a later time when water seeps into joints and seams or the load of upper and lower construction leads to cracks, holes or ruts. Uneven subsidence and deficient connections of the asphalt surface are especially found in the sector of pipeline route. Regardless of the aesthetic disfigurement of the road appearance this circumstance leads to long term damage patterns. A correlation between road damages and damages to the cars seems to be evident (wear of tires, paint damage from stone chips). A study from the USA estimated that car repairs attributed to poor roads costs $400 per vehicle annually (AGL 2005).

4 CONCLUSION

Generally there are many good and comprehensible reasons to consider the trenchless technologies in all disciplines of piping as the first alternative to the open coverage construction type. In times of tight financial control and growing transparency claims from the public view the process finding and the justification become more and more relevant. Therefore it is a good idea to discuss all the possible construction methods as early as possible and to evaluate accordingly. A reasonable and transparent decision will avoid later discussions.

Although the motives for the use of trenchless operations are different, they have to observe the same quality standards as the open construction types. The trenchless technologies promise much quality, only if they are used in the correct way. To avoid pitfalls engineers should know the limitations for each method. Pitfalls and negative experience might be a reason to reject trenchless building. Therefor it is very important to build trust by a high quality assurance. There are many established technical rules (cf. DVGW-/DWA-Standards). The material technology (multi-layer pipes, drag force resistible connections) matches the execution requirements. This underlines the claim of trenchless building: At least equal to the open cut construction with the same or even higher lifetime. In everyday life many trenchless jobsites reveal perfect results.

All arguments confirm the author in his points of view:

1. Priority: Trenchless, instead of open coverage type, if no technical limits prohibit an efficient execution,
2. Priority: Renewing or replacement before renovation and repair, if the financial circumstances allow better quality.

In addition to the technical selection standards the economic analysis becomes increasingly important. "Profitability" has different meanings for the planning engineer, the awarding authority, the construction companies and even more the society. Economic aspects vary from the private to the public sector with certainly many different interests. Relating to its definition, the sustainability theory covers all aspects of life. According to the construction tendering and contract regulations the evaluation of tenders by the contracting authority the most economical offer catches the construction contract. However, not only the lowest bid price has to be considered for the decision making. For this reason, it seems essential to highlight non-cash aspects.

Because of the difficult monetary valuation, the unclear pollution liability and the so far low socio-political significance soft criteria were underestimated or even ignored in the past. Indirect costs are one instrument to quantify soft criteria. Future sustainability aspects will influence the actions and decisions. Certainly it requires great effort to implement the theory of sustainability in everyday life. For line building the paths of the so called electricity highways, international gas pipelines, composite water network and tight sewerages have to be regarded. The technical innovations of the recent years have contributed to the increase of the application security.

The insistence of a holistic view accompanies the modern environmental engineering. Now it is necessary to point out the sustainability of the outstanding benefits, characterizing the trenchless technologies. The main intention lies in smaller or at least less visible and less disturbing and more resident-friendly construction sites. When facing the future challenges for society soft criteria may become more relevant. If climate protection is meant serious, the hazardous greenhouse gas emissions, which are responsible for the global warming, must be reduced as much as possible, in each branch of daily life. Trenchless technologies do not affect the environment and the climate as much as the open cut type. In compliance with the economic success these soft criteria might argue for a broad use.

REFERENCES

AGL, Atlanta Gas Light 2005. AGL unveils new pipeline repair technology to City of Atlanta Business Wire. Jul.2005.

Bayer, H.-J. 2011. Erdkabel aufwendiger als Fernleitungen? Tracto-Technik GmbH, http://www.pressebox.de. Apr.2011.

Berger, C. & Falk, C. 2009. Zustand der Kanalisation"; Results of the DWA-Survey 2009. In Korrespondenz Abwasser KA, Hennef, 1/2011.

Brundtland Report, 1987. Our Common Future. UN Documents Gathering a Body of Global Agreements. http://www.un-documents.net/our-common-future.pdf.

Diederich, F. & Rehling R. 2005. Status Quo der Grundstücksentwässerung. In TIS 12/2005.

DVGW 2006. W 403-3Technische Regeln Wasserverteilungsanlagen TRWWV. Teil 3: Betrieb und Instandhaltung. Deutscher Verein des Gas- und Wasserfaches, 2006.

DWA, 2012. Leitlinien zur Durchführung dynamischer Kostenvergleichsrechnungen (KVR-Leitlinien). ed. 8, Hennef. 2012.

DWA, 2014. DWA-Merkblatt M 143. Sanierung von Entwässerungssystemen außerhalb von Gebäuden. Teil 1–17; Deutsche Vereinigung für Wasserwirtschaft, Abwasser und Abfall e. V., Hennef, till 2014.

Görg, H., Krüger, A. & Hiemann, P. 2013. Auswirkungen unterschiedlicher Wirtschaftskriterien auf die Bewertung der grabenlosen Leitungserneuerung. In Bauportal, 02/2013. 26–34. Erich Schmidt Verlag GmbH & Co. KG. Berlin. Feb.2013. ISSN: 1866–0207.

GSTT, 2011. GSTT-Information Nr. 11. Vergleich offener und grabenloser Bauweisen—direkte und indirekte Kosten im Leitungsbau. German Society for Trenchless Technology e.V. ed. 3. Berlin. Dec.2011.

Hölterhoff, J. 2011. Kostenvergleichsrechnung als Entscheidungshilfe zum Einsatz von geschlossener und offener Bauweise im Kanal- und Leitungsbau. Presentation. Berliner Sanierungstage, 2011.

Krüger, A., Birbaum, J., Görg, H. & Zander, U. 2011. Zusammenhang von Straßenzustand und grabenlose Leitungserneuerung im Hinblick auf lange Nutzungsdauern und optimale Wertschöpfung. Conference proc. 6. Dt. Symposium für grabenlose Leitungserneuerung SgL. Oct.2011. Universität Siegen.

Rameil, M. 2006. Handbook of Pipe Bursting Practice. Vulkan-Verlag. Essen. 2006.

Reineck, T. 2006. Berstlining-Verfahren in der Großstadt, mit Sicherheit kein Widerspruch. In 3R-International, 9/2006, RSV-Sonderdruck. 25 Jahre Rohrleitungserneuerung mit Berstverfahren". 2006.

RSV, 2011. RSV-Information Nr. 11. Vorteile grabenloser Bauverfahren für die Erhaltung und Erneuerung von Wasser-, Gas- und Abwasserleitungen. Rohrleitungssanierungsverband RSV. Lingen. 2011.

Scheuble, 2012. Direkte und indirekte Kosten als Entscheidungshilfe für die Verfahrensauswahl im Leitungsbau. proc. 7. Dt. Symposium für grabenlose Leitungserneuerung SgL. Sep.2012. Universität Siegen.

Schmidt, T. 2010. Nachhaltigkeit der Leitungssanierung. proc. 5. Dt. Symposium für grabenlose Leitungserneuerung SgL. Oct.2010. Universität Siegen.

Schöller, G. & Kitzberger, J. 2011.Emissionsbetrachtungen im grabenlosen Leitungsbau—Einsparmöglichkeiten gegenüber der offenen Bauweise. In Wiener Mitteilungen (2011), Bd .223. Universität für Bodenkultur Wien.

TT, 2013. Company Information by Tracto-Technik GmbH. http://www.tracto-technik.de. Lennestadt. 2014.

WAVIN, 2006. Pipeline Rehabilitation for gas, water, sewer and industrial applications. Technical manual. Wavin GmbH. 2006.

Underground Infrastructure of Urban Areas 3 – Madryas et al. (Eds)
© 2015 Taylor & Francis Group, London, ISBN 978-1-138-02652-0

Image processing and analysis as a diagnostic tool for an underground infrastructure technical condition monitoring

H. Kleta & A. Heyduk
Technical University of Silesia, Gliwice, Poland

ABSTRACT: Difficult operating conditions of underground infrastructure create the need for regular inspection and monitoring of damage growth. In many cases, direct access to the surface of the lining can be very difficult. A separate problem is the large surface area and time of inspection, often exceeding the capabilities of human perception. In this situation, the easy way can be in the form of a video stream surface image recording and its automatic analysis in an on-line or off-line mode. Diverse structure shapes, spatial conditions and lighting conditions require the use of separate algorithms adapted to the specificities of detected problems. The paper presents an overview of possible solutions, algorithms and criteria for detecting different damage kinds and proposals for numerical description of their size. There have been shown examples for concrete and for brick lining surface analysis. Dripstones as the first symptoms of the initial phase of the damage can be identified on the basis of the color difference (related to the mineral contents). Cracks can be identified as elongated dark lines of the mostly irregular character (distinct from man-made components with a relatively simple and regular shape). Larger cavities can be identified on the basis of various textural parameters (detached fragments have typically much larger granularity than the remaining concrete lining; in the case of brick lining there is a breach of dimensional regularity of bricks and joints pattern). The most convenient way of detected damage presentation is a color enhancement and their overlay onto the true monochrome image plane and automatic measurement of such damage parameters as length, width or area. This—facilitating human interpretation—enhancement of the true image is called Augmented Reality (unlike Virtual Reality—generated entirely by the computer system). In many cases, processing and analysis of the image can be significantly hampered by the poor quality of the input video stream (poor and uneven illumination, glare on the wet surface, falling droplets, haze). Therefore, it is important to use appropriate methods of initial image processing to significantly reduce influence of these disturbances. Frequent registration and fast analysis of the video streams can assist fast necessary decision making and support planning of repair procedures, significantly lowering the maintenance costs and increasing the operational safety of underground infrastructure.

1 INTRODUCTION

Underground infrastructure (traffic and railway tunnels, water and sewage ducts, garages, and subways) is essential for urbanized areas, as they fulfill an important role in the transportation of people, energy, communication and water. Lining stability is one of the most important factors deciding on successful and efficient activity of the structure. Technical state of the lining is a subject of continuous aging and deterioration. The most important threats for the underground lining are:

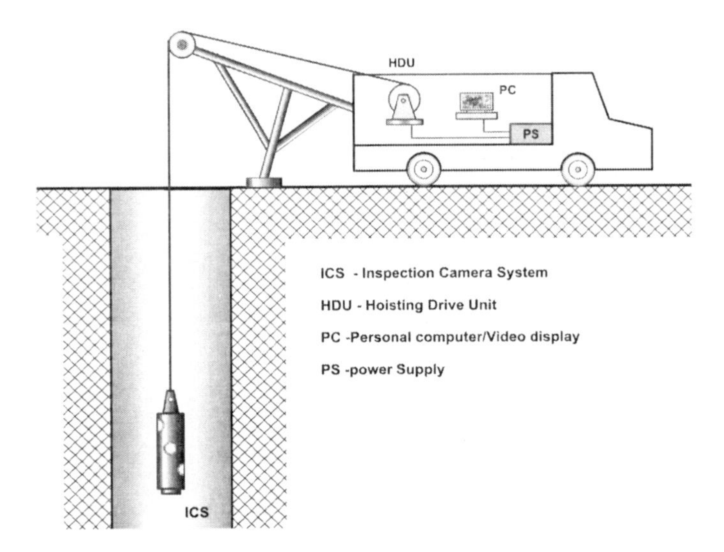

Figure 1. Schematic of a mobile video inspection system for shafts without hoisting machines (Kleta 2013).

- Geological and hydrogeological conditions (often an aggressive salty water)
- Rock mass pressure
- Thermal conditions—seasonal temperature changes
- Vibrations caused by winding machines and other devices (traffic, railways etc).

In active mining areas, the influence of primary geological structures can be modified by influence of underground excavations. Often the lining degradation is manifested by concrete cracking. Crack existence, size and progress can be used for prediction of damage development and for proper maintenance and repair decisions. Therefore it is important to inspect regularly a surface of the lining. In some cases there is practically no possibility of direct access to the lining surface. In these cases damage assessment and integrity monitoring can be performed only by periodical video recording of lining surface images by a system of vertically moving video cameras (Heyduk & Kleta 2009, Kleta & Heyduk 2011, Kleta 2013).

Therefore safety assessment can be very subjective, depending mainly on experience and perception of the inspection personnel. Because of the large surface area of the underground lining and a certain vertical speed and framerate of the camera there is usually a very long stream of video material. Therefore it is necessary to apply some methods of computer image processing and analysis because of:

- Limited perceptional and attention focusing abilities of a human inspector
- Automated comparison of films recorded in longer time intervals
- Quantitative diagnostic criteria of local-state assessment based on image analysis and measurement
- Poor image quality because of dust and steam contents in the air.

Visual inspection of concrete structure elements is one of the basic methods of their technical condition evaluation. It is connected to the fact that it can deliver a great amount of valuable information supporting diagnosis of roots and consequences of noticed structure damage. In some cases it is even regarded as a dominant and most important method. Unfortunately, visual analysis is limited to the visible part of the lining surface, so internal damages and failures cannot be detected. But visual testing can be used as a preliminary examination, monitoring earlier damage progress and limiting the scope of other much more time-consuming and expensive methods (non-destructive or low-destructive).

2 AUGMENTED REALITY CONCEPT

Augmented Reality (AR) is a modern sophisticated method based on dynamic overlaying of additional visual information (text or drawing) on the real image or video stream. It should be distinguished from Virtual Reality (VR) where the whole image (possibly 3D) is artificially generated by computer. In Augmented Reality the output image or video stream is always based on real input image or video. Augmented Reality has been successfully applied to medicine, education, machine diagnostics and military target location (Azuma 1997, Kamat & El-Tawil 2007, Klein 2006).

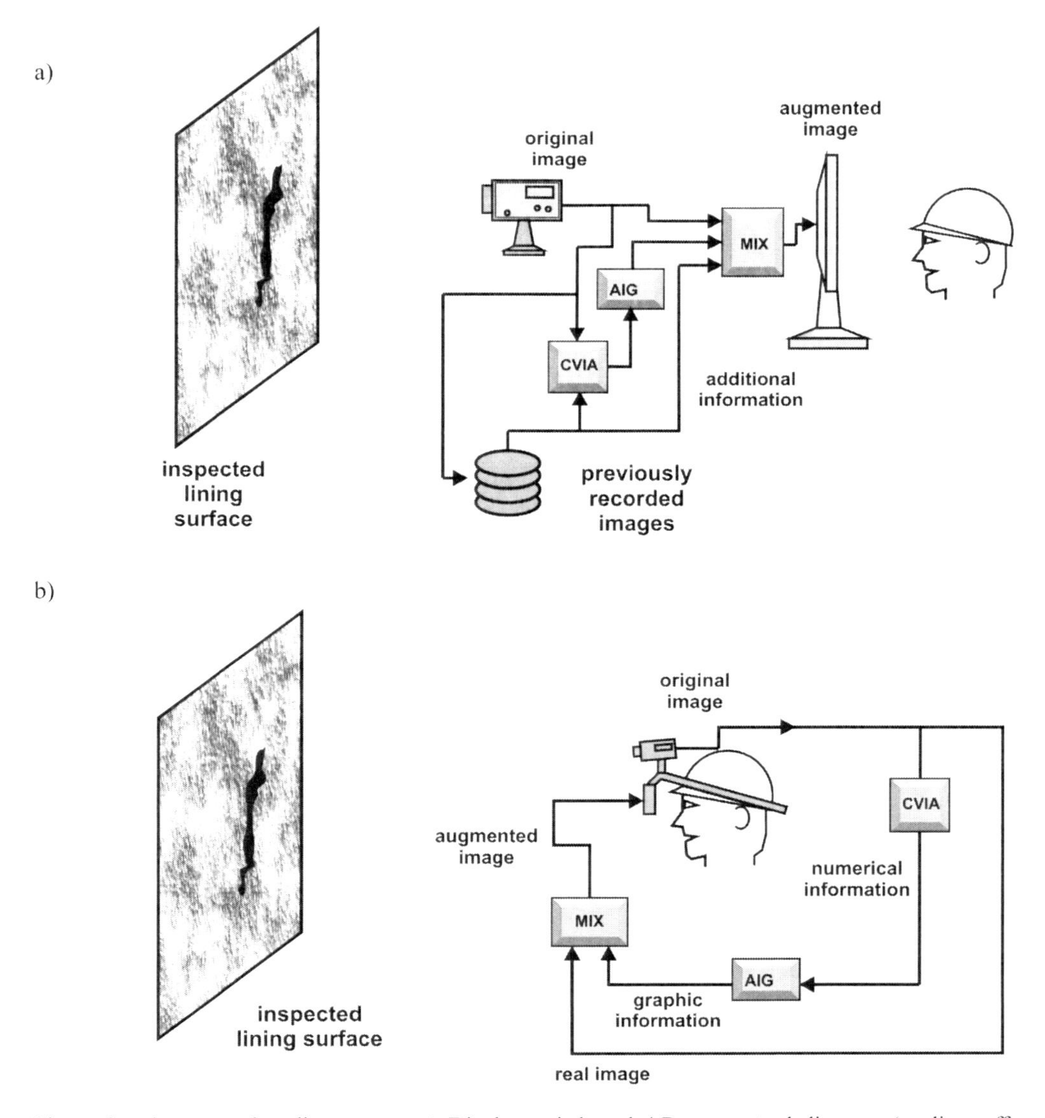

Figure 2. Augmented reality system. a) Display unit-based AR conceptual diagram (on-line, off-line or mixed), b) Head mounted display on-line AR conceptual diagram. IMX—Image Mixer, AIG—Additional information Generator, CVIA—Computer Vision Image Analyser.

The most important stages in these applications of augmented reality are

- Image preprocessing (uneven lighting equalization, contrast stretching)
- Automatic damage (crack, leakage etc.) detection, location and quantitative measurement
- Automatic quantitative crack development monitoring (comparative).

These three stages (Fig. 3) are described in detail in the following sections of this paper.

The original image (gray level or colour) has to be binarized—usually to black and white—because only binary (two logical values—true or false) images can be easily used for direct measurements. An image pixel either belongs to the wanted diagnostic image subset eg. damaged area or not. The thresholding should be performed according to application-specific principle. It can be simply done according to grayscale intensity value, or hue (location in colour space), surface roughness, or two-images difference value. Results of these thresholding operations (detected scratches, cracks, holes, leakages etc.) representing the most valuable and interesting from the diagnostic point of view information can be therefore overlayed

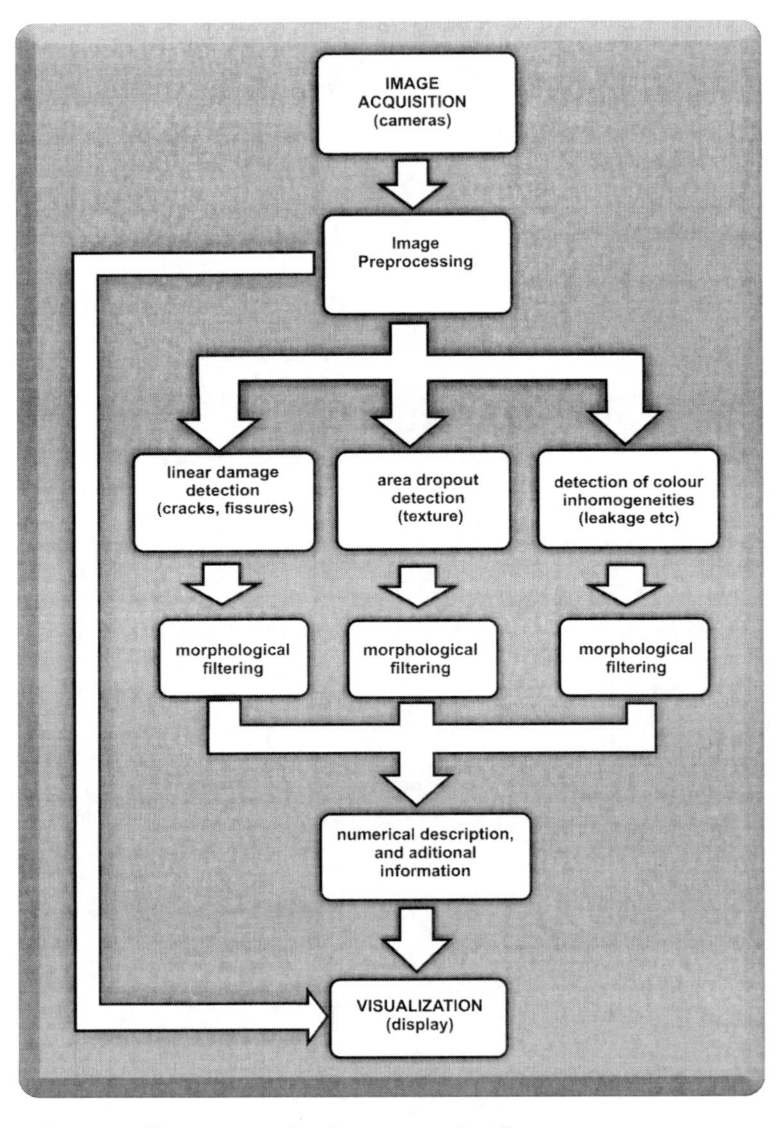

Figure 3. Basic stages of image processing for augmented reality system.

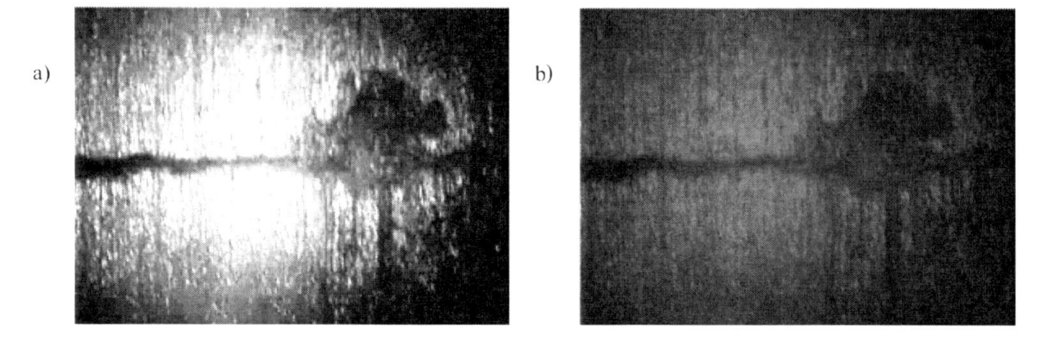

Figure 4. Image correction. a) Original image, b) Image after correction (water reflections removed).

(usually in a quite different colour or brightness) onto the original image in order to focus the inspector's attention to the possible damages or to highlight the image changes (damage progress).

3 IMAGE PREPROCESSING

Sometimes recorded by structure lining surface image is of very weak quality, because of dust and steam contents in the air and because of uneven lighting, so it makes difficult to find and assess damage location and severity. Therefore it can be useful to correct these flaws. There are several methods of these corrections (Heyduk & Kleta 2011, Shih 2010):

- Histogram equalisation
- Homomorphic processing
- Noise reduction (linear on nonlinear filtering).

Image after these corrections is therefore much easier for human inspector (or for a computer algorithm) to analyse, because signal-to-noise ratio (SNR) has a much higher value. An example has been presented in Figure 4.

4 AUTOMATIC DAMAGE (E.G. CRACK OR LEAKAGE) DETECTION

In underground structure lining inspection there is need of fast damage (e.g. crack or leakage) detection. This can be done by special edge (valley) detection methods or color space transformation and thresholding algorithms. These methods should help a human inspector to quickly focus his attention to the most suspicious areas. First example is shown in Figure 5. Cracks are detected as special types of edges—valleys (darker than its surrounding). Crack lines are usually irregular, curved and can be distinguished against a brighter background. Straight, rectangular or circular lines are usually construction details of the lining.

Second example is shown in Figure 6. Area damage (cavity) is detected as an irregular coarse stain on a much smoother concrete surface.

5 AUTOMATIC CRACK DEVELOPMENT MONITORING

One of the most important problems for the evaluation of the load-carrying capability and stability of the concrete lining is a monitoring of temporal failure development process. This is a 3-d spatial process, but for the human eye and for the vision system there are available only surface symptoms. Because of the failure progress dynamics, it is very important to visually evaluate failure development process i.e. to perform automatic image comparison

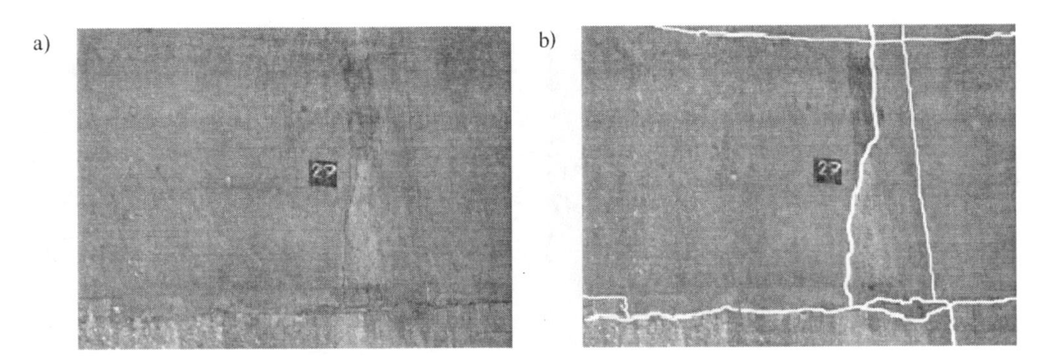

Figure 5. Automatic crack (discontinuity, valley, edge) detection. a) Original image, b) Highlighted cracks and other discontinuities.

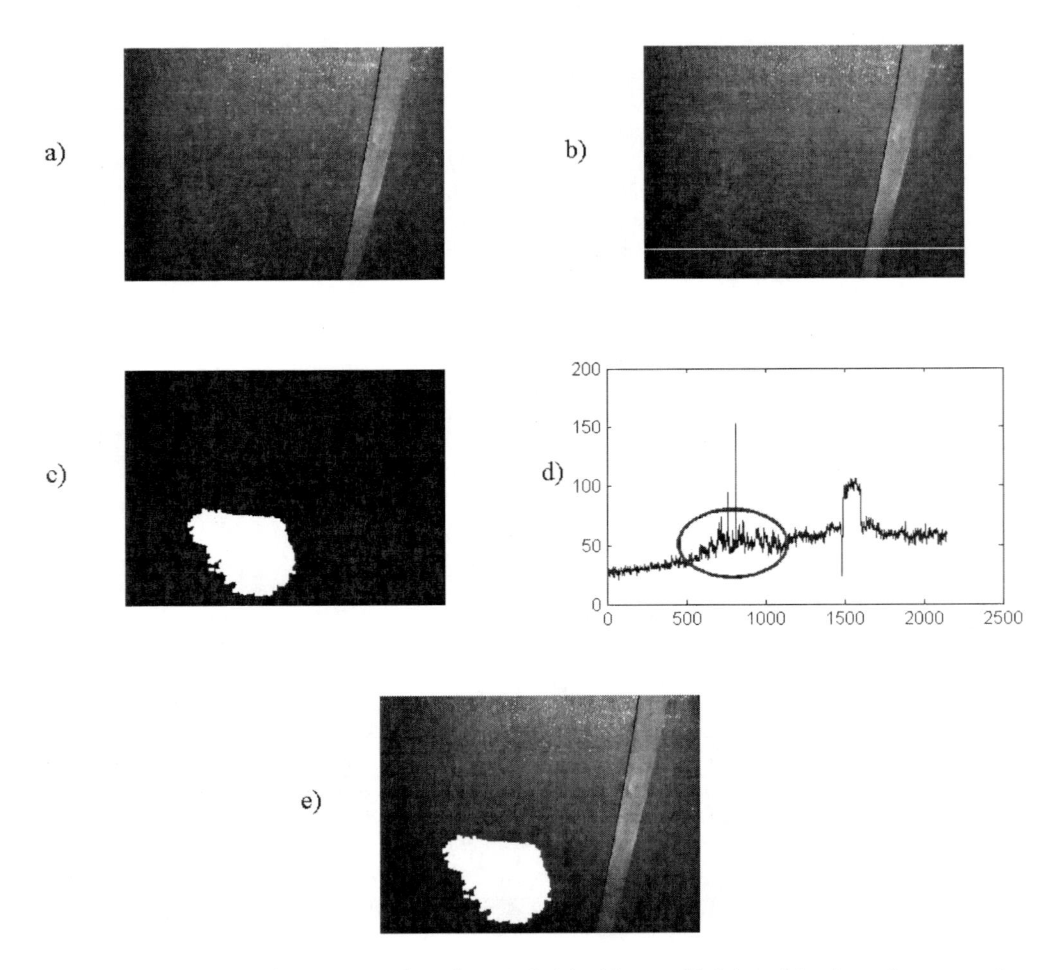

Figure 6. Automatic damage area detection. a) Original image, b) Selected horizontal cross-section line, c) Detected damaged surface, d) Intensity profile along the selected cross-section line (elipse denotes the "suspected" area), e) Detected dropout overlayed onto original image.

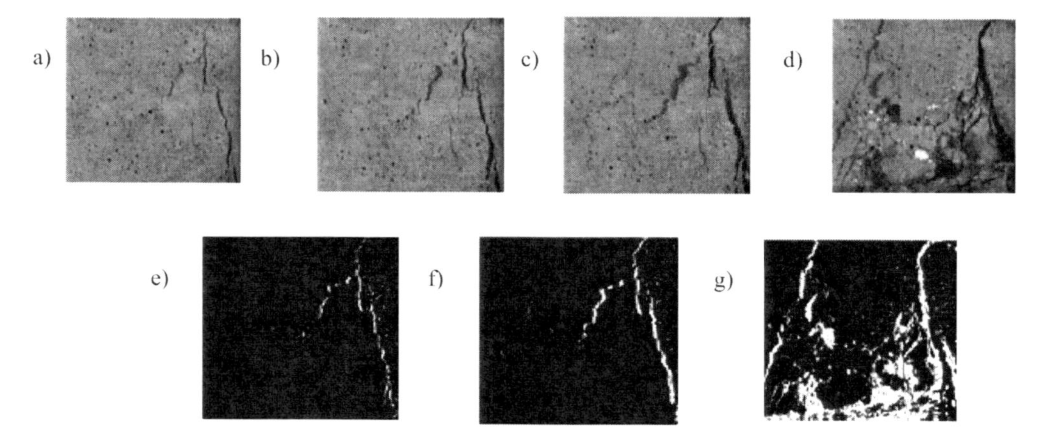

Figure 7. Automatic crack development monitoring. *Upper row*: real grayscale images: (a) Initial stage of cracking; (b), (c) Intermediate stages; (d) final stage of cracking; *Lower row*: Difference images (new crack developments): (e) = (b) − (a), (f) = (c) − (b), (g) = (d) − (c).

of the current state to the previous one. Image processing methods can be also applied to change detection and damage speed evaluation. When periodic records of the same area are taken, it is possible to automatically detect new damages (eg. increasing crack length, width or surface). Automatically detected and highlighted changes can be very small, and therefore hard to notice by a human inspector. A simple example of this method has been presented in Figure 7.

The simplest numerical expression of crack state or development can be by total surface crack area (e.g. total number of white pixels—after thresholding—in the whole investigated area) Cracks can be also described by other parameters like total length of crack branches and maximum or average crack line width. Crack direction can also provide information on the direction of external forces and strains acting on the lining.

6 CONCLUSIONS

Augmented Reality systems can facilitate and speed up human-based inspection of underground structure lining, and help in proper repair and maintenance decision making. They need fast graphic processing hardware and special (application-tailored) analysis methods. These methods are particularly useful in the case when there is no direct access to the lining, and only long streams of video recordings are available. Automatic detection, description and highlighting (detected damages are overlayed onto the original image) can speed-up human analysis and make it more detailed and accurate. This method is particularly helpful for automatic comparison of corresponding images from different video streams recorded during consecutive inspections in longer time periods.

REFERENCES

Azuma R.T. 1997. A Survey of Augmented Reality. *Presence—Teleoperators and Virtual Environments*. 6(4) 355–385.

Heyduk A. & Kleta H., 2009. Wykorzystanie cyfrowej analizy obrazu do wspomagania oceny stanu technicznego i bezpieczeństwa obudowy szybu. *Budownictwo Górnicze i Tunelowe* 15(3) 18–23.

Heyduk A. & Kleta H., 2011. Metody wstępnego przetwarzania obrazu w wizyjnym systemie monitoringu stanu technicznego obudowy szybowej. *Mechanizacja i Automatyzacja Górnictwa* vol. 49(10): 40–46.

Kamat R.V., & El-Tawil S. 2007. Evaluation of Augmented Reality for Rapid Assessment of Earthquake-Induced Building Damage. *Journal Of Computing In Civil Engineering*. 21(5) 303–310.

Klein G. 2006. *Visual Tracking for Augmented Reality*. PhD thesis. Cambridge University Press.

Kleta H. & Heyduk A. 2011. Analiza obrazu w zastosowaniu do stanu obudowy szybów. *Budownictwo Górnicze i Tunelowe* 17(1) 40–46.

Kleta H. (ed.). 2013. *Wizualizacyjna metoda wspomagania oceny stanu technicznego i bezpieczeństwa obudowy szybu z wykorzystaniem cyfrowej analizy obrazu*. Wyd. Pol. Śl., Gliwice.

Shih F.Y. 2010. *Image Processing and Pattern Recognition. Fundamentals and Techniques*. IEEE-Wiley, New Jersey.

Underground Infrastructure of Urban Areas 3 – Madryas et al. (Eds)
© 2015 Taylor & Francis Group, London, ISBN 978-1-138-02652-0

Variable-diameter conduits for water grids

A. Kolonko
Wroclaw University of Technology, Wroclaw, Poland

A. Kolonko
Municipal Water Supply and Sewage Company, Wroclaw, Poland

ABSTRACT: This paper presents a new technical solution in the form of a variable-diameter conduit which makes it possible to reduce the size of fluctuations in the velocity of the flowing medium at a changing rate of flow. The variable-diameter conduit could be used in water supply systems in which the velocity of flow of water fluctuates sharply depending on the water demand. As a result, water resides longer in the water grid whereby the quality of the water deteriorates. This adverse phenomenon cannot be eliminated by reducing pipe diameters since this is not allowed by the fire-control regulations.

1 INTRODUCTION

One of the EU's priorities is to improve the quality of drinking water supplied by water utilities. In Poland much has been done towards this end by upgrading water treatment plants. As a result, the water fed into the water supply system meets the highest requirements. Unfortunately, consumers have not noticed this improvement since the water fed into the corroded old conduits encrusted with hard deposits is polluted again and the cross section of the conduits further decreases. The applies mainly to water conduits built from steel or cast iron pipes without protective coatings, which predominate. Moreover, because of their higher failure rate (due to the increased pressure and consequently, higher electricity consumption) such pipe networks entail higher operating costs. From the economic point of view the operation of the worn out conduits is unprofitable, but water needs to be continuously supplied. This means that a high percentage of water conduits must be rehabilitated.

A tangible improvement in water quality can be achieved by carrying out the national programme of modernizing water-pipe networks by replacing them with new ones or by rehabilitating them. The scale of this undertaking is huge and so are the necessary expenditures.

However, there are areas in which the water quality improvement problem cannot be solved by conventional means, regardless of the financial outlays.

As noted in (Kuś, Ścieranka 2005) in the case of small settlements or areas with dispersed development the diameters of water conduits are much larger than the actual water demand would require. This is dictated by the fire-control regulations imposing minimum conduit diameters and fire hydrant pressure for firefighting purposes. As a result, the quality of the water significantly deteriorates since the latter resides in the conduits for as long as tens of hours. As water resides in water stagnation areas (network terminals, dead points) over extended periods of time, more corrosion products as well as compounds and elements washed out from the mineral deposits and from the biofilm pass into it. A solution to this problem, in the form of a variable-diameter conduit which was patented (Kolonko, Kolonko 2011), is presented further on in this paper.

2 CONDITION OF WATER GRIDS AND CHANGES IN WATER CONSUMPTION IN POLAND

According to GUS (Central Statistical Office 2009) data in 2008 the length of the distribution water grid amounted to about 263 thousand kilometres and in comparison with the previous year increased by about 2.2% (6 thousand kilometres). In the spatial system, the highest water grid density [in km per 100 km^2] still occurs in the Lower Silesian Province (156.4), the Kujavian-Pomeranian Province (118.3), the Lodz Province (116.7) and the Małopolska Province (112.4).

Countrywise, most of the water-pipe networks are located in rural areas. The percentage of rural water-pipe networks amounts to about 78% of the total water grid in Poland. The water grid in these areas increased by about 5 thousand kilometres. The largest increase [km] took place in the Mazovian Province (1275.4), the Małopolska Province (532.0), the Kujavian-Pomeranian Province (411.5) and the Pomeranian Province (350.5). In the West-Pomeranian Province, the Lubuskie Province, the Lodz Province and the Opole Province the increase was below 143 km per annum.

The highest water consumption [hm^3] by households was recorded in the Mazovian Province (191.4) and the Silesian Province (142.9) and the lowest in the Opole Province (30.4) and Lubuskie Province (31.0). The way water is managed is reflected in water consumption per inhabitant. In urban areas this index ranges from 4.0 m^3 in the Mazovian Province to 30.2 m^3 in the Podlaskie Province while in rural areas it ranges from 35.3 m^3 in the Wielkopolska Province to 15.1 m^3 in the Małopolska Province. The distribution of water consumption per inhabitant and per household for urban and rural areas is shown in Figure 1 (Central Statistical Office 2009).

Water consumption changed particularly dynamically in the years 1988–2000 when it sharply decreased—on the average by 50%—in the whole country. In some regions the decrease exceeded 65% (Kuś, Ścieranka 2005). The diagram (Figure 2) showing the average 24-hour water supply for Wrocław, prepared by (Municipal Water Supply and Sewerage Company in Wrocław 2006), illustrates well the decrease in water consumption, which has been observed for over ten years now.

The water-pipe networks in Poland, in both urban and rural areas, practically have already been built. Currently the main problem connected with the supply of the population with drinking water is to ensure its good quality. Most of the water treatment plants supply water with proper parameters. The problem begins when good quality water enters the water-pipe networks. The corroded old pipes encrusted with hard deposits cause deterioration in the quality of the water as it becomes polluted again as a result of the processes taking place inside such water conduits.

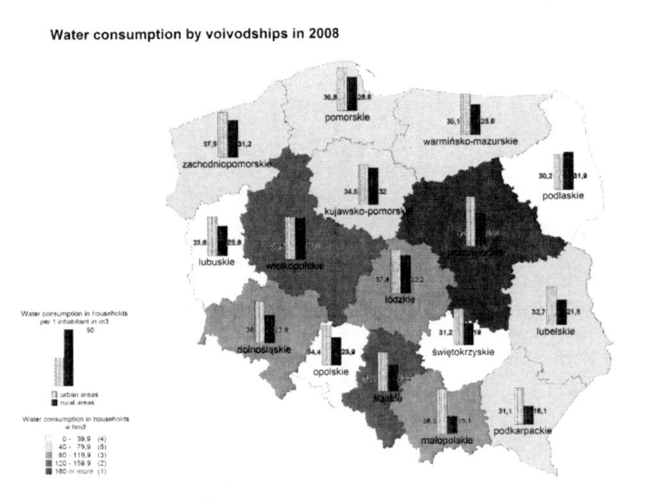

Figure 1. Distribution of water consumption in individual provinces in 2008.

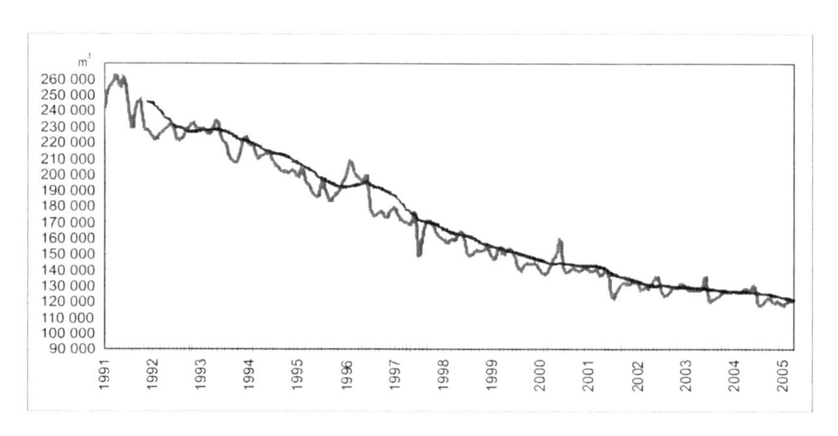

Figure 2. Average 24-hour water supply for Wrocław.

3 BIOCHEMICAL PROCESSES TAKING PLACE IN WATER GRIDS CAUSES AND EFFECTS OF SECONDARY POLLUTION

In water grids, particularly the ones built from steel or cast iron pipes without an internal protective cement mortar or plastic coating (such pipes still predominate), many chemical, microbiological and electrochemical processes adversely affecting the quality of the carried water take place. This results in various kinds of corrosion and in the formation of hard mineral deposits (incrustation). These processes often take place simultaneously and interact with each other. A survey of such processes, based on a paper published in (Szuster-Janiaczyk 2005) is presented below.

- Microbiological corrosion is connected with the activity of microorganisms present in a corroded water conduit. It is caused by mainly bacteria, but also by other living organisms, such as algae and protozoa, which may settle on the metal surfaces of the water distribution system, altering the oxygenation conditions on the surfaces being in contact with water. Bacteria do not directly damage metals, but the products of their metabolism introduce such changes into the environment which, considering the electrochemical nature of corrosion, contribute to the initiation or acceleration of corrosion processes. Corrosion stimulated by bacteria has a local and mostly pitting character.
- Chemical corrosion is caused by the action of dry gases or nonelectrolytes incapable of conducting electricity. The oxidizing action of oxygen on metals at elevated temperatures is an example here. Chemical corrosion causes damage to nonmetallic materials containing calcium compounds, i.e. concrete, asbestos cement, reinforced concrete and cement linings, through the dissolution and washing out of their components (mainly calcium and magnesium compounds).
- Electrochemical corrosion is caused by the action of electrolytes (mostly aqueous solutions). It is due to the formation of a large number of local galvanic cells on metal surfaces, between chemically different metal components or structurally heterogeneous parts of the metallic phase. As regards corrosion damage size and appearance, local and general corrosion are distinguished. Local corrosion is the most harmful and widespread corrosion affecting water distribution piping materials. It occurs selectively in places on the surface of a metal subjected to the action of a corrosion environment. It is characterized by local corrosion pits nonuniformly distributed over the whole surface of the material being damaged. It includes such types of corrosion as: deposit corrosion, pitting corrosion and crevice corrosion.

General corrosion is characterized by uniform damage to the whole metal surface being in contact with an aggressive environment. It is caused by the action of corrosion microcells distributed over the whole surface of the corroding metal. It is the least harmful from the piping durability point of view. But it can affect the quality of water flowing through corroding pipes, as the corrosion products pass into the water and as the microorganisms forming a biofilm in the layers of corrosion products (particularly iron corrosion products) proliferate. Uniform corrosion in sanitary systems can be reduced by adhering to the principles of the use of structural materials proper for the water with which they are to be in contact.

Corrosion may cause:

- a change in water colour (because of rust the latter changes to brown),
- staining of sanitary facilities and washing,
- secondary pollution of water by corrosion products,
- a metallic aftertaste,
- water toxicity,
- intensified mineral deposit accumulation and biofilm formation.

Formation of deposits on inner surface of water conduits. The main consequence of feeding chemically unstable water into the water supply system is the incrustation of its conduits with calcium carbonate deposits. The latter, in turn, contribute to the so-called deposit corrosion. As a result of the build up of deposits an oxygen deficit area develops on the pipe's surface under the layer of deposits. This area becomes an anode and the rate of corrosion there is the higher, the larger is the difference in oxygen concentration between the flowing water and the water filling the deposit's pores. Crystallization takes place, accelerating the increase in temperature and Ca^{2+} ion content and in the amount of inorganic sedimentary solids. The presence of organic colloids and polyphosphates inhibits the rate of crystallization whereby $CaCO_3$ nuclei get trapped. Depending on the chemical composition of water, the $CaCO_3$ precipitate may contain other impurities.

Formation of biofilm. Biofilm is a major factor in the microbiological assessment of water in conduits. There are many places in the water grid which favour the development of a biofilm. In steel and cast iron pipe networks these are mainly areas in which intensive corrosion and incrustation processes take place. This is further compounded by the low flow velocity whereby there are no adequate hydraulic conditions which would prevent microorganisms from forming settlements. Microbiological corrosion is initiated in such places. The corrosion affected areas quickly become larger. The aggressiveness of the water additionally contributes to this process. Characteristic blisters, which shelter the growing colonies inside, form at this stage in the areas affected by microbiological corrosion.

Microorganisms which settle conduits as the first initially are able to survive in certain unfavourable conditions. As the colonies grow, they modify their immediate environment to optimize their living conditions. As time passes, the colonies attract other more demanding organisms, e.g. Legionella, which usually feed on the metabolites of the organisms forming the pioneer colony. In this way the biofilm's structure becomes even more varied. Various bacteria, protozoa, nematodes and snails can settle in the biofilm. The biofilm ensures parallel development for many environments. It is settled by both aerobic and anaerobic organisms. The place of settlement in this case depends on, among other things, the presence/absence of oxygen. The different organisms within this complex structure which the biofilm is communicate with each other via the tangle of extracellular organic polymers interwoven with water channels.

In the case of pipes made of concrete or PVC, which are not susceptible to corrosion of this type, the biofilm forms in a similar way, except that the process is initiated in a different way. Often settlement by microorganisms starts in places where the pipes are damaged. Each small crack or crevice in the material creates conditions favourable to the development of a colony of microorganisms.

One should note here that not all the pipes (even the ones after many years of service) will undergo the above processes. To a large extent this depends on the quality of the water, the flow velocities and the water residence times. The biofilm accumulating in the water-pipe network poses a sanitary hazard since it contributes to the destruction of the conduit's structural material and when strains of anaerobic bacteria develop within its structure, the water has an unpleasant smell and taste.

4 PRESSURE IN WATER-PIPE NETWORK AND FREQUENCY OF NETWORK FAILURES

Today water conduits are rationally operated, by adjusting the pressure to the current demand, in order to extend their service life. Thanks to this the strain of the structural material is reduced. As a result of the reduction in the tensile stresses generated by the internal pressure the failure frequency decreases. But the changes of pressure have their drawback. Changes of pressure, particularly abrupt ones, cause sudden changes in flow velocity and directions, resulting in the lifting of loose sediments and in the tearing off of the outer layers of the mineral deposits and the biofilm.

Flow velocity fluctuations are sometimes unavoidable and can be due to random causes, e.g. they can result from a sharp increase in water consumption during firefighting.

5 CONSTRAINS STEMMING FROM FIRE-CONTROL REGULATIONS

One can expect that the water-pipe networks will be replaced or rehabilitated in the future. But this will not completely solve the problem of water quality, particularly at network terminals in dispersed development areas. Water residence times as long as 100 hours (Kuś et al. 2003) are unacceptable. At the same time the current fire-control regulations make it impossible to alleviate this problem by reducing the diameter of water conduits (Kuś, Ścieranka, 2005). According to the regulations, the minimum diameter conduits on which external fire hydrants are to be installed is DN 100 and DN 125 for respectively the looped water distribution network and the branched water distribution network. In such cases the demand for firefighting water can considerably exceed the demand for domestic water.

6 VARIABLE-DIAMETER CONDUIT—THE SOLUTION TO THE PROBLEM

The solution to the problem of water conduits whose relatively large diameter is dictated not by the demand for domestic water but by the fire-control regulations is a variable-diameter conduit. The diameter size would be adjusted by the pressure of the water carried by the conduit.

The idea of the variable-diameter conduit is illustrated in Figure 3. As already mentioned, a patent for this solution has been filed (Kolonko, Kolonko 2011).

The conduit shown in Figure 3 consists of two layers. The outer layer is a pipe made of a conventional structural material (e.g. PEHD, PVC, steel, cast iron) and the inner layer is a tubular shell made of a flexible material. In periods of low water demand (low pressure) the diameter of the inner elastic conduit is small. When there is a need to increase the flow capacity (e.g. during firefighting), water pressure is increased which results in an increase in the diameter of the inner flexible conduit. It is shown in Figure 4.

The variable-diameter conduit has this advantage that an even flow velocity can be maintained at a varying quantity of the utility being carried. This is of major significance for water grids in which the quantity and velocity of the flowing water depends on the size of its consumption by users. The variable-diameter conduit is particularly suitable for water-pipe networks in sparsely populated areas where because of the low water consumption, water often resides in the conduit for much longer periods, which leads to secondary pollution and bacterial contamination.

The new variable-diameter conduit makes it possible to strike a balance between water residence time reduction and the fire-control regulations specifying the minimum water supply system diameters.

Thanks to water presence detecting sensors placed inside the outer pipe it will be possible to quickly locate failures and minimize water losses.

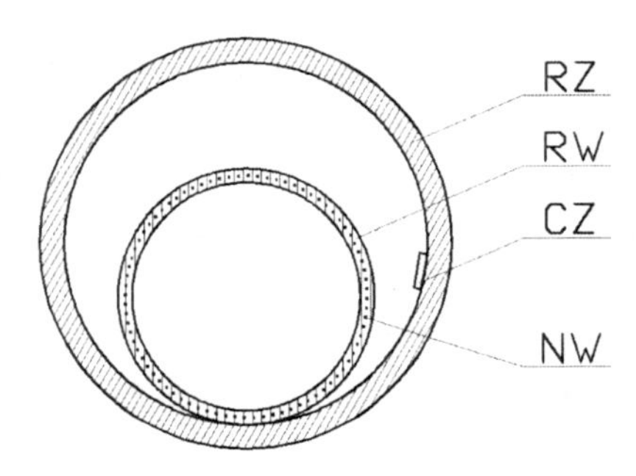

Figure 3. Variable-diameter conduit design (low pressure). RZ—outer pipe, RW—inner pipe, CZ—sensor, NW—reinforcing fibres.

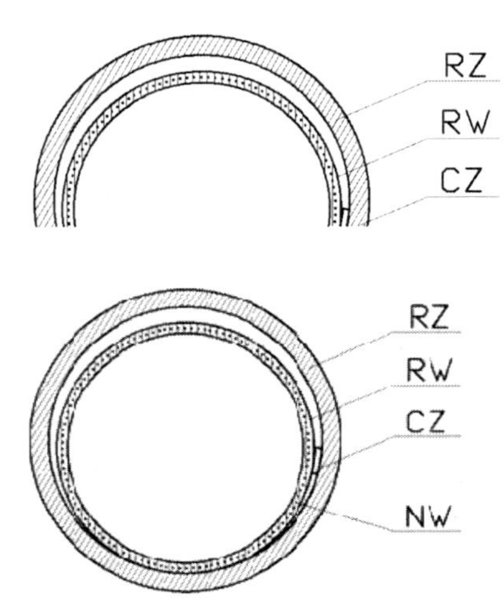

Figure 4. Variable-diameter conduit (high pressure).

7 CONCLUSION

In some situations the use variable-diameter pressure conduits can bring substantial benefits. In the case of water grids, the benefit will be improved water quality. Experimental studies need to precede the production of pipes for variable-diameter pressure conduits.

REFERENCES

Central Statistical Office in Poland: http://www.stat.gov.pl/cps/rde/xbcr/gus/PUBL_MI municipal_infrastructure_in_2008.pdf.

Information material from Municipal Water Supply and Sewerage Company Wroclaw.

Kolonko, A., Kolonko A., 2011. Wroclaw University of Technology. Patent No. PL209351.

Kuś, K., Grajper, P., Ścieranka, G., Wyczańska-Kokot, J., Zakrzewska, A., 2003. Effect of Decrease in the Consumption of Water on its Quality in the Water Distribution System of Upper Silesia, Ochrona Środowiska. Vol. 25. No. 3. pp. 29–34.

Kuś, K., Ścieranka, G., 2005. Effect of Material Parameters and Service Conditions of the Water Distribution System on the Quality of Transported Water. Ochrona Środowiska, Vol. 27. No. 4. pp. 31–33.

Szuster-Janiaczyk, A., 2005. Zjawiska i procesy zachodzące w sieci dystrybucji wody—przyczyny, skutki, przeciwdziałanie. Forum eksploatatora. No 3–4/2005, pp. 28–35.

Underground Infrastructure of Urban Areas 3 – Madryas et al. (Eds)
© 2015 Taylor & Francis Group, London, ISBN 978-1-138-02652-0

Improved road embankment loess substrate under earthquake hazards

J. Kozubal & A. Szot
Faculty of Civil Engineering, Wroclaw University of Technology, Wroclaw, Poland

D. Steshenko
Building, Transport and Mechanical Engineering, North Caucasus Federal University, Stavropol, Russia

ABSTRACT: In the paper was examined the approach to the problem of estimating the safety of vulnerable road embankments and engineering facilities constructed on collapsing loessial substrate on area a with earthquake risk, improved by technology of soil loessial columns. The main idea of the article is presented the universal method of estimating the probability of shocks using the open databases of seismic events and their impact on the projected objects for recognised geological conditions on the site. The problem was presented an issue of 3D for the proposed method with assumptions of axisymetric state. Article also allows you to solve problems widespread in East European and wide parts of Asia loessial collapsing thick layer soils.

1 INTRODUCTION

The problem of constructing road embankments on structural collapsible soils is important on widespread seismic activity terrain from an civil engineering point of view, particularly in centre of towns dominated by collapsible homogeneous layers in South Russia (Caucasus region) presented on Fig. 1. The requirements for embankments include a limitation of surface settlement (i.e., roads, deposit embankments), independence of the values of saturation for the weak layers, and embankment protection against destabilising effect of critical saturation. The implicit consideration additional forces from earthquake were done, in presented technology of substrate improved for designing on predicted safety level. The seismic threats were analyzed in the first part of the text.

The soil columns under earthquake impact discussed in this paper, which increase the stiffness of the sub-layers covered with the embankment, along with their performance with a specific technology, are an attractive alternative due to the low demand for materials, installation time and specialised equipment. The method of reinforcement using soil columns has been developed successfully by Galay and Steshenko (2012) in collapsible homogeneous loess since 2005. The scale of the problem shows the wide range of thicknesses in the loess layers in Eurasia.

In this solution an arching phenomenon is playing an important role in the estimation of the cooperation between the columns and the embankment, as this factor affects the extent of soil improvement. The literature on the subject offers a variety of solutions, such as that proposed by Zhou, Chen, Zhao, Xu and Chen (2012) for a single-layer, one-dimensional rigid column in the state of consolidation, that do not fully take into account the effects of flexible soil columns. A more complete model was developed by Kousik and Mohapatra (2013), who analysed the impact of geosynthetic membranes on the distribution of stress and settlement of a pile in one-dimensional computational space with friction between the ground and the pile.

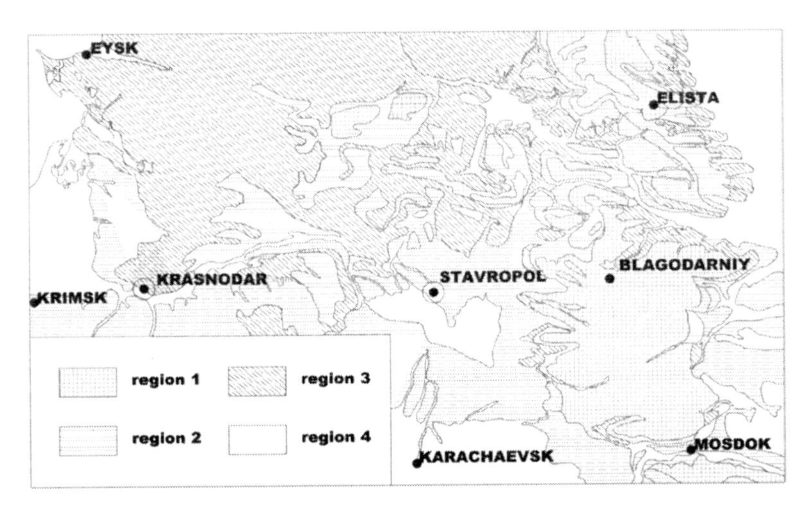

Figure 1. Marine and alluvial sediments in North Caucasus where: region 1) thickness of the layer H_{layer} above 20 m, the thickness of collapsing soils H_{loess} above 20 m and predicted collapsible settlements S_{col} above 50 cm; region 2) $H_{layer} > 20$ m, $H_{loess} > 10$ m, $S_{col} = 15$–50 cm; region 3) $H_{layer} > 20$ m, $H_{loess} = 10$–20 m, $S_{col} = 5$–15 cm; region 4) $H_{layer} < 20$ m, $H_{loess} < 10$ m, $S_{col} < 5$ cm.

The one-dimensional approach makes it difficult to estimate the bearing capacity of the column in terms of the 3D arching phenomena, and it prevents the correct determination of stress within the column core and jacket layers. The arching impact of concrete piles has been investigated by numerical experiments by many researchers, such as Han and Gabr (2002); Tan, Tjahyono and Oo (2008) and the model experiments by Chen, Cao and Chen (2007).

2 EARTHQUAKE HAZARD IN THE NORTH CAUCASUS REGION

The Eurasia-Arabia region is the place of collision of Arabia plateau with Caucasus Anatolian and Iranian block, the Arabian plate moving northerly with a rate about 18 mm/year, whereas implies a shortening along Caucasus about 6–10 mm/year. However presented data have not given a proof to increase or initiate seismically activity into region. The axis of maximum compression and shortening can be traced from eastern Turkey to the Transcaucasia (Trifonov et al. 1996).

The data selection procedure to assessment probability of earthquake extents was done in two stages. The main part of data were taken for Global Earthquake Search (Beta) USGS for 1973 until 2013 year. Other sources of data were used for comparison extremes values after Gupta (2009) The Centre for Research on the Epidemiology of Disasters (CRED) maintains the EM-DAT global database on disasters, which is source of data available in the public domain and The National Geophysical Data Centre (NGDC) which is complete database on earthquake events since 1900 for most countries in the world.

The analyzed information about earthquakes were selected in geographical limits for their epicenters by squared filter where it dimensions were: latitudes [46.86–54.975], longitudes [35.012–38.00] and events were taken only with magnitude more than 3M. Results of selection are presented on Fig. 2 where intense of colors inform about depth of epicenters and the linearly scale is stretched from darkest blue near to lightest blue and consequently is correlated with deep of events for 0.2 km to 164.9 km.

For an assessment of the influence earthquakes in investigated points was used the mathematical model after Tonouchi & Kaneko (1984) and Sultanova (1986) describing values of peak ground acceleration:

$$\log(PGA(R,M,H)) = 0.5M + 0.0043H - \log(R + 0.0055\ 10^{0.5M}) - 0.003R + 0.83, \quad (1)$$

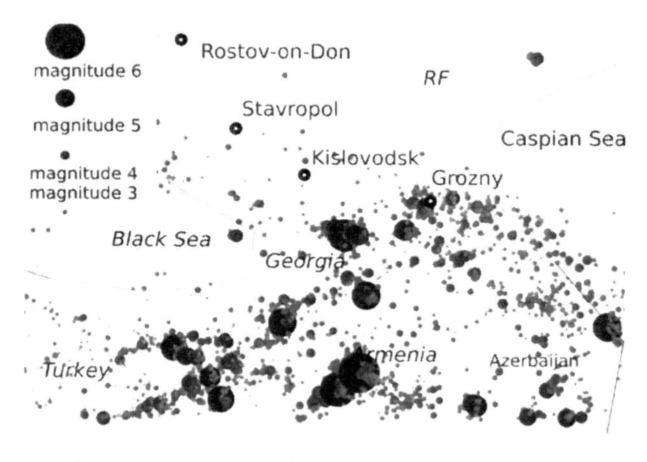

Figure 2. Seismicity of Caucasus zone with epicenters for magnitude M > 3 earthquakes the occurred during period 1973–2013. The intense of colors inform about depth of epicenters: the darkest blue near means depth equal 0.2 km; the lightest—164.9 km.

Table 1. Amplification ratios for eleven groups of geomorphologic by Yamazuki, Wakamatsu, Onishi & Yamauchi (1999).

Digital national land information		Classification after Yamazuki et al.	Amplification ratio	
Geomorphologic classification	Subsurface geology		PGA	PGV
Reclaimed land, polder	Sand, sand soil	Reclaimed land	1.31	2.12
Natural levee, sand bar, lowland, between sand dune, sand dune covered with vegetation	Sand, sandy soil, dune sand	Sand bar, sand dune	1.40	2.12
Reclaimed land, delta, flood plain	Mud, muddy soil, silt, clay, peat	Delta mud clay	1.54	2.92
Reclaimed land, delta, flood plain	Sand, sandy soil, sand and mud, alternation of sand and mud	Delta sandy soil	1.37	2.39
Alluvial fan, volcanic fan	Gravel, gravelly soil, sand and gravel	Alluvial fan	0.87	1.48
Loam terrace, shirasu terrace, volcanic sand terrace	Volcanic ash, loam, pumice flow deposit, shirasu	Terrace—volcanic ash	2.05	2.50
Sand and gravel terrace	Gravel, gravelly soil, sand and gravel	Terrace—sand and gravel	1.26	1.62
Rock terrace, limestone terrace	Rock	Terrace—sand and gravel	0.95	1.34
High relief hills, low relief hills, volcanic hills	Rock	Hill	1.45	1.71
Volcanic footslope, lava flow field, lava plateau	Volcanic material, lava, mud flow deposit	Volcanic footslope	1.80	1.91
Hig relief mountain, midle relief mountain, low relief mountain, foot of mountain, volcano	Rock, volcanic rock	Mountain	1.00	1.00

where R is a distance between the epicenter and the examined point (on a sphere) [km], H is a depth of epicenter [km] and M is a magnitude.

In assessment of PGA connected to local geomorphologic conditions were used the amplification ratios, which are correlated witch structure of substrate and bedrock, they are presented with short description in Tab. 1.

The calculated values of PGA were sorted to strings of annual maxims in each examined city center points and analyzed under regime of the scale presented by the United States Geological Survey. This scale was developed into Instrumental Intensity scale, which is presented on Tab. 2. The peak ground acceleration PGA and peak ground velocity PGV values on an intensity scale are similar to the felt Mercalli proposition.

The probability of earthquake in selected cities is presented on Fig. 3 in scale of probability (Fig. 3c, d) and in useful scale of the disaster occurrence index – γ (Fig. 3a, b). The γ index

Table 2. The Instrumental Intensity scale by The United States Geological Survey.

Intensity	Acceleration (g)	Perceived shaking	Potential damage
I	<0.0017	Not felt	None
II–III	0.0017–0.014	Weak	None
IV	0.014–0.039	Light	None
V	0.039–0.092	Moderate	Very light
VI	0.092–0.18	Strong	Light
VII	0.18–0.34	Very strong	Moderate
VIII	0.34–0.65	Severe	Moderate to heavy
IX	0.65–1.24	Violent	Heavy
X+	>1.24	Extreme	Very heavy

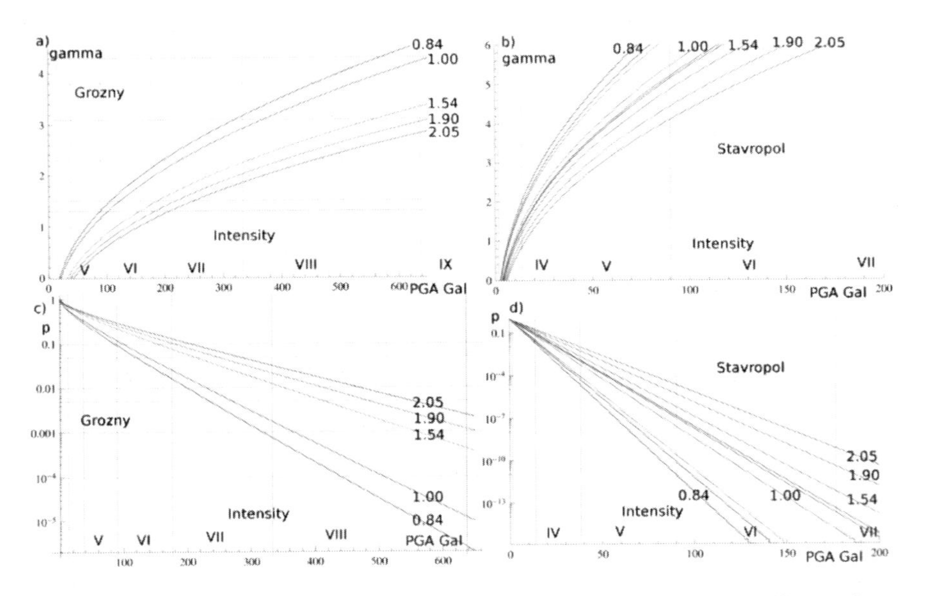

Figure 3. Results of PGA γ index for two cities in the North Caucasus: a) Grozny in γ scale, b) Stavropol in γ scale, c) Grozny in probability scale, d) Stavropol in probability scale, all graphs present the set of amplification ratios {2.05, 1.80, 1.54, 1.00, 0.87}, which are according to values in Table 1.

is associated with probability of an expected earthquake occurrence p by means of the following relationship $p = \Phi_0(-\gamma)$, where Φ is the standard normal cumulative distribution function. For the best fitting results of probability occurrence earthquake events, the type 3 of Pearson's distribution was used.

3 CHARACTERISTIC OF SHAKES

We used as the source of detailed shake resources from the project database of ground motion recordings and supporting information PEER Next-Generation Attenuation (PEER NGA). This database was developed as the principal resource for the development of updated attenuation relationships in the NGA research project coordinated by PEER-Lifelines Program (PEER-LL), in partnership with the United State Geological Survey (USGS) and the Southern California Earthquake Center (SCEC). For a presentation of analytical method were used data from long distance station, however the characteristic of wave is similar to smaller events near to interesting areas. For deterministic analyze was used proposed by Baker's (2007) a method of decomposition to a wavelet form. The wavelet basis function at time t is defined in general:

$$\Phi_{s,l}(t) = s^{-0.5}\Phi\left(\frac{t-l}{s}\right), \tag{2}$$

where Φ is the base wavelet function, s is the scale parameter that dilates the wavelet, and l is the location parameter that that translates a wavelet in function of t time. Any measured signal depend from time $f(t)$ can be represented, in this decomposition approach, as a linear combination of basic functions (2). Their linear combination is determined by the following Fourier transform, presented for analytical form of base wavelet function:

$$C_{s,l} = \int_{-\infty}^{+\infty} f(t)\Phi_{s,l}(t)dt = \int_{-\infty}^{+\infty} f(t)s^{-0.5}\Phi\left(\frac{t-l}{s}\right)dt, \tag{3}$$

where s—scale, l—position. The base function must fulfilled a some constrains (Percival and Walden 2000), and possible predefined analytical functions to decomposition are: Mexican Hat, Morlet, DGaussian, Meyer, Gabor and Poul. The original ground motions and residuals after pulse extraction can be used to predict whether the given ground motion has pulse characteristic. To perform this classification, a variety of potential predictor variables and functions were computed and evaluated. The results of decomposition for an example selected impact with pulse characteristics are presented on Figure 4. This original wave is based on PEER recording data for its both parts: horizontals (S-Wave) and a vertical (P-Wave).

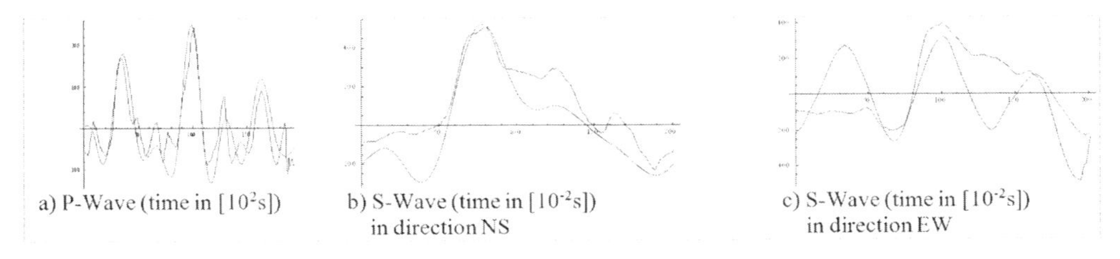

Figure 4. Decomposition of original data of accelerations (PGA scale) in set for time period [0;200] 10^{-2} s (it is the peak episode) for earthquake in Erzincan 1992, blue color of graphs—original recorded data, red color curves are effects of decompositions, a) P-Wave, b) S-Wave in direction North-South, c) S-Wave in direction East-West.

4 SHEAR MODULUS OF SOILS WITH DEPENDENCIES FROM EARTHQUAKES

On observable frameworks of soil behavior based on strain regimes presented by Vucetic (1994), Santamarina (2001) and stress-strain response the following points are illustrated on Figure 5 and described below:

1. The experimental results for most cohesive soils described the very small strain elastic regime is a region referred by approximately constant stiffness, where very small energy losses are presented and there is not changes in pore pressure. The limitation for the linear shear strain behavior γ_{tl} predominantly are strain values for 0.001% to 0.005%.
2. For strains more than γ_{tl} soil stress-strain relation is changes to non-linear dependencies; there is not accumulation of additional pore pressure during undrained oscillated loading or volume change for drained conditions. The volumetric cyclic threshold strain γ_{tv} is the border for small strain regime, with fully elasticity behavior where the next range is medium strain regime, with strength degradation. The limitation γ_{tv} separated strain and stress described and controlled conditions. The modulus reduction G/G_{max} is in this case between 0.60 and 0.85 relates from value of γ_{tv}. Silts and clays with their plastic index from 14 to 30 have got γ_{tv} in range [0.024%; 0.06%] by Hsu and Vucetic (2004, 2006).
3. The boundary of the medium strain regime with a degradation strain threshold described by γ_{td} represents the strain values which have more destructive effects of the specimen and separates the medium strain regime from the large strain regime. The degradation strain threshold γ_{td} is adapted by Santamarina (2001).

In work of Isao Ishibashi Xinjian Zhang (1993) the dependency between shear modulus G and cyclic shear strain amplitude is:

$$G = K(\gamma) f(e)\, \sigma_0^{-m(\gamma)} \tag{4}$$

where f is function of e void ratio, $m(\gamma)$ is an increasing function of γ and σ'_0 (in kPa) is the mean effective confining pressure, γ cyclic shear strain amplitude, K is decreasing function of γ, G_{max}—maximum dynamic shear modulus is usually obtained at $\gamma = 10^{-6}$ or less. The appropriate constrains are:

$$G_{max} = K_0(e)\, \sigma_0'^{m_0}, \tag{5}$$

$$K_0 = K(\gamma \leq 10^{-6}) = 1.0, \tag{6}$$

$$m_0 = m(\gamma \leq 10^{-6}), \tag{7}$$

The K function originally had been constructed for sandy soils only, but later was developed to form incorporated cohesive soil also, descriptions of fitting curves to experimental

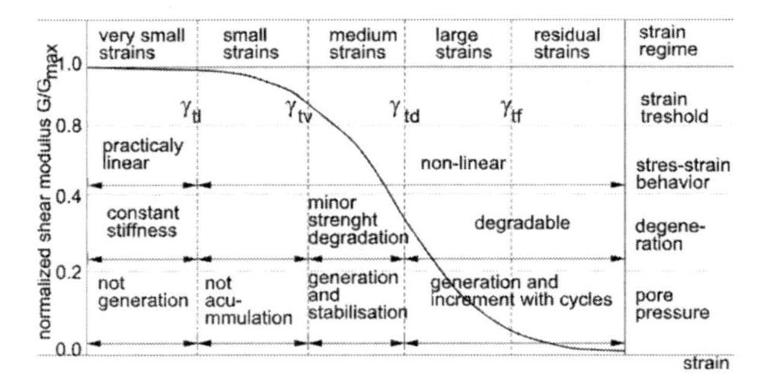

Figure 5. Soil behavior between proposed by Díaz-Rodríguez and López-Molina (2008) strain thresholds for saturated cohesive soils versus cyclic shear strain amplitude γ in logarithmic scale.

data are prepared for ranges of I_p (plasticity index) values as follow:

$$K(\gamma, 0 \leq I_p < 15) = 0.5(1.0 + Tanh(Ln(1.02\times10^{-4} + 3.37\times10^{-6}\, \overline{I_p\,\gamma}^{\,1.404})^{0.492} \tag{8}$$

$$K(\gamma, 15 \leq I_p < 70) = 0.5(1.0 + Tanh(Ln(1.02\times10^{-4} + 7.0\times10^{-7}\, \overline{I_p\,\gamma}^{\,1.976})^{0.492} \tag{9}$$

$$K(\gamma, 70 \leq I_p) = 0.5(1.0 + Tanh(Ln(1.02\times10^{-4} + 2.7\times10^{-5}\, \overline{I_p\,\gamma}^{\,1.115})^{0.492} \tag{10}$$

and finally shear modulus G has a form:

$$\frac{G}{G_{max}} = K(\gamma, I_p)\, \sigma_0^{m(\gamma, I_p)-m_0} \tag{11}$$

where $m(\gamma, I_p)$ function is:

$$m(\gamma, I_p) - m_0 = 0.272(1.0 - Tanh(Ln(5.56\times10^{-4}/\gamma))\, e^{-1.455\times10^{-2} I_p^{1.3}} \tag{12}$$

On Fig. 6 the dependencies between $K(\gamma)$ and $m(\gamma)-m_0$ are presented for different soils types. After integrity of analyticaly decomposed function of acceleration in both cases (vertical and horizontal) values of velocities displacements were reached with assumption of zero start values as boundary conditions for velocity and position (for $t = 0$). Integration results are presented on Fig. 7 for all cases of waves and directions.

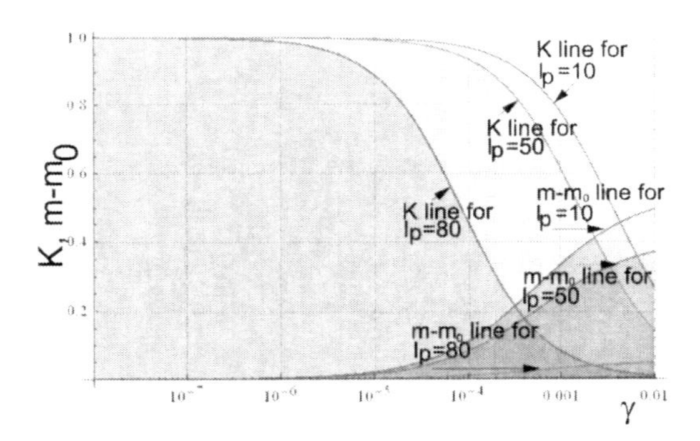

Figure 6. Graphs of $K(\gamma)$ and $m(\gamma) - m_0$ functions for soils with different I_p values equal {10, 50, 80} versus cyclic shear strain amplitude γ in logarithmic scale.

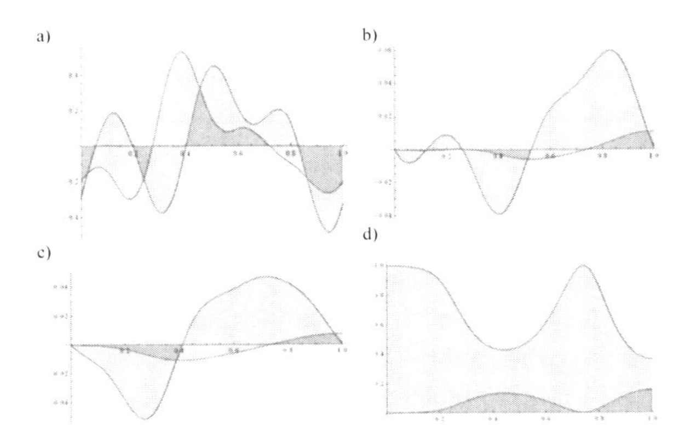

Figure 7. Results of integration of Wavelet which represented measured values of accelerations in selected peak time: a) sources accelerate for both S-Wave WE and S-Wave NS, b) velocity (blue), displacement (purple) for S-Wave EW, c) velocity (blue) and displacement (orange) S-Wave NS, unity of displacement [m], and velocity [m/s] d) $K(\gamma,t)$ and $m(\gamma,t) - m_0$.

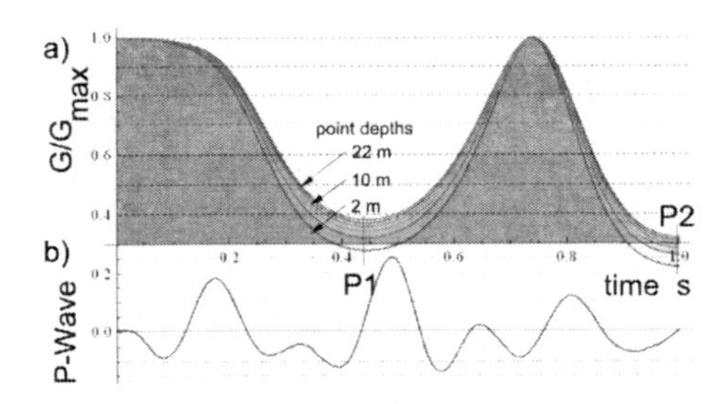

Figure 8. Results G/G_{max} for depths $z = \{2, 6, ..., 22\}$m, horizontal axis—time [s], vertical axis G/G_{max} [ ] correlated inversely with σ_0', below results P-Wave is presented [m/s^2].

The modulus reduction G/G_{max} in the until wave is going in a place of road embankment is showed by graphical representation on Fig. 8. A distance from earthquake to tested site was artificially, by authors, reduced for better clarity of results.

5 LOESS IMPROVED BY SOIL COLUMNS—THE DIRECT SOLUTION FOR ROAD EMBANKMENTS

The problems associated with collapsible loess are substantial in the economies (Delage and Cui (2005)) of China, North America and Russia. Loess has been well recognised and described in many articles with respect to its geology (structure, history of deposition, morphology). The specific geotechnical problems linked with loess have been addressed in some areas of Eurasia, where loess deposits appeared to be sensitive to water content changes and susceptible to collapse. Collapsible unsaturated loess deposits have a low plasticity index and high porosity. The sensitivity of loess to changes in water content appears to be particularly important in the case of railways, roads and other types of embankments. The general aim is to reduce the collapsibility settlement risk by improving the mechanical properties of the soil. The typical potential sources of water that can cause wetting are described by Houston (1995), Borecka and Olek (2013) and include the following: broken water lines, canals and landscape irrigation, roof run-off and poor surface drainage, a rising ground water table, unintentional and intentional recharge, moisture migration and protection from the sun.

The technology described in our article was developed by Galay and Steshenko (2012) over 8 years in the south of Russia (particularly in the North Caucasus Region).

These researchers used simple equipment, such as a common truck (4 to 8 ton total weight) with a typical 125 mm hydraulic drilling pad on the bed of the truck.

Stages of improvement:

1. the introduction of the drill head into loess for the design depth,
2. the downward upward movement while rotating the drill head counter-clockwise,
3. the addition of supplementary material in the hole next to the local soil by the workers while still rotating the drill head counter-clockwise and the uplift of the head by 0.30–0.40 m.
5. Steps 2–4 are sequentially repeated in a loop until the level of the drilling head reaches the ground level. In step 5, the result of the presented loess improvement method is a two-layered soil column with the same material composition that existed before improvement but with a different structure.

6 THE DETERMINISTIC RESULT FROM ANALYTICAL MODEL OF IMPROVED SUBSTRATA

Due to its greater efficiency in strengthening the weak layer, we selected the optimal triangular mesh for further example calculations. Other possible variants are rectangular and dense

Table 3. Mechanical parameters for all soil column and tubes—in a steady state task.

Soil column/tube	Mechanical parameters	
Description	Young's modulus [MPa]	Poisson's coefficient [ ]
'*a*' high compacted loess	30.0	0.3
'*b*' compacted loess	20.1	0.3
'*c*' intact dry loess/saturated loess	18.7/ 4.0	0.2

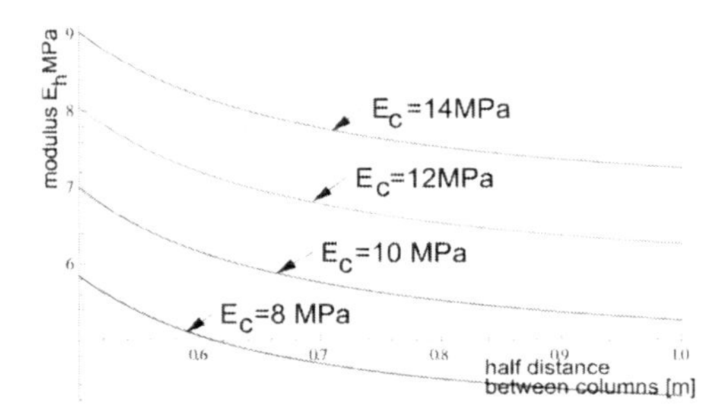

Figure 9. The influence S-Wave results for simple homogenized E_h [MPa] of improved substrate under road embankment for weak layer Young's modulus equal: {8, 10, 12, 14} Mpa.

packing mesh. The construction of the column corresponding to the technology used. The described core (full cylinder '*a*') is a cylinder with the compacted soil zone containing possible chemical or gravel additives, however in general th method is used without any additives. The next layer—jacket '*b*'—is compacted only with local materials, and the third layer represents the weak intact soil '*c*'. The last two layers are treated as tubes.

The deterministic result were done in 3D analysis under axisymmetric terms (for vertical axis of symmetry, along depth of column '*a*' which was went by center of its). In spite of fact that γ values are greater than γ_{tv}, the elastic soil model was used to calculation, the reason was a simple model used to approximate of improved substrate stiffness under earthquake influence. The table 3 presents mechanical parameters for all elements of substrate.

The influence S-Wave results for simple homogenized E_h [MPa] of improved substrate under road embankment for weak layer Young's modulus equal: {8, 10, 12, 14} MPa are presented on Fig. 9.

The modulus reductions were calculated on transposition values from minimal shear modulus G/G_{max} proportion within selected weave passes by construction site (Fig. 8—the point P2 on the sketch). The simple homogenization were done as proportional to cross areas ('*a*','*b*','*c*') participation in total stiffness of substrate. The individual columns stiffness were taken as result of axisymmetric constrains for a flexible embankment.

7 REMARKS

Following problems have been described and resolved by simple numerical task in the paper:

- The procedure of assessment and determination probability of occurrences the seismic event based on freely data bases published by universities and government institutions. The universal method of representation seismic S-wave and P-wave with the use of wavelet functions.

- Parameters concerning seismicity influence on mechanical parameters of substrate for both cases: sandy and clayey soils with dependence to, plasticity index.
- The technology of strengthening loess by direct soil columns.
- Decreasing of geotechnical homogenized improved substrate parameters under the embankment with reference to seismicity activity of region.

The paper gives a base under further reliability analysis for geotechnical structures subject to seismic load in the areas of occurrence homogenous ground of eolic origin, such like loess, with the view of their dynamic response.

REFERENCES

Baker, J.W. (2007). Quantitative classification of near-fault ground motions using wavelet analysis. *Bulletin of the Seismological Society of America, v. 97, no. 5, pp. 1486–1501.*

Borecka, A., & Olek, B. (2013). Loesses near Kraków in light of geological-engineering research. *Studia Geotechnica Et Mechanica, Vol. 35, No. 1, pp. 41–57.*

Chen, Y., Cao, W., & Chen, R. (2008). An experimental investigation of soil arching within basal reinforced and unreinforced piled embankments. *Geotextiles and Geomembranes 26, pp. 164–174.*

Ching, J. (2011). *Practical Monte Carlo Based Reliability Analysis and Design Methods for Geotechnical Problems.* Applications of Monte Carlo Method in Science and Engineering, Prof. Shaul Mordechai.

Delage, P., Cui, Y., & Antoine, P. (2005). *Geotechnical problems related with loess deposits in Northern France, Proceedings of International Conference on Problematic Soils*, Eastern Mediterranean University, Famagusta, N. Cyprus.

Díaz-Rodríguez, J. & López-Molina J. (2008). Strain thresholds in soil dynamics. *The 14 World Conference on Earthquake Engineering, Beijing, China.*

Galay, B., & Steshenko, D. (2012) *Recommendations on designing and arrangement of bored ground piles made a screw way in subsidence and soft soils.* Stavropol (in Russian).

Houston, S. (1995) Foundations and pavements on unsaturated soils—*Part 1: Collapsible soils. Proc. 1st Int. Conf. Unsaturated Soils UNSAT'95 (3), pp. 1421–1439.*

Hsu, C. & Vucetic, M. (2004). Volumetric threshold shear strain for cyclic settlement. *ASCE Journal of the Geotechnical and Geoenviromental Engineering, 130 (1), pp. 58–70.*

Hsu, C. & Vucetic, M. (2006). Threshold shear strain for cyclic pore-water pressure in cohesive soils. ASCE Journal of the Geotechnical and Geoenviromental Engineering, 132 (10): 1325–1335.

Ishibashi, I. & Zhang, X. (1993). Unified dynamic shear moduli and damping ratios of sand and clay soils and foundations. *Vol. 33, No. 1, pp. 182–191, 199,. Japanese Society of Soil Mechanics and Foundation Engineering.*

Kousik, D., Mohapatra S. Analysis of stone column-supported geosynthetic-reinforced embankments Applied Mathematical Modelling 37, *pp.* 2943–2960, 2013.

PEER Ground Motion Database *http://peer.berkeley.edu/peer_ground_motion_database/site/tutorials.*

Percival, D., Walden A. (2000). Wavelet Methods for Time Series Analysis, Cambridge University Press.

Santamarina, J.C. (2001). Soils and Waves. Wiley, New York. *Geotechnical Engineering Division, 106 (6): pp. 691–712.*

Sultanova, Z. (1986). Earthquakes of Azerbaijan for the Period of 1966–1982. *Elm, Baku, 96. (in Russian).*

Tan, S.; Tjahyono, S.; & Oo, K. (2008). Simplified Plane-Strain Modeling of Stone-Column Reinforced Ground. *Journal of Geotechnical and Geoenvironmental Engineering, pp. 185–194.*

Tonouchi, K. & Kaneko, F. (1984). Methods of evaluating seismic motion at base layer. *OYO Technical Report No. 6, Japan, pp. 12–14.*

Trifonov, V.O., Karakhanian, A.S., Berberian, M., Ivanova, T.P., Kazmin, V.G., Kopp, M.L., Kozurin, A.I., Kuloshvili, S.I., Lukina, N.V., Mahmud, S.M., Vostrikov, G.A., Svedan, A. & Abdeen, M. (1996). Active faults of the Arabian plate bounds in Caucasus and Middle East. *Journal of Earthquake Prediction Research, v. 5, pp. 363–374.*

Vucetic, M. (1994b). Cyclic threshold shear strains in soils. ASCE Journal of Geotechnical Engineering 120: *pp.* 2208–2228.

Yamazuki, F., Wakamatsu, K., Onishi, J. & Yamauchi, H. (1999). Relationship between Geomorphological Land Clasification and Soil Amplification Ratio Based on JMA Strong Motions Record Bull. *ERS No. 32.*

Zhou, W., Chen, R., Zhao, L., Xu, Z. & Chen, Y. (2012). A semi-analytical method for the analysis of pile-supported embankments, *J Zhejiang Univ-Sci A (Appl Phys & Eng) 13(11), pp. 888–894.*

Underground Infrastructure of Urban Areas 3 – Madryas et al. (Eds)
© 2015 Taylor & Francis Group, London, ISBN 978-1-138-02652-0

Influence of PVC pipe deflection on the thickness of CIPP rehabilitation liners

E. Kuliczkowska
Kielce University of Technology, Kielce, Poland

ABSTRACT: Differences in recommendations on permissible deflection of sewer plastic pipes were indicted. The results of field measurements of the magnitude of PVC pipe deflections were presented. The inspections were conducted both with the CCTV method and a special device equipped with a deflection gauge. The influence of PVC pipe deflections on the thicknesses of CIPP rehabilitation liners was assessed. The analysis was performed separately for non- and fully structural liners.

1 INTRODUCTION

Trenchless rehabilitation of sewers can be carried out using numerous technologies, described, among others, in (Kuliczkowski A. et al. 2010) Cured-in-Place Pipe (CIPP) technology is of major importance. It falls into the category of Close-Fit Lining technologies, which include those using PE pipes with mechanically or thermo-mechanically reduced cross sections, for example Swage-lining, or reshaped (folded or convoluted) ones, for example Compact Pipe. Those technologies are widely applicable as they only very slightly reduce cross sectional area of sewers.

Close-Fit Lining technologies can be used to install non-structural liners, when a sewer has a proper load capacity but it needs to be sealed, protected from corrosion or abrasion, or its hydraulic parameters have to be improved. The technologies are also applicable to semi- or fully structural liners, when in addition to the effects mentioned above, a liner is also required to have semi- or full load capacity.

The thickness of CIPP, PE or others liners is determined for non-semi- or fully structural options by conducting appropriate static calculations (ATV-A 127 P 2000, ATV-M 127 P 2000, ASTM International, Kuliczkowski 2004, Madryas, Kolonko, Wysocki 2002, Per Aarsleff Polska 1996). The thickness is a function of, among others, deformation of cracked rigid pipes or deflection of plastic pipes.

2 DEFLECTIONS OF THERMOPLASTIC PIPES

Deformability constitutes a major advantage of plastic pipes (Kuliczkowski, Kuliczkowska 2002), especially when errors are found in designs, or disadvantageous conditions of pipe bedding occur. In such cases, pipes do not crack, but they only deflect too much when compared with the deflection values stated in the code.

Different guidelines on permissible deflection of thermoplastic plastic pipes were developed. At the design stage, for instance, in accordance with the guidelines most frequently applied in Poland (ATV-A 127 P), it is assumed that permissible long-term deflection of pipes should not exceed 6%. According to recommendations in Scandinavia (Madryas, Kolonko, Wysocki 2002), the values should not be higher than 8% for short-term deflection of PVC pipes, 9% for short term deflection of PE-HD pipes, and 15% for long-term deflection. The value of 6% given in the guidelines (ATV-A 127 P) refers to a majority of typical sewage systems. The designer, following the guidelines, may, however, decrease the above value of permissible deflection when the type of road pavement and type of vehicles in the traffic flow are accounted for. For sewers laid under railway tracks, permissible deflection may not exceed 2%, occasionally the values may have to be even lower if the condition $\delta_v \leq 10$ mm is not satisfied.

In accordance with the guidelines stated in (DWA-M-149-3 2007), however, sewers with deflections $2\% \leq \delta_v < 6\%$ are intended for long–term rehabilitation, which seems to suggest that the value of pipe long-term deflection should not exceed 2%. Thus an apparent contradiction between recommendations in different guidelines is found.

3 RESULTS OF FIELD MEASUREMENTS OF PVC PIPE DEFLECTIONS

3.1 *Results of CCTV inspections*

The analysis concerned deflections of newly laid PVC sewer pipes. The investigations were conducted with CCTV camera during the pre-commissioning inspection, before the pipes were put into service. The present study summarizes results from the analysis of 11,388.6 m PVC sewers randomly selected in eleven Polish cities. Forty three inspections were conducted, diverse in terms of the city, street and pipe diameter. 336 sections of the pipelines were inspected. The diameters of the sewer pipes varied from 200 to 630 mm. On the basis of German guidelines, ATV-A 149, deflections of PVC pipes were categorized into five classes.

The investigations demonstrated that deflections of pipes were found along 49% of the total length of the sewers of concern (Kuliczkowska 2005). Figure 1 presents percentages of the sewer lengths with observed pipe deflections, grouped into five classes. Pipe deflections denoted as δ_v, given as percentage values, in classes I–V, ranged as follows: $\delta_v \geq 40$, $25 \leq \delta_v < 40$, $15 \leq \delta_v < 25$, $6 \leq \delta_v < 15$, $\delta_v < 6$. Class V includes 86.6% of the observed pipe deflections, which were small, lower than 6%. Class IV comprises 13.3% of the sewers with deflections higher than 6% and lower than 15%. According to the classification adopted in the study, only 0.1% of sewers, the deflections of which exceeded 15%, fell into Class III.

Photographs show pipes with deflection slightly below 6% (Figure 2), and that higher than 6% (Figure 3).

3.2 *Results of field measurements with a device equipped with a deflection gauge*

As in section 3.1, the analysis covered the results of field measurements of newly laid PVC pipes that have a total length of 3498.9 m. Investigations, in which a device equipped with

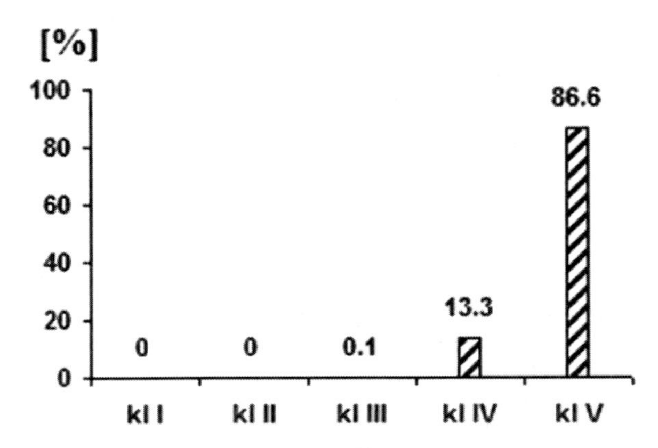

Figure 1. Lengths of sewers (percentage) with deflections categorised into five classes.

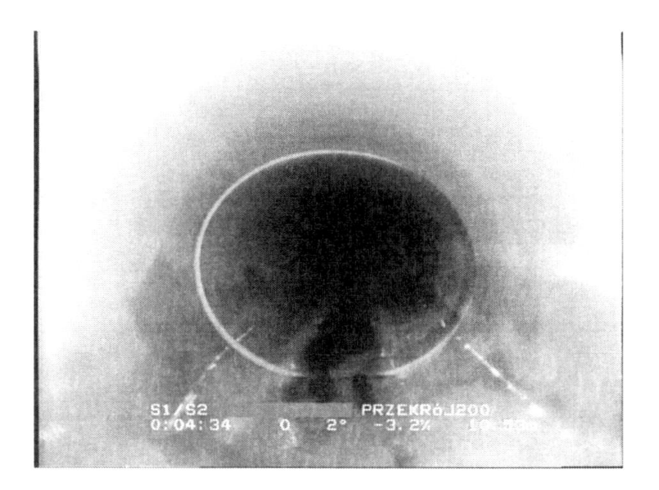

Figure 2. PVC pipe deflection slightly below 6% (the author's photograph).

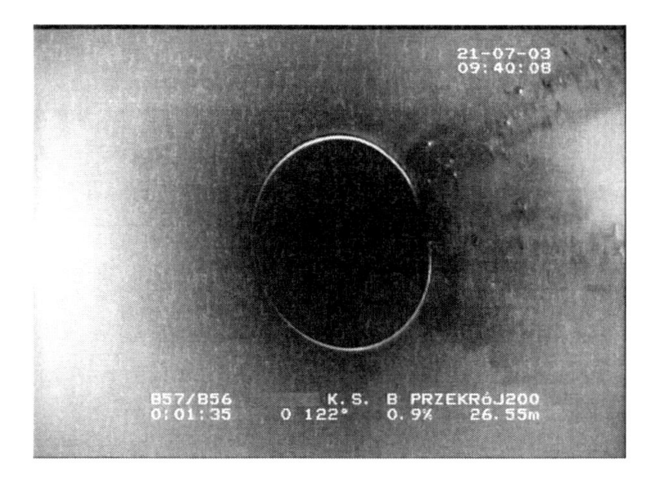

Figure 3. PVC pipe deflection over 6% (the author's photograph taken with the camera rotated 90°).

a deflection gauge was used, were performed in eight Polish cities. Eighty two surveys were conducted in pipes with diameters 200, 225, 315, 400 and 500 mm and length from 15 to 84 m. The deflections were recorded on continuous basis. The research apparatus, namely the Delphine Systeme with Titanic 2 and Titanic 3 systems was described, in detail, in (Kubicka 2000).

On the basis of graphs showing continuous deflection measurements, lengths of sewers with deflections categorised into five classes proposed in (DWA-M-149-3 2007) were shown (Figure 4). Deflections, denoted as δ_v, are given as percentage values. In the successive classes from 0 to 4, they are as follows: $\delta_v \geq 15$, $10 \leq \delta_v < 15$, $6 \leq \delta_v < 10$, $2 \leq \delta_v < 6$, $\delta_v < 2$.

3.3 *Discussion of results*

The surveys described in sections 3.1 and 3.2 concerned various PVC sewers laid in different regions of the country, yet the results are consistent. It is disturbing that already at the pre-commissioning stage pipes show deflections higher than those recommended in the code. Pipe deflection occurrence can result from the following:

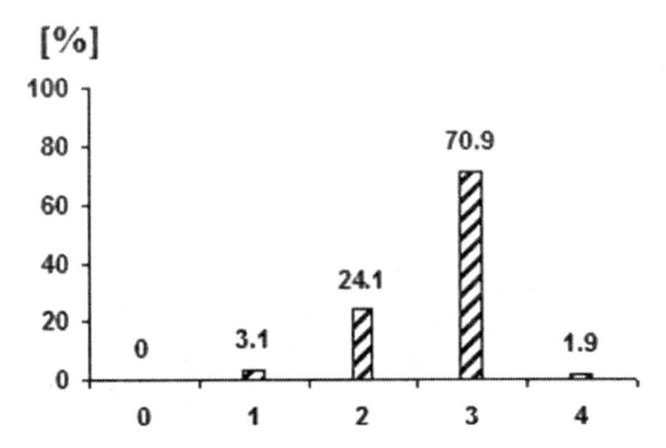

Figure 4. Lengths of sewers (percentage) with deflections categorised into five classes.

- failure in static calculations, e.g. those concerning the assumed load capacity of the pipes,
- embedment of either damaged pipes, or other than those stated in the design,
- pipe embedment which does not conform to the design, e.g. inappropriate soil compaction, or pipe embedment other than that stated in the design,
- thermal impact.

Pipe deflections exceeding the values given in the code only slightly affect the hydraulic parameters. At the deflection of 10%, the reduction in flow is only 1% (Stein 1999). Higher pipe inflections can produce more negative results, including the unsealing of pipe joints and a higher hazard of pipe buckling.

Another major disadvantage caused by higher pipe deflections, as discussed further on, is that it will be necessary to increase the thickness of liners that will have to be deployed for sewer rehabilitation. For pipe deflections greater than 10%, it is not possible to design CIPP sewer rehabilitation liners in accordance with the binding guidelines.

4 DEPENDENCE BETWEEN PLASTIC PIPE DEFLECTIONS AND REQUIRED THICKNESSES OF CIPP REHABILITATION LINERS

In accordance with the American National Standard (ASTM International) on the design of CIPP rehabilitation liners, the thickness t of non-structural CIPP liners can be calculated from formula:

$$t = \frac{D}{\left[\dfrac{2K \cdot E_L \cdot C}{P \cdot N \cdot (1-\nu^2)}\right]^{\frac{1}{3}} + 1} \tag{1}$$

where:

D—outside diameter of liner, [mm]
K—enhancement factor of the soil and of the existing pipe adjacent to the new one (a minimum value of 7.0 is recommended where there is full support of the existing pipe), [–]
E_L—long-term (time corrected) modulus of elasticity for CIPP, [N/mm²]
C—ovality reduction factor, [–]
P—groundwater load measured from the invert of the pipe, [N/mm²]
N—factor of safety, [–]
ν—Poisson's ratio (0.3 average), [–]

Table 1. Factor C dependence on the deflection of rehabilitated sewer (Per Aarsleff Polska 1996).

Deflections of the sewer [%]	0	1.0	2.0	4.0	5.0	6.0	8.0	10.0
Factor C	1.00	0.91	0.84	0.70	0.64	0.59	0.49	0.41

In the formula, the deflection of rehabilitated sewer is accounted for by means of factor C, which is taken from Table 1.

Using the standard above, it is possible to compute the thickness of fully structural CIPP liner from formula:

$$t = 0{,}721 \cdot D \cdot \left(\frac{\left(\frac{N \cdot q_t}{C} \right)^2}{E_L \cdot R_W \cdot B' \cdot E'_s} \right)^{\frac{1}{3}} \tag{2}$$

Some of the notations are explained above, the remaining ones represent the following:

q_t—total external pressure on pipe, [N/mm²]
R_w—water buoyancy factor, [–]
B′—coefficient of elastic support, [–]
E_s—modulus of soil reaction, [N/mm²]

When the pipe deflection is higher than 10%, the thickness of CIPP liner is large enough to choose trenchless replacement, trench replacement, or rehabilitation with lower cross-section pipes. Depending on the sewer diameter and location, that may prove a more economical solution. Rehabilitation with lower cross-section pipes is acceptable provided that it is possible to reduce the sewer capacity.

5 INFLUENCE OF PVC PIPE DEFLECTION ON THE THICKNESS OF NON-STRUCTURAL CIPP LINERS

Analysis of the impact of deflection of PVC pipes on the thickness of non-structural CIPP liners was performed for the following parameters:

a. material of pipes—PVC,
b. deflections—1, 2, 4, 6, 8, 10%,
c. diameters of PVC pipes—200, 500, 630 mm
d. technical state of the sewers—partially deteriorated condition,
e. depth of cover—1.5 m; 3.0 m,
f. short-term modulus of elasticity for CIPP—E_K = 2950 N/mm² (Per Aarsleff Polska 1996),
g. long-term (time-corrected) modulus of elasticity for CIPP—E_L = 1650 N/mm² (Per Aarsleff Polska 1996),
h. long-term (time-corrected) flexural strength for CIPP – σ_{dop} = 31.5 N/mm² (Per Aarsleff Polska 1996).

Minimal thicknesses of non-structural CIPP liners for different deflections and the depth of cover 1.5 m are shown in Figure 5, and for depth of cover 3.0 m—in Figure 6.

The thicknesses of non-structural CIPP liners in the pipes with diameters ø200, ø500 and ø630 mm increase with a growth in their deflections. For instance, for a deflection of 6%, the increase, relative to the deflection of 1%, amounts to:

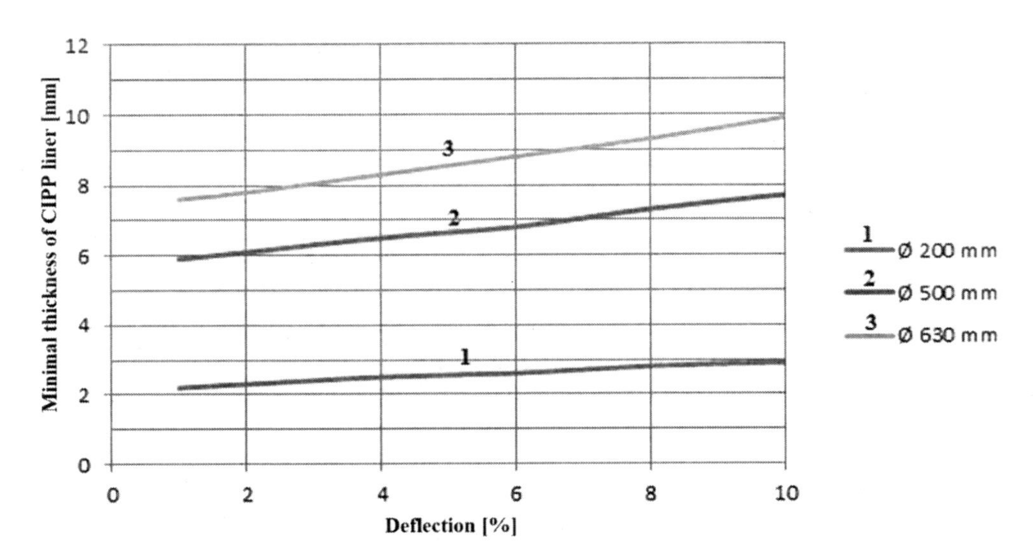

Figure 5. Minimal thicknesses of CIPP liners for the sewer depth cover 1.5 m and for different deflections (Beznosik 2012).

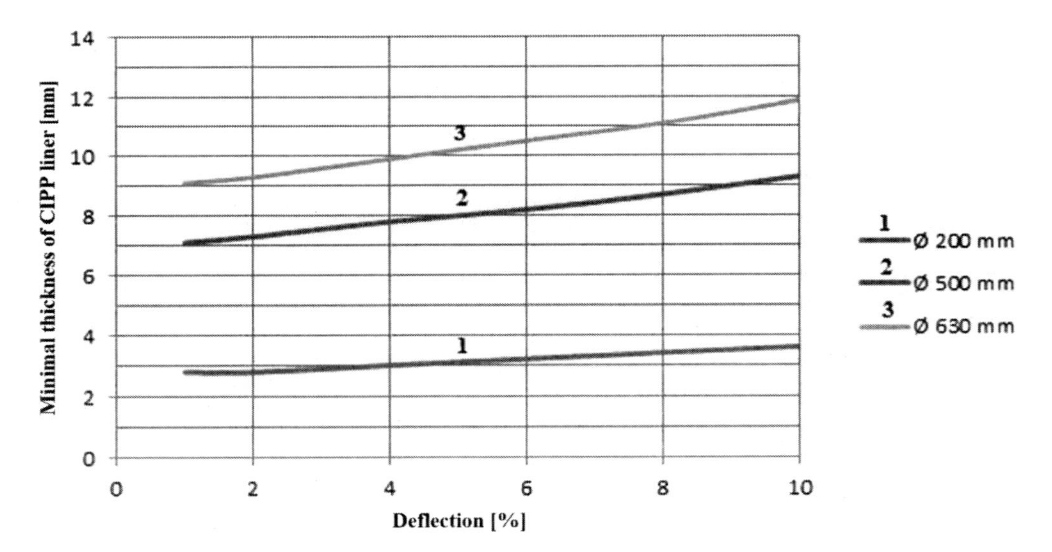

Figure 6. Minimal thicknesses of CIPP liners for the sewer depth cover 3.0 m and for different deflections (Beznosik 2012).

Table 2. Thicknesses of non-structural CIPP liners for the sewer with diameter ø500 mm and h = 3.0 m.

Deflection [%]	Minimal thickness of CIPP liner [mm]	Design thickness of CIPP liner [mm]
1	7.1	7.5
2	7.3	7.5
4	7.8	9.0
6	8.2	9.0
8	8.7	9.0
10	9.3	10.5

– for the cover depth h = 1.5 m, successively 18.2%, 15.3%, and 15.8%
– for the cover depth h = 3.0 m, successively 14.3%, 15.5%, and 15.4%.

For a deflection of 10%, the increase, relative to the deflection of 1%, amounts to:

– for the cover depth h = 1.5 m, successively 31.8%, 30.5%, and 30.3%
– for the cover depth h = 3.0 m, successively 28.6%, 31.0%, and 30.8%.

Design thickness of liners is selected from the catalogues issued by the companies that offer those technologies. Exemplary thicknesses of CIPP liners selected from the table provided in AT-15-7988/2009 for the PVC sewer with diameter 500 mm and h = 3 m are shown in Table 2.

In the analysed case (ø500 mm and h = 3 m), an increase in the design thickness of CIPP liner, when the deflection grows from 1% (for the thickness of 7.5 mm) to 10% (for the thickness of 10.5 mm), is equal to 40.0%.

6 INFLUENCE OF PVC PIPE DEFLECTION ON THE THICKNESS OF FULLY STRUCTURAL CIPP LINERS

The analysis of the impact of PVC pipe deflection on the thickness of fully structural liners was performed for the same parameters as before. One change was, however, introduced, which concerned point d), namely the technical state of the sewer. It was assumed that the sewer condition badly deteriorated. Minimal thicknesses of fully structural CIPP liners for different deflections and the sewer cover depth of 1.5 m is shown in Figure 7, and for the cover depth of 3.0 m in Figure 8.

Like before, thicknesses of fully structural liners in pipes with diameters ø200, ø500 and ø630 mm increase, as pipe inflections grow. For instance, for a deflection of 6%, the increase, relative to the deflection of 1%, amounts to:

– for the cover depth h = 1.5 m, successively 10.7%, 11.6%, and 11.5%
– for the cover depth h = 3.0 m, successively 36.7%, 32.9%, and 33.3%.

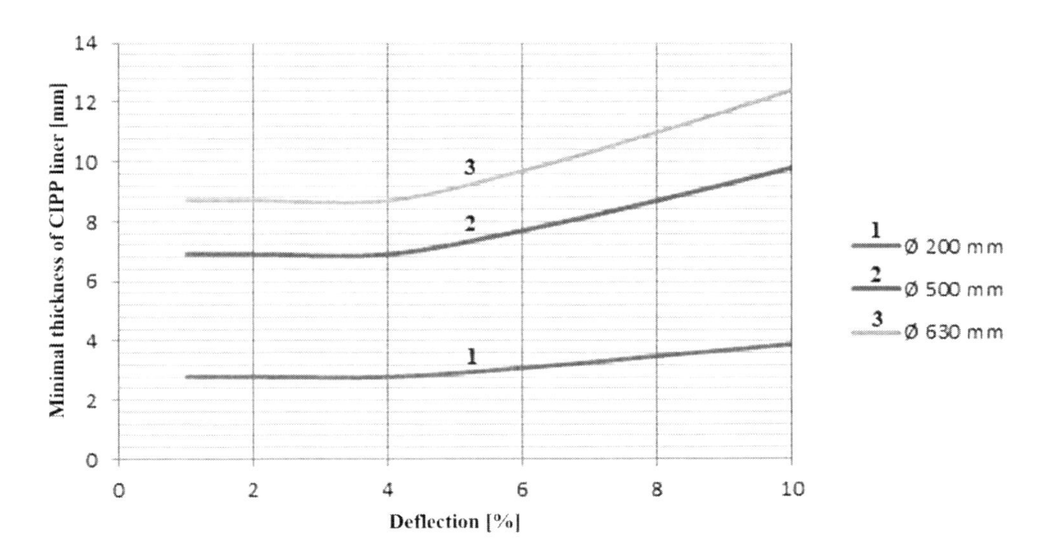

Figure 7. Minimal thicknesses of fully structural CIPP liners for the depth cover of 1.5 m and different deflections (Beznosik 2012).

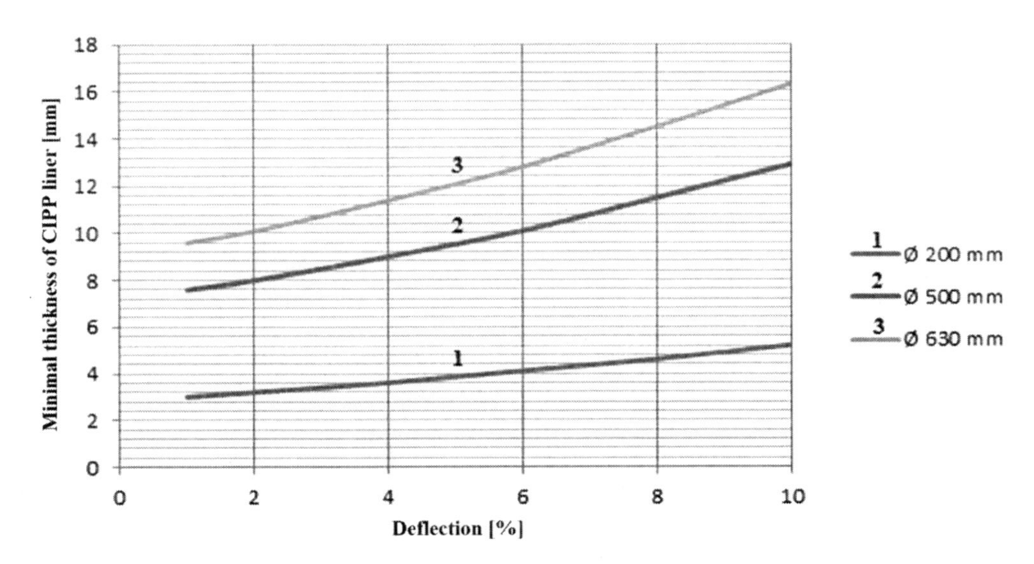

Figure 8. Minimal thicknesses of fully structural CIPP liners for the depth cover of 3.0 m and different deflections (Beznosik 2012).

Table 3. Thicknesses of fully structural CIPP liners for the sewer with diameter ø630 mm and h = 3 m.

Deflection [%]	Minimal thickness of CIPP liner [mm]	Design thickness of CIPP liner [mm]
1	9.6	10.5
2	10.1	10.5
4	11.4	12.0
6	12.8	13.5
8	14.5	15.0
10	16.3	18.0

For a deflection of 10%, the increase, relative to the deflection of 1%, amounts to:

- for the cover depth $h = 1.5$ m, successively 39.2%, 42.0%, and 42.5%
- for the cover depth $h = 3.0$ m, successively 73.3%, 69.7%, and 69.8%.

Design thicknesses of fully structural CIPP liners selected from the table provided in AT-15-7988/2009 for the PVC sewer with diameter 630 mm and $h = 3$ m are shown in Table 3.

In the analysed case (ø630 mm and $h = 3$ m), increase in the design thickness of CIPP liner, when the deflection grows from 1% (for the thickness of 10.5 mm) to 10% (for the thickness of 18.0 mm), is equal to 71.4%.

7 CONCLUSIONS

Field measurements of newly built PVC sewers, taken by means of the CCTV method and also with a device with a deflection gauge, indicate that along certain sections, pipe deflections are higher than the permissible ones. As regards deflections surveyed with the CCTV method, it was found that:

- along 13.3% of the length of pipes, deflections were greater than 6% and smaller than 15%,
- along 0.1% of the length of pipes, deflections were greater than 15% and smaller than 25%.

When deflections were inspected using a device equipped with a deflection gauge, the following results were obtained:

- along 70.9% of the length of pipes, deflections were greater than 2% and smaller than 6%,
- along 24.1% of the length of pipes, deflections were greater than 6% and smaller than 10%,
- along 3.1% of the length of pipes, deflections were greater than 10% and smaller than 15%.

The results obtained from field tests indicate that PVC pipes should be more carefully embedded to avoid excessive deflections.

In the guidelines (ATV-A 127 P 2000), it is acceptable to design sewer PVC pipes with deflections up to 6%, whereas in accordance with the guidelines (DWA-M-149-3 2007), sewer PVC pipes with deflections ranging from 2% to 6% are categorised, on the basis of the safety criterion, as belonging to Class 3—partially deteriorated sewers (classification of the technical state of sewers from Class 0—heavily deteriorated, to Class 4—without defects). A clash of recommendations resulting from the guidelines can be clearly seen.

In the future, thermoplastic pipes will need rehabilitation, just as presently it is the case with many brick, vitrified clay or concrete sewers that have been in service for a long time. Trenchless rehabilitation of plastic pipes with deflections will be, however, much more costly than the rehabilitation of rigid pipes, which frequently maintain their circular cross-section shape. The analysis presented in the paper indicates that increase in the thickness of CIPP rehabilitation liners for the sewers not laid deep in the ground (1.5–3.0 m), when deflection grows from 1.0% to 6.0%, ranges 10.7%–36.7%. For the deflection increase from 1.0% to 10%, liner thickness increase amounts to 28.6%–73.3%, in relation to liner thickness determined for the deflection of 1%. The data were obtained for the thickness of liners obtained by computations. The data depend on whether the liner is non- or fully structural, also on the diameter of the sewer, how deep it is buried in the ground, and the ground water table level relative to the sewer bottom.

Taking into account that all sewers built from plastic pipes, including PVC pipes, that are presently constructed will need to be rehabilitated in the future, it seems reasonable to seek changes in the code. The changes should involve lowering the values of permissible deflection of plastic pipes. The guidelines stated in (DWA-M-149-3 2007), which assume that the correct value of deflection should not exceed 2%, seem to show an advantageous trend.

REFERENCES

ASTM Standard: F 1216-09. 2009. Standard Practice for Rehabilitation of Existing Pipelines and Conduits by the Inversion and Curing of a Resin – Impregnated Tube, ASTM International, West Conshohocken, PA.

ATV–DVWK–A 127 P. 2000: Obliczenia statyczno-wytrzymałościowe kanałów i przewodów kanalizacyjnych, Wydawnictwo Seidel-Przywecki, s. 92.

ATV–DVWK–M127P–część 2. 2000: Obliczenia statyczno-wytrzymałościowe dla rehabilitacji technicznej przewodów kanalizacyjnych przez wprowadzenie linerów lub metodą montażową, Wydawnictwo Seidel-Przywecki, s. 86.

Beznosik M. 2012: Analiza wpływu ugięć rur PVC na grubość powłok renowacyjnych i rekonstrukcyjnych żywicznych oraz polietylenowych, praca magisterska, promotor: dr inż. E. Kuliczkowska, Politechnika Świętokrzyska, Kielce, s.64.

DWA-M-149-3.2007: Zustanderfassung und Beurteilung von Entwässerungssystemen außerhalb von Gebäuden, Deutsche Vereinigung für Wasserwirtschaft, Abwasser und Abfall e.V. Hennef, s. 66.

Kubicka U. 2000: Analiza ugięć rur z PVC ułożonych w gruncie, praca doktorska, Politechnika Świętokrzyska, Kielce, s. 191.

Kuliczkowska E. 2005: Wyniki badań nowo wybudowanych przewodów kanalizacyjnych z rur PVC, Gaz, Woda i Technika Sanitarna, nr 10, s. 16–20.

Kuliczkowski A. 2004: Rury kanalizacyjne, t. II. Projektowanie konstrukcji, Wydawnictwo Politechniki Świętokrzyskiej, Kielce, s. 507.

Kuliczkowski A. i in. 2010: Technologie bezwykopowe w Inżynierii Środowiska, Wydawnictwo Seidel-Przywecki, Warszawa, s. 735.

Kuliczkowski A., Kuliczkowska E. 2002: Odkształcalność—istotną zaletą rur z tworzyw sztucznych, Gaz, Woda i Technika Sanitarna, nr 5, s. 178–179.
Madryas C., Kolonko A., Wysocki L. 2002: Konstrukcje przewodów kanalizacyjnych, Oficyna Wydawnicza Politechniki Wrocławskiej, Wrocław, s. 377.
Per Aarsleff Polska. 1996: Informator, Projektowanie w technologii Insituform, Kraków, s. 32.
Stein D. 1999: Instandhaltung von Kanalisationen, 3 Auflage, Ernst & Sohn, Berlin, s. 941.

Underground Infrastructure of Urban Areas 3 – Madryas et al. (Eds)
 ISBN 978-1-138-02652-0

The effect of glazing of vitrified clay pipes to load capacity growth of them after rehabilitation with CIPP liners

A. Kuliczkowski & K. Mogielski
Department of Water Supply and Sanitary Installations, Kielce University of Technology, Kielce, Poland

ABSTRACT: The article describes tests of specimens made of vitrified clay DN 200 pipes with CIPP liners. A method of preparation of them is presented. Specimens were made of vitrified clay pipes with and without glazing. Liners installed inside them were 3.0, 4.5 and 6.0 mm thick and made of epoxy resin and polyethylene felt. Some of the specimens tested were without liners. Load capacity test procedure for all specimens made of pipes is described together with ring stiffness test procedure of specimens made of liners only. There are also described results of adhesion level (pull-off) tests of installed liners and roughness tests of pipes with and without glazing. Load capacities of both kinds of pipes with liners are compared. Differences in load capacity growth are shown. Obtained results were referred to roughness and pull-off tests. Factors reducing the thickness of the liner were determined for the case of pipes without glazing.

1 INTRODUCTION

Cured-In-Place Pipe (CIPP) lining is one of the most common trenchless technology used for underground pipeline rehabilitation (Kuliczkowski, 2000). Its growing popularity results from the fact that this technology can be used for non-, semi- and fully structural rehabilitation and for the rehabilitation of pipes conveying any type of fluid (Kuliczkowski et al., 2010). Its advantage over loose fit technologies lies—because of a significant reduction in hydraulic resistance—in potential for increasing the flow capacity of the rehabilitated pipelines (Kuliczkowski & Dańczuk, 2009).

Appropriate resistance of the lining to loads has a considerable influence on its long failure-free operation and reliability (Kuliczkowska, 2012; Kuliczkowska & Gierczak, 2013). The load that is primarily considered while choosing the thickness of the rehabilitation liner is groundwater pressure and, additionally, dead and live loads in the case of fully structural rehabilitation lining (Kuliczkowski, 2010). The authors of (Szot, 2002) add also under- and overpressure created by the flows inside the pipes as well as thermal loads.

The tests conducted on the liners laid in the pipelines 4 to 30 years earlier confirm (Allouche et al., 2014; Alzraiee et al., 2013; Bosseler & Schluter, 2002) that, in most cases, they meet the requirements in terms of leak-tightness. Their strength parameters are appropriately high to ensure failure-free operation time longer than 50 years design life. From the tests (Falter, 2007; Law & Moore, 2002; Kapasi & Hall, 2002) it follows that the liner thickness is a result of adopting simplified assumptions in the design methods (ASTM, 2009; Danish Standards, 2001; DWA-M127P part 2, 2000). According to (MC-Bauchemie, 2002), the thickness of the liner wall well bonded to the pipeline can be limited depending on the type of the resin used, up to 35.5% or 26.6% in relation to the thickness of the liner that is not bonded to the pipeline. This may evidence a considerable over dimensioning of certain liners in regard to the actual need, thus increasing the rehabilitation costs.

This paper deals with preparation and implementation of the experimental programme designed to find the influence of glazing on the increase in vitrified-clay pipe load capacity following the installation of resin CIPP liner inside the pipes. Other findings presented in this paper include the measurement results related to pipe inner surface roughness and the level of adhesion determined by pulling-off the liner from the pipes. The load capacity tests results were used to determine the factors that reduce the thickness of liners installed in non-glazed vitrified clay pipes.

2 EXPERIMENTAL PROGRAMME AND PREPARATION OF TEST SPECIMENS

The specimens were made from six DN200 vitrified clay pipes—three glazed and three non-glazed. Prior to installation, a 30 cm long reference specimen was cut out of each pipe. The reference specimens were used to determine base load capacity of each of the six pipes. The remaining parts of the pipes were fitted with CIPP liners with nominal thicknesses of 3.0, 4.5 and 6.0 mm, made of epoxy resin and polyester felt. The liners were inserted into the brand-new, clean and dry pipes using an inversion drum pressurized with compressed air and cured via the use of hot water. Then the pipes with the cured liners were cut into approximately 30 cm long sections. Each pipe provided five specimens including four with liners inside and one without the liner. Altogether 30 specimens were subjected to load capacity tests.

Resin liners were slightly longer than the total pipe length. Having been cured, the liners were used to produce specimens. Six CIPP liner specimens were made and subjected to ring stiffness tests. Two of the specimens had the nominal thickness of 3.0 mm, other two specimens were 4.5 mm thick and the remaining two were 6.0 mm in thickness. Table 1 presents the number of specimens and their characteristics (liner thickness, glazed or non-glazed pipes).

The tests, the results of which are discussed further in the paper, were carried out in the Kielce University of Technology as part of the research programme devised to determine the effect of the CIPP resin liners and pipes bonding on the load capacity of rehabilitated sewage pipes. The load capacity, ring stiffness and adhesion tests were conducted in the Department of Water Supply and Sanitary Installations. Surface roughness was measured in the Department of Manufacturing Engineering and Metrology. The influence of groundwater pressure on the external surface of the liner, which may reduce its adhesion to the pipeline, was not taken into account (Artillan et al., 2000).

3 LOAD CAPACITY AND RING STIFFNESS TESTS

The specimens with and without liners were tested in accordance with the requirements laid down in (PN-EN 295-3, 1999). The specimens were not soaked before the tests. The pipes were placed in the Zwick/Roell Z 100 universal testing machine on the steel, V-shaped lower bearing element with spread angle of 170°. The load was applied from the top and distributed evenly using a duralumin upper bearing element. Elastomeric belts with a hardness of 50 IRHD were inserted between the specimen and upper and lower bearing elements. Distribution of the bearing elements with the elastomeric belts is shown in Figure 1, and their dimensions are summarized in Table 2. The four pipe specimens with liners after the load capacity test are shown in Figures 2 and 3.

Table 1. Number of tested specimens.

Type of specimen	3.0 mm thick liner	4.5 mm thick liner	6.0 mm thick liner
Glazed vitrified clay pipe + liner	4 + 1*	4 + 1*	4 + 1*
Non-glazed vitrified clay pipe + liner	4 + 1*	4 + 1*	4 + 1*
Liner only	2	2	2

* Each pipe was used to produce four specimens with and one without liner (reference specimen).

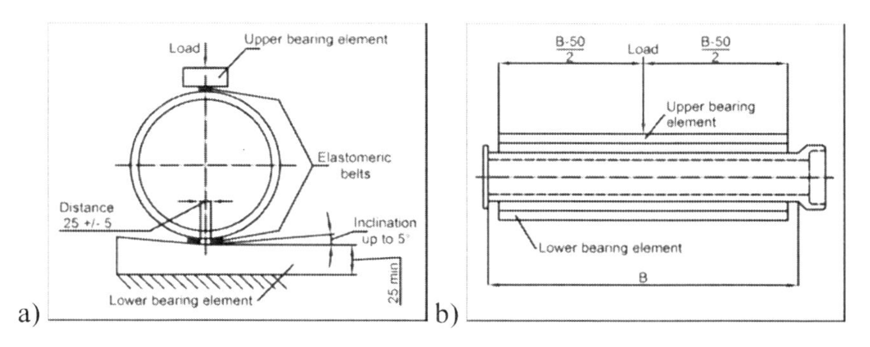

Figure 1. Placing vitrified clay pipe specimens in universal testing machine: a) front, b) side view (PN-EN 295-3, 1999).

Table 2. Bearing elements and elastomeric belts measurements.

Name of element	Height [mm]	Width [mm]	Length [mm]
Upper bearing element	30	30	400
Lower bearing element	12 ÷ 21*	200	400
Elastomeric belt	32	50	400

* Minimum and maximum height of V-shaped lower bearing element.

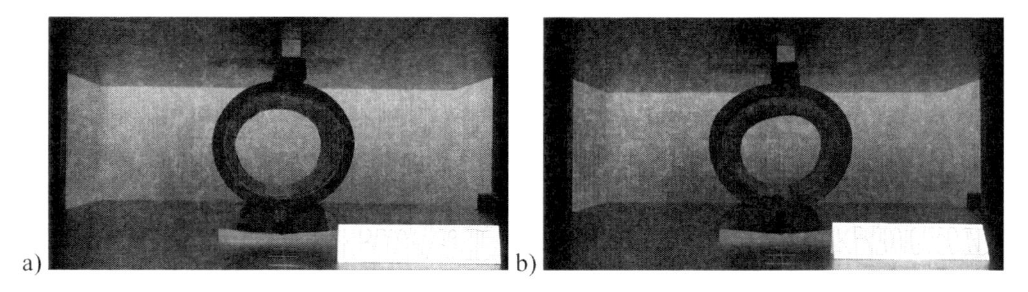

Figure 2. Specimens made of non-glazed vitrified clay pipes with liners: a) 4.5 mm thick and b) 6.0 mm thick after load capacity test [pictures by authors].

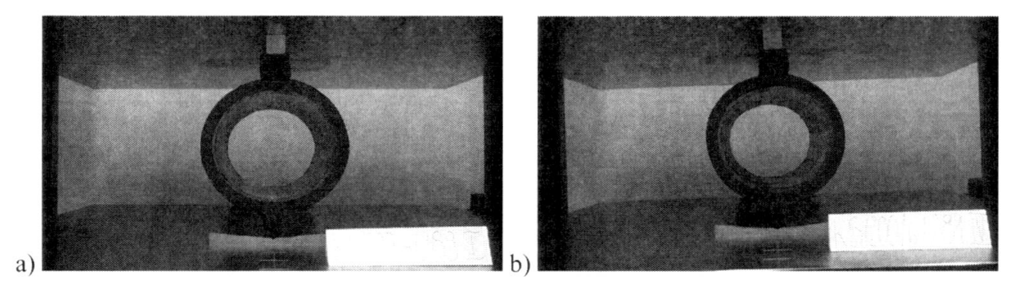

Figure 3. Specimens made of glazed vitrified clay pipes with liners: a) 4.5 mm thick and b) 6.0 mm thick after load capacity test [pictures by authors].

During the load capacity tests, the liners with thickness of 3.0 and 4.5 mm did not detach from the inner walls of the pipes after fracture (Fig. 2a and Fig. 3a), unlike the stiffer liners with 6.0 mm in thickness (Figs 2b and 3b). This was associated with the deformation of the vitrified clay pipe after its cracking at four points (at the crown, invert and springings). The cross-section of the pipe specimen loaded from the top turned from circular into more or less oval. Thinner liners with lower ring stiffness deflected with the pipe. Thicker liners with higher ring stiffness did not deflect, which resulted in their detachment from the pipe wall.

The ring stiffness tests were performed on the Zwick/Roell Z 20 universal testing machine as per standard (PN-EN 1228, 1999). The test pieces of the liner were placed on the lower compression plate, 40 × 40 cm, and compressed from the top with a beam, 10 × 40 cm. The lower compression plate and the upper beam were made of steel. Six specimens of approximately 30 cm in length, cut out from the cured liners, were subjected to the test (each two specimens had nominal thickness of 3.0, 4.5 and 6.0 mm). Each resin liner specimen was tested in accordance with (PN-EN 1228, 1999) at three positions every 120° around its circumference.

4 TEST RESULTS

The load capacity of the specimens without lining (six pieces) varied considerably (in the range from 36.87 to 51.79 kN/m²). For that reason, while comparing the test results with those for the pipes with lining, the base load capacity should be referred to, that is, the load capacity of the specimen without the liner. Base load capacity (100%) used to determine the normalized load capacity of each specimen with the liner is expressed in percentage and represents the ratio of the load capacity of the given specimen with the liner to the load capacity of the specimen without the liner, both produced from the same pipe. For example, if the specimen manufactured from the non-glazed pipe with 3.0 mm thick liner has the load capacity of 81.67 kN/m² and the specimen without the liner made from the same pipe has the load capacity of 51.79 kN/m², then the normalized load capacity of the specimen with the liner is equal to 81.67/51.79 * 100% = 157.68%.

Figures 4–8 and Table 3 show the load capacity test results of 30 pipe specimens with and without liners. Figures 4–6 compare normalized load capacities of glazed and non-glazed pipes with liners of nominal thickness 3.0 mm (Fig. 4), 4.5 mm (Fig. 5) and 6.0 mm (Fig. 6). Figures 7 and 8 compare normalized load capacities of pipe specimens with liners of nominal thickness 3.0, 4.5 and 6.0 mm, made from glazed (Fig. 7) and non-glazed (Fig. 8) pipes. Table 3 shows the results of load capacity tests for each specimen separately. Actual thickness of the liner without the inliner (column 4) is the mean value of 12 measurements of the liner thickness in each specimen (see item 6). Measured load capacities correspond to the results of measurements described in item 3, whereas the method of determining normalized load capacity was described at the beginning of item 4. Mean normalized load capacity were computed for a group of specimens with the same characteristics in terms of liner nominal thickness and presence or absence of glazing. Table 4 summarizes the results from ring stiffness tests of the samples made of liners only.

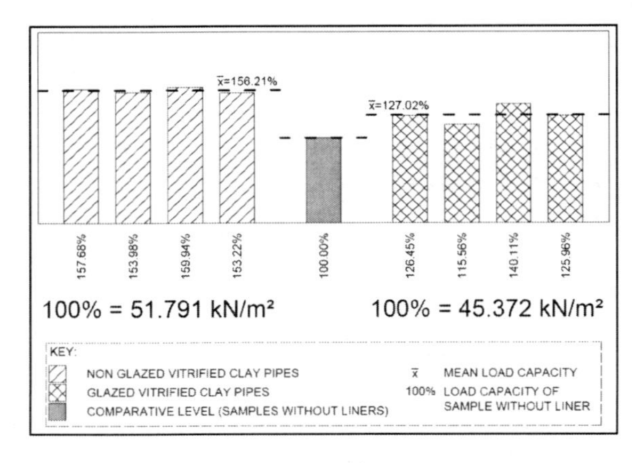

Figure 4. Normalized load capacity of vitrified clay DN 200 pipes with 3.0 mm thick liner.

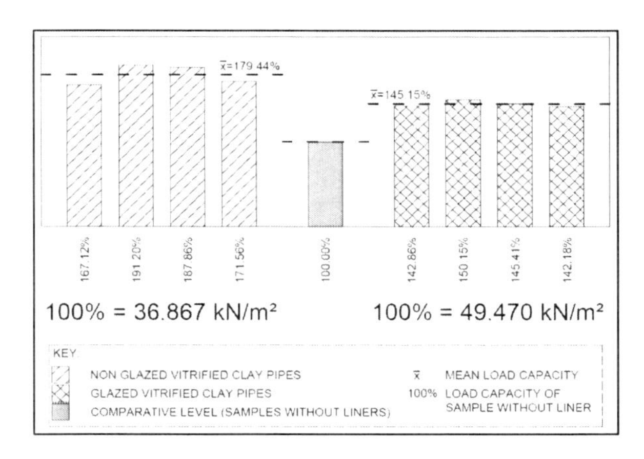

Figure 5. Normalized load capacity of vitrified clay DN 200 pipes with 4.5 mm thick liner.

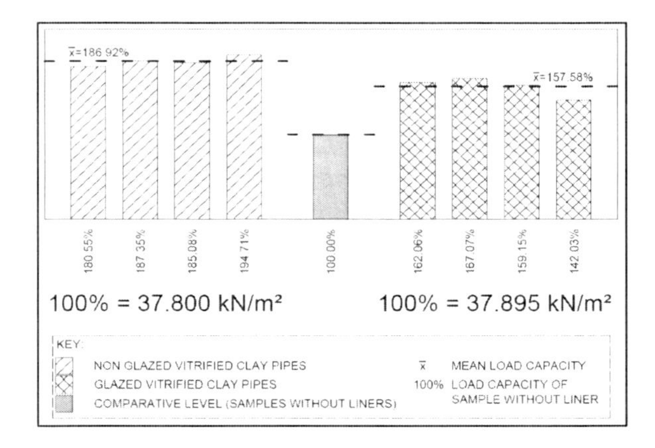

Figure 6. Normalized load capacity of vitrified clay DN 200 pipes with 6.0 mm thick liner.

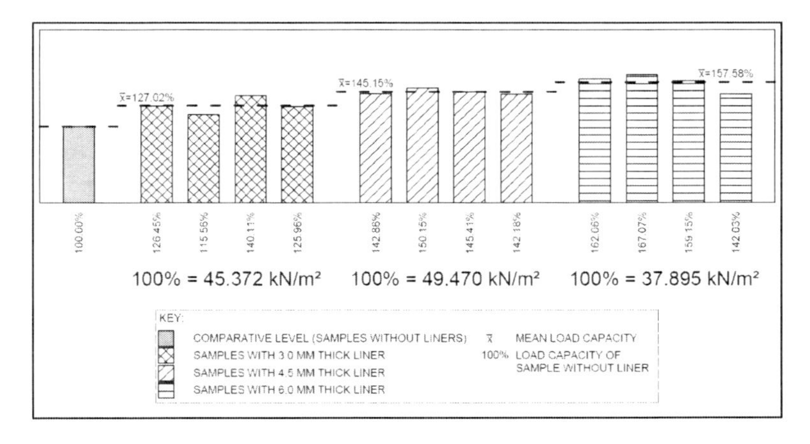

Figure 7. Normalized load capacity of glazed vitrified clay DN 200 pipes.

Compared with the load capacity of the pipes with liners, ring stiffness of the liners only is low and equals maximum 5.71% of the load capacity of the pipe with the liner in which it was installed (ring stiffness of the liner specimen L_200_1_60 II versus load capacity of specimen

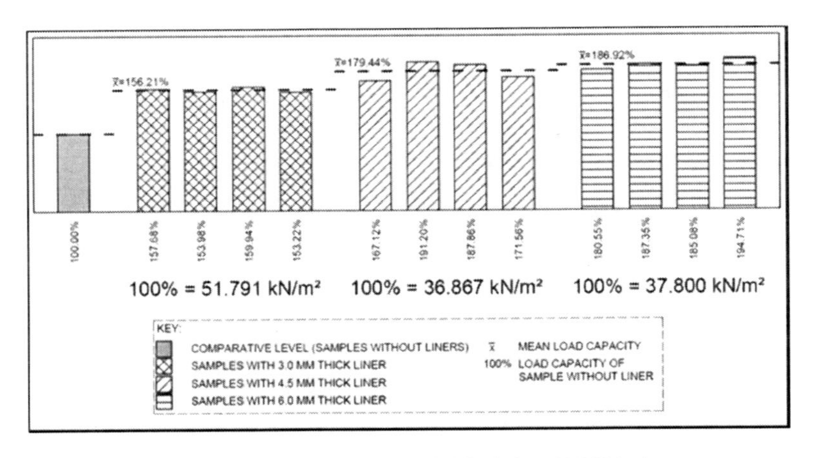

Figure 8. Normalized load capacity of non-glazed vitrified clay DN 200 pipes.

Table 3. Load capacity tests results.

Symbol of the specimen	Glazed (B) or non-glazed (S) pipe/nominal liner thickness [–]	Mean measured liner without inliner thickness [mm]	Measured load capacity [kN/m²]	Normalized load capacity [%]	Mean normalized load capacity [%]
KB_200_0_30 K	B/0.0	0.00	51.791	100.00	100.00
KB_200_0_30 I	B/3.0	3.24	81.665	157.68	156.21
KB_200_0_30 II	B/3.0	3.21	79.749	153.98	
KB_200_0_30 III	B/3.0	3.30	82.832	159.94	
KB_200_0_30 IV	B/3.0	3.19	79.353	153.22	
KB_200_0_45 K	B/0.0	0.00	36.867	100.00	100.00
KB_200_0_45 I	B/4.5	4.34	61.612	167.12	179.44
KB_200_0_45 II	B/4.5	4.13	70.491	191.20	
KB_200_0_45 III	B/4.5	4.48	69.258	187.86	
KB_200_0_45 IV	B/4.5	4.34	63.249	171.56	
KB_200_0_60 K	B/0.0	0.00	37.800	100.00	100.00
KB_200_0_60 I	B/6.0	5.59	68.247	180.55	186.92
KB_200_0_60 II	B/6.0	5.64	70.817	187.35	
KB_200_0_60 III	B/6.0	5.93	69.959	185.08	
KB_200_0_60 IV	B/6.0	5.90	73.600	194.71	
KS_200_0_30 K	S/0.0	0.00	45.372	100.00	100.00
KS_200_0_30 I	S/3.0	3.56	57.371	126.45	127.02
KS_200_0_30 II	S/3.0	3.47	52.434	115.56	
KS_200_0_30 III	S/3.0	3.49	63.570	140.11	
KS_200_0_30 IV	S/3.0	3.10	57.148	125.96	
KS_200_0_45 K	S/0.0	0.00	49.470	100.00	100.00
KS_200_0_45 I	S/4.5	4.21	70.671	142.86	145.15
KS_200_0_45 II	S/4.5	4.13	74.279	150.15	
KS_200_0_45 III	S/4.5	4.17	71.933	145.41	
KS/200_0_45 IV	S/4.5	4.39	70.334	142.18	
KS_200_0_60 K	S/0.0	0.00	37.895	100.00	100.00
KS_200_0_60 I	S/6.0	5.32	61.412	162.06	157.58
KS_200_0_60 II	S/6.0	5.57	63.311	167.07	
KS_200_0_60 III	S/6.0	5.55	60.309	159.15	
KS_200_0_60 IV	S/6.0	5.43	53.822	142.03	

Table 4. Ring stiffness tests results.

Symbol of the specimen [–]	Nominal liner thickness [mm]	Ring stiffness [kN/m²]	Mean ring stiffness [kN/m²]
L_200_1_30 I	3.0	0.708	0.782
L_200_1_30 II	3.0	0.855	
L_200_1_45 I	4.5	1.859	1.941
L_200_1_45 II	4.5	2.022	
L_200_1_60 I	6.0	2.569	2.822
L_200_1_60 II	6.0	3.074	

KS_200_0_60 IV is 3.07/53.82 * 100% = 5.71%). It suggests that ring stiffness should not have a great influence on the increase in load capacity of the pipe after rehabilitation.

5 ROUGHNESS MEASUREMENTS AND TEAR-OFF TESTS

Roughness of inner surfaces of the vitrified clay pipes, glazed and non-glazed, was assessed on a Form Talysurf PGI 1200 profiler. Roughness parameters were determined using a Gaussian filter with a wavelength of λc = 0.8 mm (glazed pipes) and λc = 2.5 mm (non-glazed pipes). The measurements were performed on four pipe specimens without liners: two glazed and two non-glazed. Their roughness was studied on five measurement segments each of which had a length of five elementary segments equal to λc. This means that the measurement segment was 4.0 mm long on the glazed pipe and 12.5 mm long on the non-glazed pipe. Altogether 20 surface roughness measurements were made along the pipe (in the flow direction). Value Ra, that is, the arithmetic mean of the given profile ordinates was taken to represent the surface roughness (Adamczak, 2008). One of the specimens under test is shown in Figure 9. A summary of the roughness tests results is presented in Table 5.

The pull-off tests were performed on six specimens. Three tests were carried out on each specimen, thus the total number of measurements made was 18. The testing was conducted in compliance with the requirements laid down in (PN-EN 1542, 2000). To minimize the influence of the pipe curve on the measurement results, test discs 20 mm in diameter were used during the test. Owing to this, the point furthest away from the liner surface was located 0.5 mm from the test disc. For that reason the liner surface was treated as approximately flat. Table 6 summarizes the results from the pull-off tests.

6 COEFFICIENTS REDUCING THE THICKNESS OF LINERS INSTALLED IN NON-GLAZED VITRIFIED CLAY PIPES

Actual thicknesses of the liners were different from their nominal thicknesses. This is associated with the specific characteristics of the CIPP technology, in which the liner is formed in-situ, not in the plant.

The thicknesses of the liners were measured to determine the correlation between the actual liner thickness and the force needed to destroy the specimen. The measurements were performed to (PN-EN 1228, 1999) for each specimen at 12 sites distributed every 60° around the circumference on both sides. To this end, a digital calliper was used. The arithmetic mean from 12 measurements was taken to be the actual thickness of each liner. All the liners installed had 0.6 mm thick polyurethane inliners fixed on their inner surfaces. The inliner prevented the resin from blending with hot water used for curing. Structural action of the membrane (inliner) is negligibly small compared with the action of cured resin (Boot et al., 2004); therefore its thickness was not accounted for in calculating coefficients reducing the thickness of the liner. To graphically determine the coefficients, graphs were prepared to

Figure 9. Form Talysurf PGI 1200 profilometer during specimen roughness measurement [picture by authors].

Table 5. Roughness measurements results.

		Roughness R_a [μm]		
Specimen material	Specimen symbol	Mean	Minimum	Maximum
Non-glazed vitrified clay	KB_200_0_30 K	8.8501	7.7065	10.2632
	KB_200_0_60 K	6.8421	5.7166	8.0378
	Mean	7.8461	–	–
Glazed vitrified clay	KS_200_0_30 K	0.2471	0.1619	0.3682
	KS_200_0_60 K	0.3618	0.2877	0.5139
	Mean	0.3045	–	–

Table 6. Pull-off tests results.

		Adhesion f_h [MPa]			
Specimen material	Specimen symbol	Measurement 1	Measurement 2	Measurement 3	Mean
Non-glazed vitrified clay	KB_200_0_30 III	1.455	1.688	1.367	1.503
	KB_200_0_45 III	2.105	2.179	2.501	2.262
	KB_200_0_60 III	2.128	2.207	1.995	2.110
	Mean	–	–	–	1.958
Glazed vitrified clay	KS_200_0_30 III	0.416	0.872	1.036	0.774
	KS_200_0_45 IV	1.329	0.571	2.016	1.305
	KS_200_0_60 I	0.606	1.463	0.838	0.969
	Mean	–	–	–	1.016

show the dependence of the actual thickness of the liner on the normalized load capacity of the specimens. One such graph is presented in Figure 10a. Figure 10b shows polynomial regression curves describing the relationship between the specimen load capacity and the actual thickness of the liner, separately for glazed and non-glazed pipes. Figure 10c describes graphically the method of determining liner thickness-reducing coefficients for six different values of normalized load capacity.

Dividing the thicknesses (read from Fig.10c) of liners installed in the glazed pipe specimens by the thicknesses of the liners installed in the non-glazed pipe specimens of the same normalized load capacity values gave the liner thickness-reducing coefficients expressed by formulas 1–6. Liner thickness calculated for the glazed vitrified clay pipe and multiplied

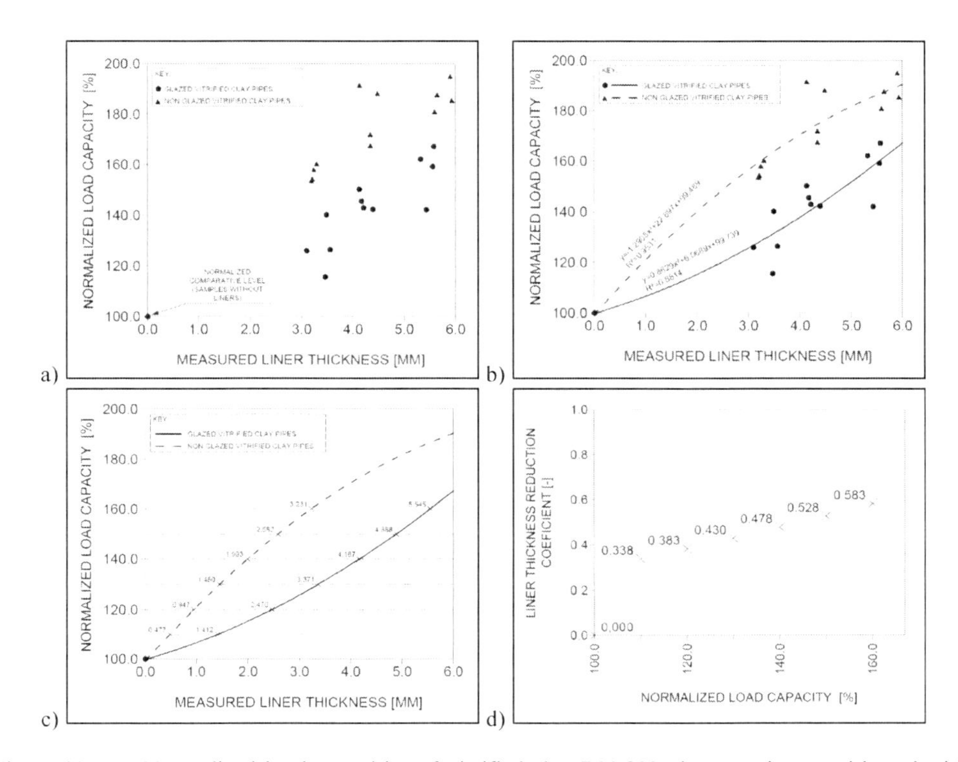

Figure 10. a) Normalized load capacities of vitrified clay DN 200 pipes specimens with and without glazing, b) regression drawings: normalized load capacity of the specimens made of vitrified clay DN 200 pipes with and without glazing in a function of measured liner thickness, c) normalized load capacity of vitrified clay DN 200 pipes specimens with and without glazing. Intersections of regression drawings and vertical lines show thicknesses of the liners with the same normalized load capacity (from 110% to 160% with 10% step), d) values of the liner thickness reduction coefficients or specimens with the same normalized load capacities (see Eq. 1–6).

by that coefficient will give a smaller thickness of the liner which might be installed in the non-glazed pipe at the same detrimental force increase. These coefficients are also shown in Figure 10d.

$$W_{KB.D200.\%110} = \frac{0.477}{1.412} = 0.338[-] \quad (1)$$

$$W_{KB.D200.\%120} = \frac{0.947}{2.470} = 0.383[-] \quad (2)$$

$$W_{KB.D200.\%130} = \frac{1.450}{3.371} = 0.430[-] \quad (3)$$

$$W_{KB.D200.\%140} = \frac{1.993}{4.167} = 0.478[-] \quad (4)$$

$$W_{KB.D200.\%150} = \frac{2.582}{4.888} = 0.528[-] \quad (5)$$

$$W_{KB.D200.\%160} = \frac{3.231}{5.545} = 0.583[-] \quad (6)$$

The coefficients determined in Equations 1–6 indicate that the thickness of the liners installed in non-glazed vitrified clay pipes might be 33.8%; 38.3%, 43.0%, 47.8%, 52.8% and

58.3% of the thickness of the liners installed in the glazed pipes at the same level of the load capacity increase. Figure 10d shows that the value of the liner thickness-reducing coefficient increases with increasing normalized load capacity. This means that at very thick liner the coefficient can reach the value of 1. This would suggest that the load capacity increase in both pipe types is the same because ring stiffness of the liner (closely related to its thickness) is high enough to render the adhesion effect minor.

Coefficients reducing the liner thickness mentioned earlier (35.5% and 26.6%) are similar only to the result from the tests conducted on specimens whose normalized load capacity is low (110.0 and 120.0%). This indicates that in the case of glazed and non-glazed vitrified clay pipes these coefficients are correct only for a narrow application range of the liners (the liners only slightly increasing load capacity of the pipe, that is, non-structural liners, not fully-structural ones).

The tests did not account for the presence of groundwater around the pipe. The water pressure on the external wall of the liner might result in its detachment from the pipe after certain time (Artillan et al., 2000) and reduction in pipe-liner bonding, thereby decreasing the load capacity of the pipe with liner.

7 INFLUENCE OF PIPE ROUGHNESS AND THE ADHESION LEVEL BETWEEN PIPE AND LINER ON THE INCREASE IN THE PIPE LOAD CAPACITY AFTER REHABILITATION

From the results summarized in Table 6 it follows that the mean roughness R_a of non-glazed vitrified clay pipes (7.8461 μm) is about 26 times the roughness of the glazed pipes (0.3045 μm). It directly translates into a varied level of adhesion between the liners and the pipes, which was confirmed in the pull-off tests. The results from those tests are shown in Table 6. The adhesion of the epoxy resin liner f_h to the non-glazed pipes is on average 92.7% (Tab. 5) greater than that observed in the glazed pipes.

The graph in Figure 11 shows the relationship between the normalized load capacity of the specimens and their roughness. The dependence between these two values is considerable in the case of samples with liners with nominal thickness of 3.0 and 6.0 mm.

Figure 12 presents a graph of the correlation between the normalized load capacity and the level of adhesion between the pipe and the liner. For the non-glazed pipes with greater roughness, the dependence between the increase in normalized load capacity and the adhesion level is high for three different nominal liner thicknesses. For smooth glazed pipes, no such dependence was observed (the plot is approximately steady). This suggests that the increase in the load capacity of glazed pipe specimens depends on the ring stiffness of the liner and is independent of the good adhesion of the liner to the pipe.

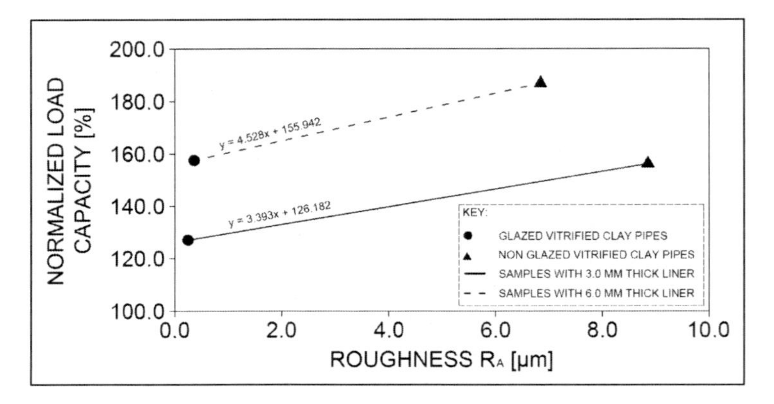

Figure 11. Correlation between mean normalized load capacity (of four specimens) and mean roughness of specimen without liner made of the same pipe.

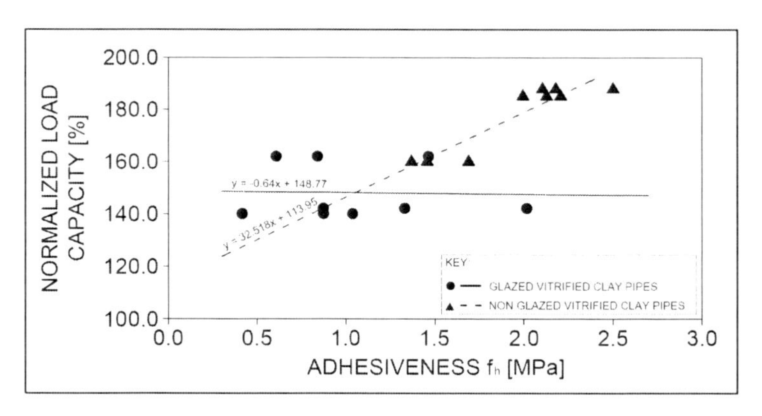

Figure 12. Correlation between mean normalized load capacity (of six specimens) and mean adhesion level between liner and pipe.

8 CONCLUSIONS

- Much lower load capacity increase was confirmed for glazed vitrified clay pipes DN 200 relative to the same non-glazed pipes after the installation of CIPP rehabilitation liners. The increase was in the range from 15.56% to 67.07% for the glazed pipes and from 53.22% to 94.71% for the non-glazed pipes (Tab. 3). This indicates that glazing has an influence on the load capacity change in the pipes.
- Lack of glazing in the vitrified clay pipe provides an opportunity to considerably reduce the thickness of the liner. This reduction varies from case to case and can reach 33.8%–58.3% of the thickness of the liner fixed in the glazed pipe (Fig. 10d), assuming that there is no groundwater near the pipeline (see item 6).
- Lack of glaze in the vitrified clay pipe and good adhesion of the liner reduce tensile strain spreading in the top and bottom parts inside the pipe, thereby increasing the load capacity of the pipe-liner system (synergy effect).
- Due to their sensitivity to deformation, liners with nominal thickness of 6.0 mm detached from the inner surfaces of the glazed and non-glazed pipes. After cracking, the liners with smaller nominal thickness did not detach but they deformed at the cracking location (Fig. 2 and Fig. 3).
- The greatest differences in load capacity increase between the glazed and unglazed pipes occurred when the liners were the thinnest. For that reason the coefficients reducing the liner thickness have the lowest values (Eq. 1–6 and Fig. 10d). The coefficients increase with increasing liner thickness, which may be associated with a relatively smaller reduction in the liner thickness after installation in a non-glazed pipe. This results from greater ring stiffness of thicker liners and from much smaller effect of the pipe-liner adhesion on the pipe load capacity increase.
- Load capacity of the vitrified clay pipes increases with increasing liner thickness, regardless of the presence or absence of glazing (Fig. 7 and Fig. 8).
- The glazed and non-glazed pipes used in the tests were supplied by the same manufacturer. The only differences between them included glazing or its absence and the resulting varied roughness of their surfaces. This suggests that different pipe-liner adhesion level (Fig. 11 and Fig. 12) is responsible for different increase in load capacity (Figs 10a–10c).
- The results from the tests described here can be used in the future to correct currently used methods of CIPP liners design.

ACKNOWLEDGMENT

The authors would like to thank the MC-Bauchemie and Unimark Sp. z o.o. for the preparation of the specimens examined in the study presented in the article and the companies

Keramo Steinzeug N.V., P.V. Prefabet Kluczbork S.A. and Pipelife Polska S.A., Marywil S.A. and Insituform Sp. z o.o. for making the test materials available.

The research was performed using equipment the purchase of which was co-financed by the European Union from the resources of the European Regional Development Fund within the framework of MOLAB project.

REFERENCES

Adamczak S. 2008. Pomiary geometryczne powierzchni: zarysy kształtu, falistość i chropowatość. Warsaw: WNT.

Allouche E. et al. 2014. A pilot study for retrospective evaluation of cured-in-place pipe (CIPP) rehabilitation of municipal gravity sewers. Tunnelling and Underground Space Technology. Vol. 39. pp. 82–93.

Alzraiee H. et al. 2013. A retrospective evaluation of cured-in-place pipe (CIPP rehabilitation technique used in municipal sewers. Proceedings. NASTT No-Dig Sacramento. pp. 1–9.

Artillan L. et al. 2000. A new design method for non-circular sewer linings, Proceedings. International No-Dig Perth. pp. 1–11.

ASTM Standard F1216–09. 2009. Standard practice for rehabilitation of existing pipelines and conduits by the inversion and curing of a resin-impregnated tube. West Conshohocken, PA. ASTM International.

Boot J.C. et al. 2004. The structural performance of polymeric linings for nominally cylindrical gravity pipes. Thin-Walled Structures. Vol. 42. pp. 1139–1160.

Bosseler B., Schluter M. 2002. Durability of liners, influence, effects and trends. Proceedings. International No-Dig 2002. Copenhagen.

Danish Standards. 2001. Static calculations of liners for the gravitational sewers. Taastrup.

DWA-M127P part 2. 2000. Obliczenia statyczno-wytrzymałościowe dla rehabilitacji technicznej przewodów kanalizacyjnych przez wprowadzanie linerów lub metodą montażową. Warsaw: Seidel—Przywecki.

Falter B. 2007. Buckling experiments on egg-shaped PE linings. Proceedings. International No-Dig Roma. pp. 1–12.

Kapasi S.I., Hall D. 2002. Monitoring deflections of pipe liners under external water pressure during liner buckling experiments. Proceedings. NASTT No-Dig Montreal.

Kuliczkowska E. 2012. Utrata stateczności rur i powłok renowacyjnych na wybranych przykładach. Gaz, Woda i Technika Sanitarna., No. 7, pp. 297–301.

Kuliczkowska E., Gierczak M. 2013. Buckling failure numerical analysis of HDPE pipes used for trenchless rehabilitation of a reinforced concrete sewer. Engineering Failure Analysis. Vol. 32. pp. 106–112.

Kuliczkowski A. 2000. Bezwykopowe technologie renowacji nie przełazowych przewodów kanalizacyjnych. Instal. No. 1, pp. 2–5, 30.

Kuliczkowski A., Dańczuk P. 2009. Hydrauliczne aspekty bezwykopowej renowacji przewodów kanalizacyjnych. Gaz, Woda i Technika Sanitarna. No. 1. pp. 16–19, 30.

Kuliczkowski A. 2010. Błędy w ustalaniu wymagań dotyczących wartości sztywności obwodowej utwardzanych in situ powłok żywicznych. Instal. No. 4. pp. 63–67.

Kuliczkowski A. et al. 2010. Technologie Bezwykopowe w Inżynierii Środowiska. Warsaw: Seidel—Przywecki.

Law M., Moore I.D. 2002. Laboratory investigation of the static response of repaired sewers. Proceedings. Pipelines. pp. 1–13.

MC-Bauchemie. 2002. Nowoczesne systemy bezwykopowej renowacji systemów kanalizacyjnych firmy—info materials of MC-Bauchemie company.

PN-EN 1228. 1999. Systemy przewodowe z tworzyw sztucznych. Rury z termoutwardzalnych tworzyw sztucznych wzmocnionych włóknem szklanym (GRP). Oznaczanie początkowej właściwej sztywności obwodowej. Warsaw: PKN.

PN-EN 1542. 2000. Wyroby i systemy do ochrony i napraw konstrukcji betonowych. Metody badań. Pomiar przyczepności przez odrywanie. Warsaw: PKN.

PN-EN 295-3. 1999. Rury i kształtki kamionkowe i ich połączenia w sieci drenażowej i kanalizacyjnej. Warsaw: PKN.

Szot A. 2002. Nośność rurociągów kanalizacyjnych wzmacnianych polimerowymi wykładzinami ciągłymi. PhD thesis. Wrocław: Wrocław University of Technology, Civil Engineering Institute.

Underground Infrastructure of Urban Areas 3 – Madryas et al. (Eds)
© 2015 Taylor & Francis Group, London, ISBN 978-1-138-02652-0

The technical and economic conditions for the construction of the central section of Metro Line II in Warsaw

J. Lejk
Metro Warszawskie Sp. z o.o., Poland

ABSTRACT: This presentation will deal with selected aspects of constructing the central section of Metro Line II in Warsaw. To fit within the conference format I have focussed on only one aspect of the investment process, which is the establishing of conditions for the safety of works, while also discussing some economic issues. Many events during construction, technologies used and results will require separate discussions and presentations.

The main premise in formulating the construction objectives was to complete the process in as short a time as possible, while retaining the appropriate economic parameters. For obvious reasons, the overriding goal was to maintain top safety levels across the investment project.

In economic terms the investment entitled *"The central section of Metro Line II in Warsaw"* involves four major projects—the construction of the central section of Metro Line II, the construction of the infrastructure for the future Metro network, the conversion and construction of a new urban infrastructure, and the modernisation of land development over and around the construction sites, along with the purchase of rolling stock for Metro Lines I and II.

1 INTRODUCTION

The harmonious development of Warsaw is dependent on the growth of a modern transport system. This is in addition to the fact that the capital city of Poland, as the country's political and economic centre, is facing challenges connected with outward transportation links—with both the immediate region and the rest of Europe—and the need to provide the residents and visitors with convenient means of commuting around the city. Consequently, we have to model our transportation system in such a way as to make the living standards in Warsaw on a par with those offered by other European cities.

One of the key elements of the transportation system which should help us achieve this objective is the designing and launching of a metro line system.

The first spatial planning decisions regarding Line II of the Warsaw metro were made in 1992. On 26 September of that year, the Municipal Council adopted "The Local General Land Development Plan for the Capital City of Warsaw".

Successive detailed studies pursued a variant providing for a 32-km line with 27 stations and a new technical holding station in the city's western part (Mory). (Figure 1). The whole project was divided into four tasks:

- the western section—8 stations, 10.2 km;
- the north-eastern section—6 stations, 7.5 km;

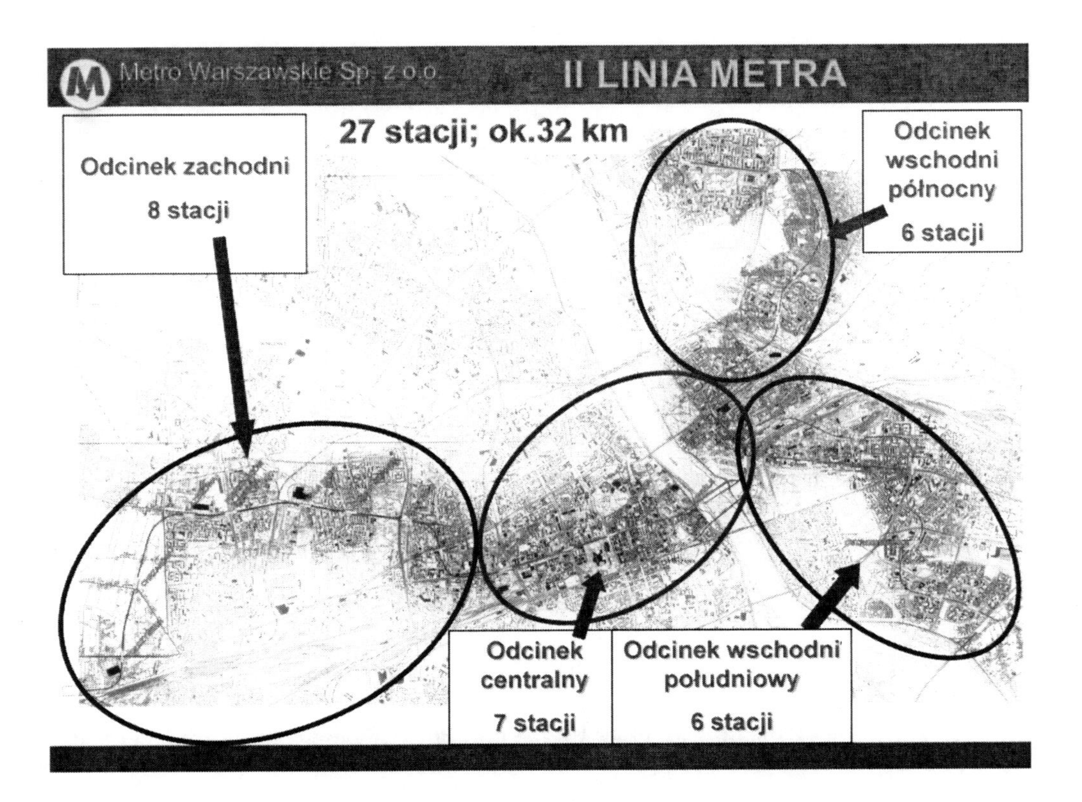

II LINIA METRA	METRO LINE II
27 stacji; ok. 32 km	27 stations; about 32 km
odcinek zachodni	western section—8 stations
odcinek wschodni-północny	north-eastern section—6 stations
odcinek wschodni-południowy	south-eastern section—6 stations
odcinek centralny	central section—7 stations

Figure 1.

- the south-eastern section—6 stations, 7.8 km;
- the central section—7 stations, 6.5 km.

The project started from the central section which consists of (Figure 2):

- seven stations, two of which are an element of the metro network (the Świętokrzyska station for passenger connections between Lines I and II and the Stadion Narodowy station for future connections with the south-eastern section),
- a one-track tunnel connecting Lines I and II,
- the redevelopment of the track layout within the area where both Lines connect,
- two track-reversing terminals on both ends of the section,
- 6308 metres of a double, one-track tunnel and a 600-m passage below the Vistula bed.

In this paper I would like to describe selected details of the construction of the central section of Metro Line II in Warsaw. To meet obvious requirements arising from the formula of the conference, I will focus on just one aspect of the investment process, i.e. the creation of the conditions ensuring safe construction and certain economic aspects. It would require

II LINIA METRA—ODCINEK CENTRALNY	METRO LINE II—THE CENTRAL SECTION
6,5 km długości	6.5 km in length
7 stacji	7 stations
Łącznik I i II linii	Connector of Lines I and II
Komora do zawracania	Track-reversing terminal
Wspólna stacja I i II linii	Joint station for Lines I and II
Węzeł Przesiadkowy „Stadion"	The "Stadion" transfer node

Figure 2.

separate studies and presentations to cover other, numerous, events accompanying the construction, along with the technologies used and the achieved results.

2 PROJECT OBJECTIVES

Working on the project objectives, we were guided mainly by the need to make the construction time as short as possible, with optimum economic parameters observed. The primary task was, obviously, to meet the strictest safety standards throughout the investment process.

The following assumptions were made:

- The stations will be built with the use of the open-pit method,
- All stations will be built simultaneously,
- The tunnels will be drilled with a tunnel-boring machine,
- There will be two one-track tunnels with prefabricated reinforced-concrete structures,
- The platforms will be arranged to resemble a central island.

3 SELECTED PREPARATORY PROCESSES

One of the first project activities was to collect data on the land and water conditions within the construction site. To this end, 650 test drillings were ordered to a total length of 14 800 metres, along with seismic probes and analyses of water-bearing strata. All these studies and analyses provided us with a picture of the stratigraphic section. (Figure 3).

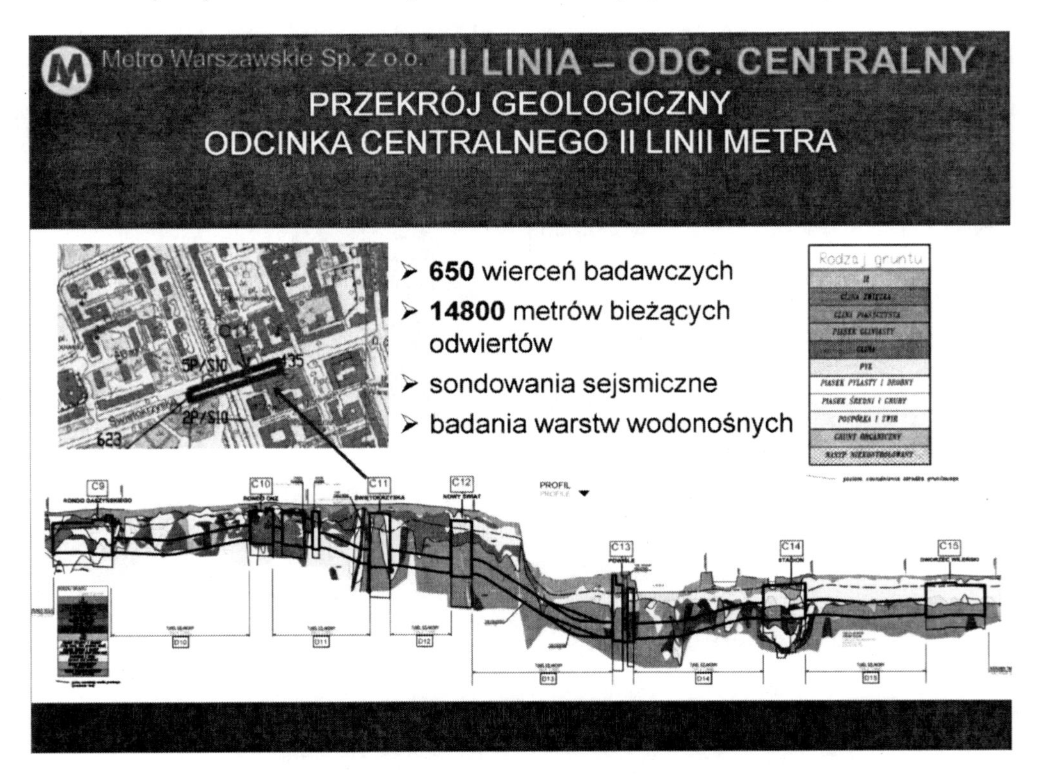

II LINIA METRA—ODCINEK CENTRALNY	METRO LINE II—THE CENTRAL SECTION
Przekrój geologiczny	The stratigraphic cross-section of the central section of Metro Line II
650 wierceń badawczych	650 test drillings
14 800 metrów bieżących odwiertów	Drillings to the length of 14 800 linear metres
sondowania sejsmiczne	Seismic probing
badania warstw wodonośnych	Analyses of water-bearing strata
Rodzaj gruntu	Soil type
Ił	Loam
Glina zwięzła	Compacted clay
Glina piaszczysta	Sandy clay
Piasek gliniasty	Clayey sand
Glina	Clay
Pył	Silt
Piasek pylasty i drobny	Silty and fine sand
Piasek średni i gruby	Medium and coarse sand
Pospółka i żwir	Sandy gravel and pebble
Grunt organiczny	Organic soil
Nasyp niekontrolowany	Uncompacted embankment
Poziom zawodnienia ośrodka gruntowego	Water accumulation in the soil

Figure 3.

Rejon stacji C-15	The vicinity of the C-15 station
Przed rozpoczęciem budowy została sporządzona dokumentacja (...)	Before the construction started, documentation had been drawn up to specify the technical condition of the buildings within the impact zone of the central section under construction. Each building adjacent to the construction site was assigned a unique catalogue card.
Informacja z kart katalogowych budynków	The information in the buildings' catalogue cards
Targowa—stan budynku w 2007 r. (...)	Targowa St.—the building's technical condition according to its catalogue card: unsatisfactory (4 out of 5 on the ITB scale); 73 measuring devices have been mounted on the building
Targowa—stan budynku w 2010 r. (...)	Targowa St.—the building's technical condition according to its catalogue card: unsatisfactory (4 out of 5 in the ITB scale); 9 measuring devices have been mounted on the building
Targowa—zarysowania (...)	Targowa St.—wall scratches between the windows
Targowa—elewacja od strony (...)	Targowa St.—wall cracks in the facade, as seen from the yard

Figure 4.

On the basis of the collected material, a boring machine was selected for the tunnels—the EPB TBP with an external diameter of 6.27 m.

In the next step, three data sets had to be compiled. The first one focussed on the extent to which the construction would affect the surrounding buildings and structures. And so, the range of the operational impact of the underground sections of the Metro Line on the buildings in the vicinity, in average land conditions, had been estimated at 40 m from the extreme wall of the nearest tunnel

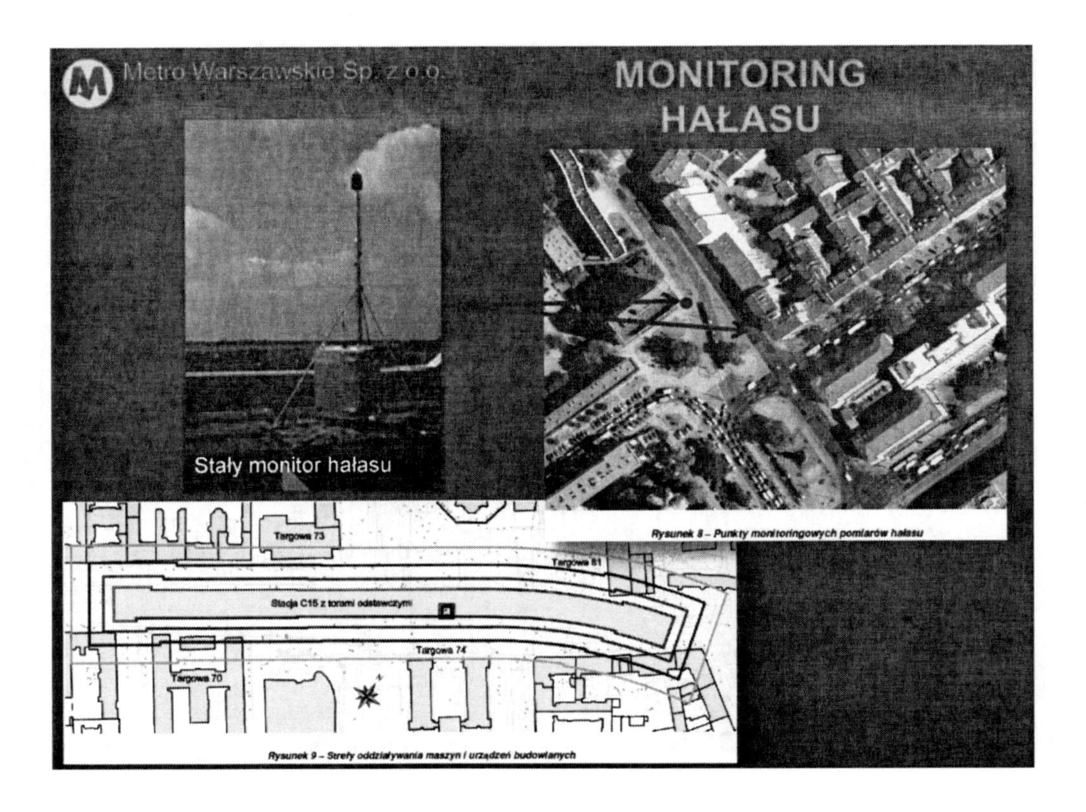

MONITORING HAŁASU	NOISE-LEVEL MONITORING
Stały monitor hałasu	Permanent noise-level monitoring device
Rysunek 8	Figure 8—Noise-level monitoring sites
Rysunek 9	Figure 9—Impact zones of construction machines and devices
Stacja C15..	C15 Station with holding tracks

Figure 5.

or station, on both sides of the Metro Line. Furthermore, for each type of construction work (a station, a tunnel), an in-depth analysis was carried out with local conditions taken into account.

According to these criteria, the construction of the central section of Metro Line II affects 204 buildings of which 33 are located within the impact zone of the TBM.

The second data set comprises documentation devoted to the technical condition of the buildings within the construction impact zone. Each building was assigned a "catalogue card" with information on its structure, any renovation or repair work conducted there, and malfunctions, together with a photographed description of its technical condition. The collected materials and on-site inspection form the basis for determining the building's technical condition, in line with the standards issued by the Building Research Institute in Warsaw. (Figure 4)

Finally, there is the third data set featuring environmental parameters such as the types, species and state of vegetation, the groundwater level, and the noise level. (Figure 5)

4 THE MONITORING SYSTEM

A very important element of supervision over the construction of Metro Line II is the automatic monitoring system which gives us a systematic overview of a specific set of parameters. The system monitors horizontal and vertical dislocations, deviations from the vertical, and

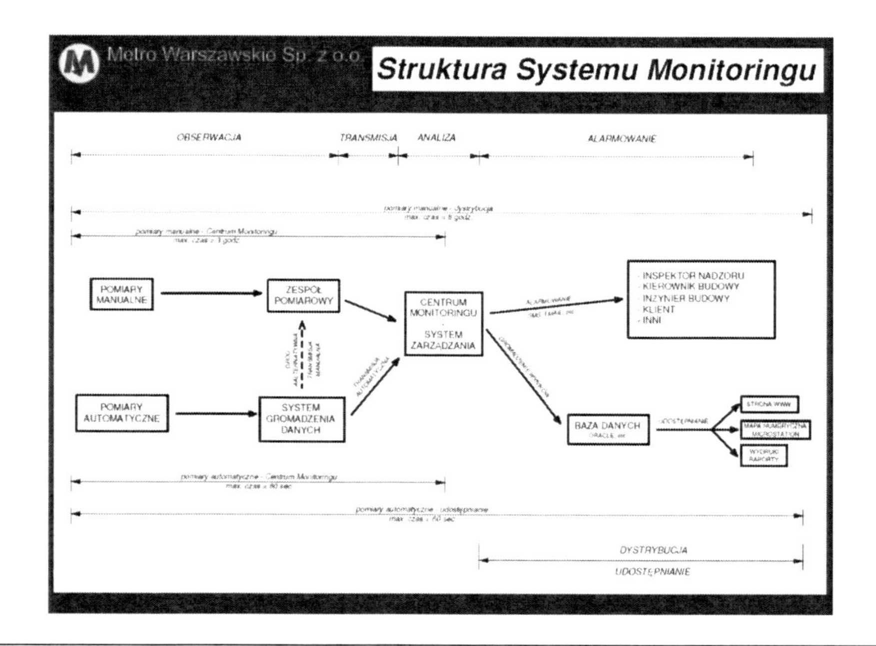

STRUKTURA SYSTEMU MONITORINGU	THE STRUCTURE OF THE MONITORING SYSTEM
OBSERWACJA	OBSERVATION
TRANSMISJA	TRANSMISSION
ANALIZA	ANALYSIS
ALARMOWANIE	ALERT
pomiary manualne—dystrybucja	manual measurement—distribution; max. 6 hours
pomiary manualne—Centrum Monitoring	manual measurement—the Monitoring Centre; max. 3 hours
POMIARY MANUALNE	MANUAL MEASUREMENT
ZESPÓŁ POMIAROWY	MEASUREMENT TEAM
CENTRUM MONITORINGU—SYSTEM ZARZĄDZANIA	MONITORING CENTRE—THE MANAGEMENT SYSTEM
ALARMOWANIE (SMS, EMAIL)	SENDING AN ALERT (A TEXT MESSAGE, EMAIL, etc.)
INSPEKTOR NADZORU, KIEROWNIK BUDOWY (...)	SUPERVISION INSPECTOR, SITE MANAGER, SITE ENGINEER, CLIENT, OTHERS
DROGA ALTERNATYWNA	ALTERNATIVE METHOD—MANUAL TRANSMISSION
POMIARY AUTOMATYCZNE	AUTOMATIC MEASUREMENT
SYSTEM GROMADZENIA DANYCH	DATA-COLLECTION SYSTEM
TRANSMISJA AUTOMATYCZNA	AUTOMATIC TRANSMISSION
GROMADZENIE WYNIKÓW	COLLECTING RESULTS
BAZA DANYCH	DATABASE (ORACLE, etc.)
UDOSTĘPNIANIE	DATA SHARING
STRONA WWW	WEBSITE
MAPA NUMERYCZNA	MICROSTATION NUMERICAL MAP
WYDRUKI, RAPORTY	PRINTOUTS, REPORTS
pomiary automatyczne—Centrum Monitoringu	automatic measurement—the Monitoring Centre; max. 60 seconds
pomiary automatyczne—udostępnianie	automatic measurement—data sharing; max. 60 seconds
DYSTRYBUCJA	DISTRIBUTION
UDOSTĘPNIANIE	DATA SHARING

Figure 6.

METRO WARSZAWSKIE SP. Z O. O.		EMERGENCY PROCEDURE	
RISK LEVEL	ALERT THRESHOLD	RECIPIENTS	RESPONSE
LOW	MESSAGE	The measurement team; the system automatically initiates the required response; the notification is recorded in the database	Increasing the frequency of measurements
MEDIUM	WARNING	CONTRACTOR: -the designer - the site manager - the contract manager SUPERVISION INSPECTOR CLIENT - the person in charge of monitoring	Introducing additional measurements (measuring horizontal dislocations, inclinometers, direct observations, etc.) and, where necessary, considering preventive measures (structure reinforcement, injections, etc.)
HIGH	ALERT	CONTRACTOR: - the designer - the site manager - the contract manager SUPERVISION INSPECTOR CLIENT - the person in charge of monitoring	Stopping the construction (station digging, TBMs, etc.). A meeting must be held to decide on follow-up actions. Preventive measures (designing structural reinforcements, additional tie beams, injections, etc.)

Figure 7.

any changes in the size of each monitored facility. The data from all measuring sites (7993 in total) are collected by 11 devices (the "total station") and sent in real time to the Monitoring Centre. The system's structure and the emergency procedure are presented in Figures 6 and 7.

The monitoring system was launched prior to the actual start of the construction to collect data which could serve as a reference point for future observations and thus allow us to assess the patterns of the monitored parameters. Additionally, this facilitates the observation of any dislocations that might be caused by other factors within the future construction site. The monitoring system will also remain in operation after the construction is complete.

5 THE IMPLEMENTATION

The contract with the general contractor for the construction of the central section of Metro Line II was signed on 28 October 2009 in line with the "design and build system", under an open tender.

Before the construction work could start with regard to the station and tunnel structures, project documentation was prepared, the relevant administrative decisions and permits were obtained, and any conflicting infrastructure underwent redevelopment. The scale of the undertaking is shown in the following tables (Figures 8 and 9).

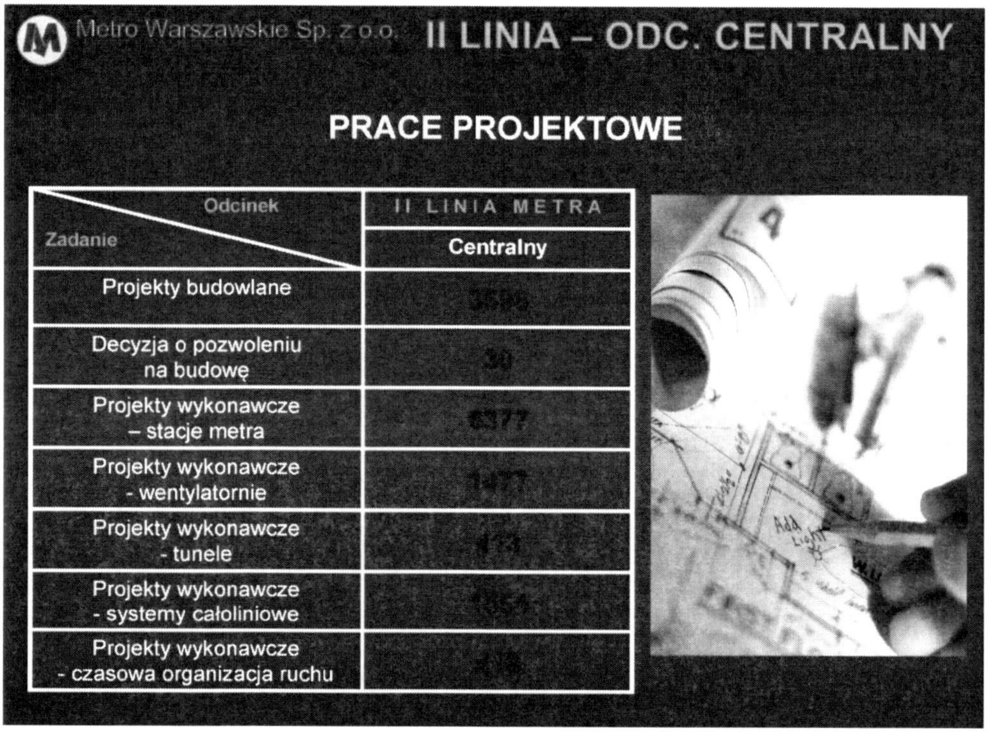

Metro Warszawskie Sp. z o.o. II LINIA – ODC. CENTRALNY

PRACE PROJEKTOWE

Zadanie / Odcinek	II LINIA METRA – Centralny
Projekty budowlane	[illegible]
Decyzja o pozwoleniu na budowę	30
Projekty wykonawcze – stacje metra	6377
Projekty wykonawcze - wentylatornie	[illegible]
Projekty wykonawcze - tunele	[illegible]
Projekty wykonawcze - systemy całoliniowe	[illegible]
Projekty wykonawcze - czasowa organizacja ruchu	[illegible]

LINE II—THE CENTRAL SECTION

DESIGN WORK

Task / Section	METRO LINE II THE CENTRAL SECTION
Construction designs	3598
Construction permits	30
Detailed designs—metro stations	6377
Detailed designs—track fan units chambers	1477
Detailed designs—tunnels	413
Detailed designs—full-line systems	1554
Detailed designs—temporary traffic control	413

Figure 8.

6 THE COSTS

The project entitled "The central section of Metro Line II in Warsaw" consists of three main tasks:

- Task 1—the construction of the central section of Metro Line II (including the stations, the track fan units chambers, tunnels with railway tracks, the turn-out chambers, the holding tracks, the underground and overground infrastructure—a cost of PLN 3000.0 mln,
- Task 2—the building of the infrastructure for the future metro line (including a connector between the lines, a turn-out chamber near the Stadion Narodowy station, interline

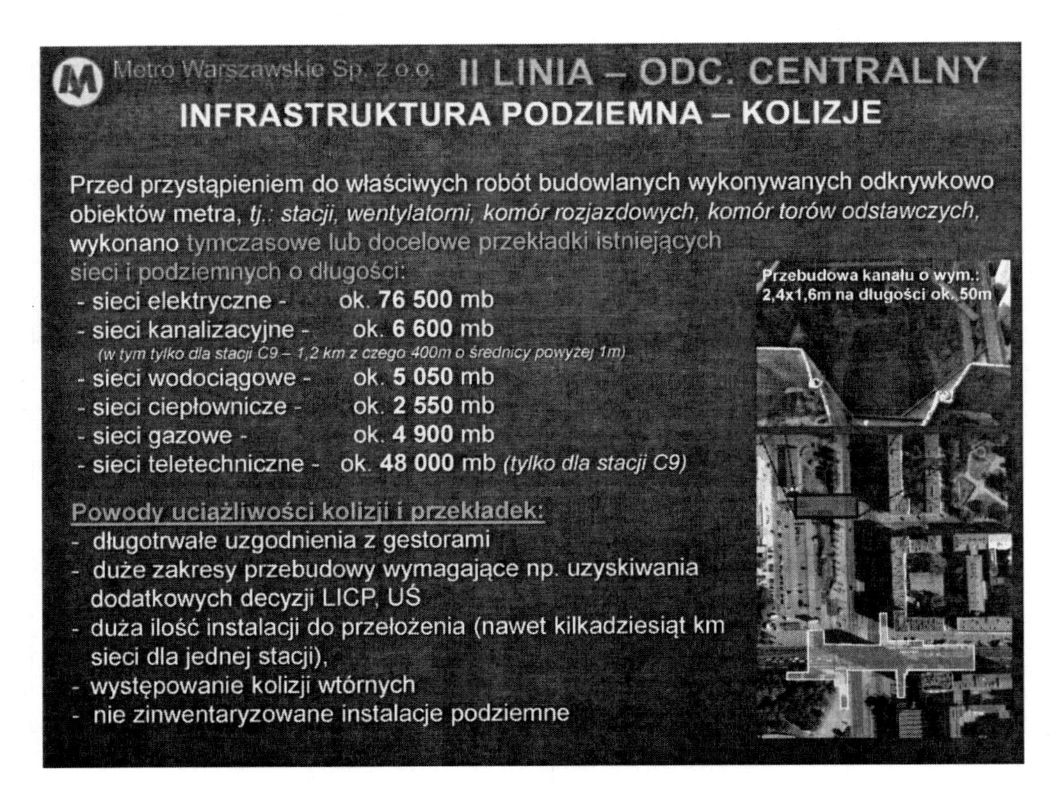

LINE II—THE CENTRAL SECTION
UNDERGROUND INFRASTRUCTURE—CONFLICTS
Prior to the actual construction, which involved the open-pit provision of metro facilities, i.e. the stations, the track fan units chambers, the track-reversing terminals, the holding tracks, temporary or final utility relocations were provided for existing underground networks with the following lengths:

- power networks—about 76 500 line metres
- sewage systems—about 6 600 line metres
  (of which for the C-9 station alone—1.2 km, including 400 m with a diameter of over 1 m)
- water supply systems—about 5 050 line metres
- heating networks—ca. 2 550 line metres
- gas-distribution systems—4 900 line metres
- telecom networks—ca. 48 000 line metres (for the C9 station alone)

The reasons for the inconvenience caused by conflicting infrastructure and utility relocations:

- prolonged negotiations with administrators
- broad scale of redevelopment work requiring additional decisions such as the Determination of the Location of a Public Project, Environmental-Impact Assessment, etc.
- many installations to be relocated (up to several tens kilometres of a network per station),
- secondary infrastructural conflicts,
- underground installations, not recorded in any inventory

Przebudowa kanału (…)—The redevelopment of a 2.4 m × 1.6 m duct along a 50 m section.

Figure 9.

passage systems, the double Stadion Narodowy station, and the technical requirements for the new rolling stock)—a cost of PLN 281.234 million.
- Task 3—the redevelopment and construction of new municipal infrastructure (including systems for city-wide zones under the ONZ Roundabout, and the Marszałkowska/

Świętokrzyska and Al. Solidarności/Targowa junctions, the redevelopment of Wileński Square and the related reconstruction of the underground infrastructure, together with overground land development.

7 SELECTED PARAMETERS—THE LARGEST, THE LONGEST...

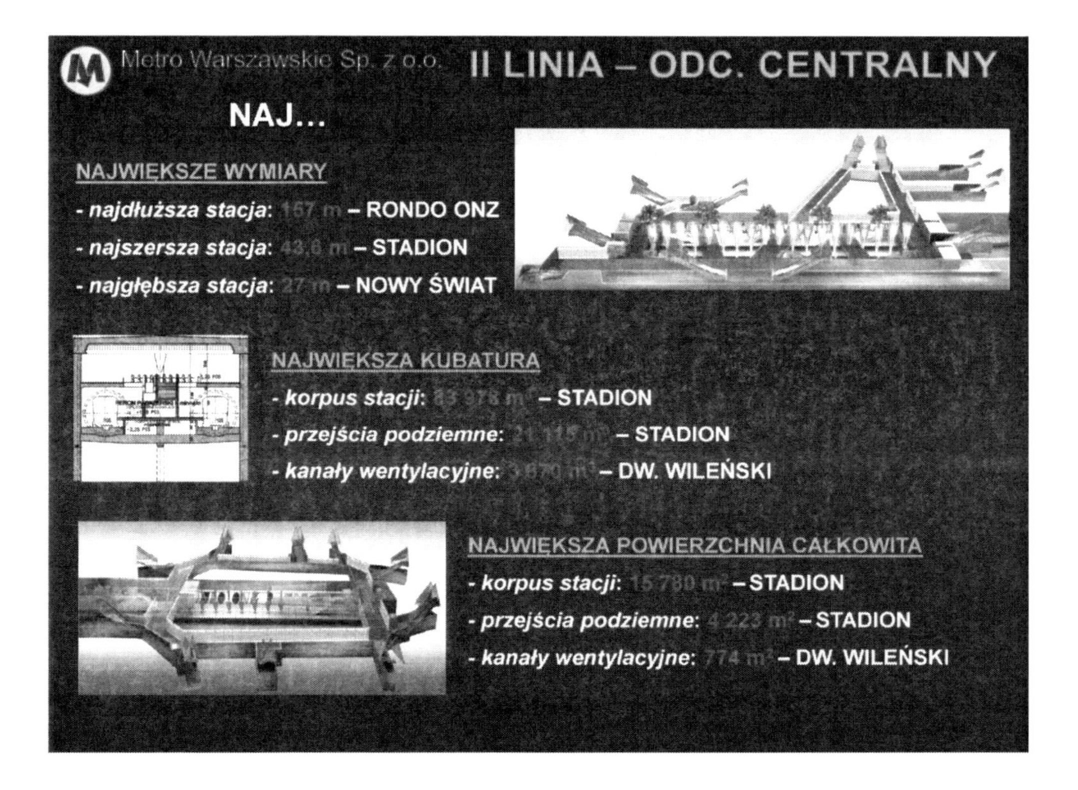

II LINIA—ODC. CENTRALNY	LINE II—THE CENTRAL SECTION
NAJ...	The largest, the longest...
NAJWIĘKSZE WYMIARY	THE LARGEST SIZE
– najdłuższa stacja (...)	– the longest station: 157 m—RONDO ONZ
	– the widest station: 43.6 m—STADION
	– the deepest station: 27 m—NOWY ŚWIAT
NAJWIĘKSZA KUBATURA	THE LARGEST CUBIC CAPACITY
– korpus stacji (...)	– station body: 83 978 m^3—STADION
	– underpasses: 21 115 m^3—STADION
	– ventilating ducts: 3 870 m^3—DWORZEC WILEŃSKI
NAJWIĘKSZA POWIERZCHNIA CAŁKOWITA	THE LARGEST TOTAL AREA
– korpus stacji (...)	– station body: 15 780 m^2—STADION
	– underpasses: 4 223 m^2—STADION
	– ventilating ducts: 774 m^2—DWORZEC WILEŃSKI

Figure 10.

Zagospodarowanie urobku—Spoil management
Ilość urobku…—Volume of spoil in m^3
Odcinek centralny—Central section
Cała II linia…—The whole Line II
Górka Szczęśliwicka—Szczęśliwicka Hill
Kopa Cwila—Cwila Hill
Piramida Cheopsa—The Great Pyramid of Giza

	ODCINEK CENTRALNY—THE CENTRAL SECTION	CAŁA II LINIA METRA—THE WHOLE LINE II
Podstawa—Base	190 × 190 m	300 × 300 m
Wysokość—Height	125 m	200 m
Objętość—Volume	1 500 000 m^3	6 000 000 m^3
Ilość wywrotek—No. of dump wagons (10 m^3/dump wagon)	150 000 pcs.	600 000 pcs.
Ilość wagonów—No. of carriages (30 m^3/carriage)	50 000 pcs.	200 000 pcs.

Figure 11.

Underground Infrastructure of Urban Areas 3 – Madryas et al. (Eds)
© 2015 Taylor & Francis Group, London, ISBN 978-1-138-02652-0

Deformations of corrugated steel tunnel during construction

C. Machelski
Wroclaw University of Technology, Wroclaw, Poland

L. Janusz
ViaCon, Rydzyna, Poland

ABSTRACT: The paper is about deformations of corrugated steel tunnel during construction. The issue is presented based on a 100 m long tunnel built in Poland in 2012. As deformations during construction are very typical and desired authors present an algorithm allowing calculation of bending stresses related to change of curvature of the structure derived from deformations. Results of the algorithm are confronted with real measurements taken during realization of the tunnel. A special feature of the tunnel is a fire-resistant layer that was applied during construction of it.

1 INTRODUCTION

The paper deals with construction road tunnel under ski-slope built in Karpacz, Poland in 2012. The tunnel was built as cut and cover operation. The main shell of the tunnel is a corrugated steel structure VBH19 shaped as a high profile arch MP 200 × 55 × 7 [mm] (a × f × t—corrugation length, pitch, gage). Top radius of the steel arch is $R = 7{,}07$ m. Basic geometrical dimensions are given on Figure 1 (span $L = 11{,}20$ m, rise $h = 6{,}50$ m). The length of the tunnel is 100 m.

The corrugated steel arch is placed on concrete foundations. The span between the support points is $L_0 = 10{,}23$ m. The inside of the tunnel is protected with fire-resistant layer of cement mix with micro-fibers. The thickness of this layer is 120 mm on top of the corrugation.

The tunnel consists of two parts: a straight part length of 60 m and a horizontal bend with 80 m radius. In the straight section the tunnel is shaped as concave arch with 1000 m radius and in the curved section with a slope of 0,5%. The cover of crown of the arch is from 1,0 to 2,25 m. The corrugated steel arch is made of steel S235 JRG2 in accordance to European Standard PN-EN 10025-2:2007 with the minimum yield stress of 235 MPa. Bolting connection is 10 bolts Φ 20 mm/m quality 8.8. The corrugated steel structure was backfilled with sand-gravel mix with SPD 98%. Above the structure and umbrella made of HDPE geo-membrane protected by geotextile layers was performed. After completion of backfilling of the corrugated steel structure the fire-resistant cement shotcrete layer was performed. Figure 2 shows the plan view of the corrugated structure prior to backfilling.

Corrugated steel structures are flexible structures and they would undergo deformations during backfilling. This is a natural behavior and promotes soil-structure interaction in carrying the loads. However one need control the magnitude of deformation in order to remain on the safe side as far as structural integrity of the corrugated steel structure is concerned. In order to monitor the behavior of the structure during backfilling as well as a long-term behavior number of deformation control sensors were installed around the periphery of the structure on the in-side.

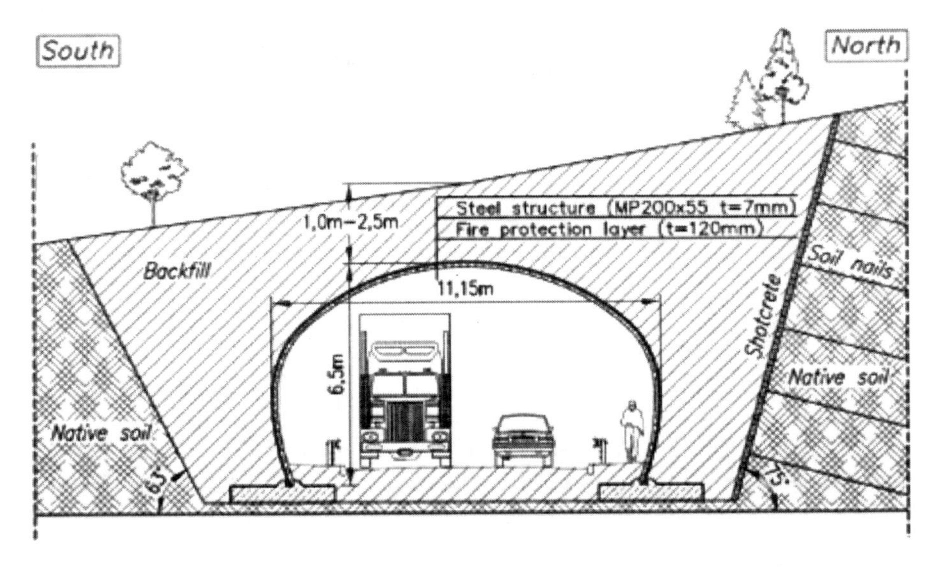

Figure 1. Cross section of the tunnel.

Figure 2. Plan view of the corrugated steel structure and inside view prior to shotcrete.

2 DEFORMATION OF FLEXIBLE SHELL DURING BACKFILLING

Theoretical deformations of a flexible shell backfilled in the soil are considered below. The case based on an arch with single-radius R. For the sake of simplicity even value of radius was assumed however it can be re-calculated to address the multi-radii situation. Figure 3 provides details of the geometry. Backfill is placed symmetrically and generates symmetric interaction with corrugated still presented through normal $p(\varphi)$ tangent $t(\varphi)$ forces. Practically this is not really the reality as there's always some un-symmetry during backfilling process. The deformation of the flexible shell is described through two characteristic values: peaking w and narrowing (shortening) $2u$. Proportions between these two vary during backfilling process (Bakht 2007), (Duncan 1977), (Machelski & Janusz 2011), (Pettersson 2007), (Vaslestad 1990). The resultant deformation of the structure is expected to happen and has a positive impact on load carrying capacity of the backfill structure during service stage.

Figure 4 provides details of deformations measured in three cross-sections during backfilling of the corugated steel structure described in the introduction chapter. Maximum values were as follows $w_I = 172$ mm, $w_{II} = 214$ mm, $w_{III} = 167$ mm. In the end cross-section located in the vicinity of the entrance portals values were respectively lower: $w_O = 130$ mm, $w_{IV} = 143$ mm. For the sake of an anlysis the middle (II) cross-section values were taken into consideration as they were the highest. The peaking factor of based on that value is:

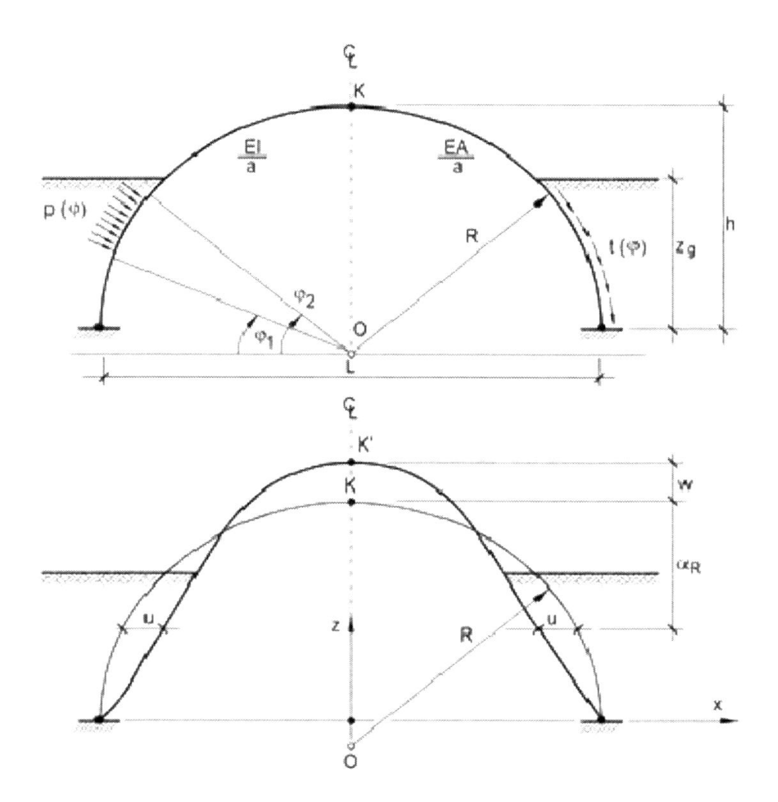

Figure 3. Backfilling behavior of a flexible structure.

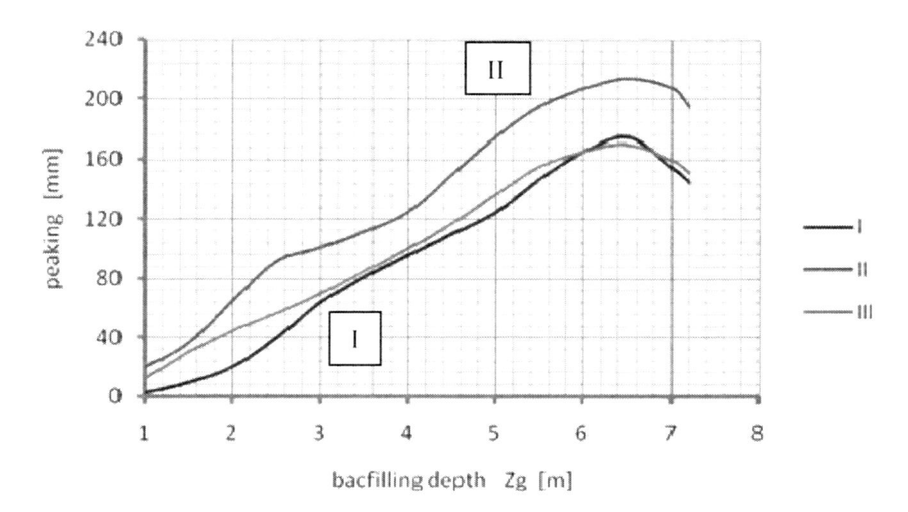

Figure 4. Peaking of the structure measured in the crown during backfilling.

$$\omega = \frac{w}{L} 100\% = \frac{214}{11200} 100\% = 1{,}91\%$$

Measurements of u were conducted at following levels related to support level (as per Fig. 3).

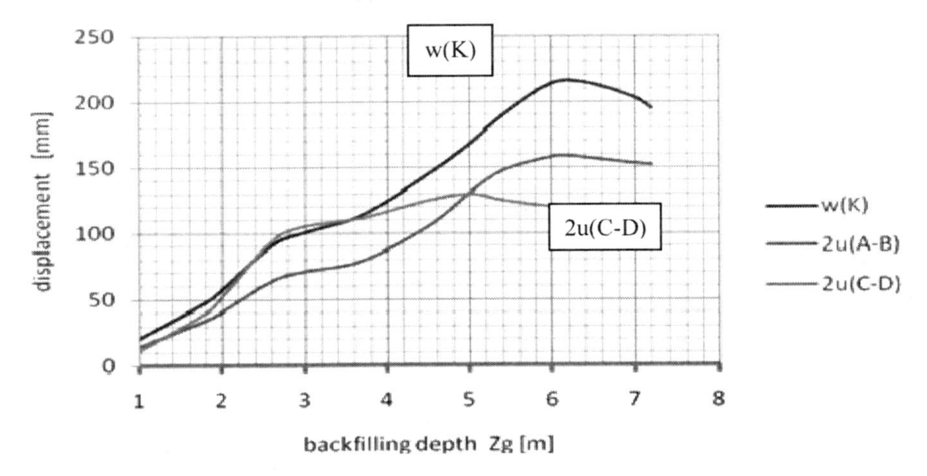

Figure 5. Characteristic deformations during backfilling.

(A–B) $h - \alpha R = 4{,}85$ thus $\alpha R = 0{,}2065 \cdot 7{,}07 = 1{,}46$;
(C–D) $h - \alpha R = 2{,}50$ thus $\alpha R = 0{,}5629 \cdot 7{,}07 = 3{,}98$.

Figure 5 shows the horizontal deformations measured at these levels. They are related to peaking w(K) in the crown. In order to simplify the further analysis data from (A–B) level was considered.

Based on the results one can conclude that narrowing is smaller than peaking. As expected the narrowing measured with lower level when backfill is at initial stage are higher. The maximum value of $2u$ is for (A–B) level which is $\alpha = 0{,}2065$ when backfill reaches the crown. The narrowing curves at different measurement levels determined by α are variable functions. During construction the peaking increases untill backfill reaches the crown. Once an overburden soil starts to load the steel shell i.e. $z_g > h$, a reduction of formerly developed deformations w and u occurs.

3 BENDING MOMENTS IN THE CROWN OF THE STEEL ARCH

Based on deformations measured during backfilling one can calculate the change of radius of the curvature which constitutes the basis of estimation bending moments in the crown. Bending moments are the main component of normal stresses in the steel. The analysis of the upper part of the arch is sufficient for the estimation of the bending moments. Unit measurements can be verified (Duncan 1977).

Important for the safety of structure performance is a construction stage presented on Figure 3 Based on that one can prognose the maximum bending monets. Useful paramete for that is a change of the radius of the curvature as follows:

$$dR_{uw} = R - R_{uw} \tag{1}$$

This change is set based on below presented algorithm and deformations w and u, as per Figure 6. The circumferentail strip is bended during the roll-forming process of the steel in production to a constant value R. Based on drawing at Figure 6 relative peaking in comparison to level αR, is calculated as follows

$$w = w_k - \frac{1}{2}\left(w_A + w_B\right) \tag{2}$$

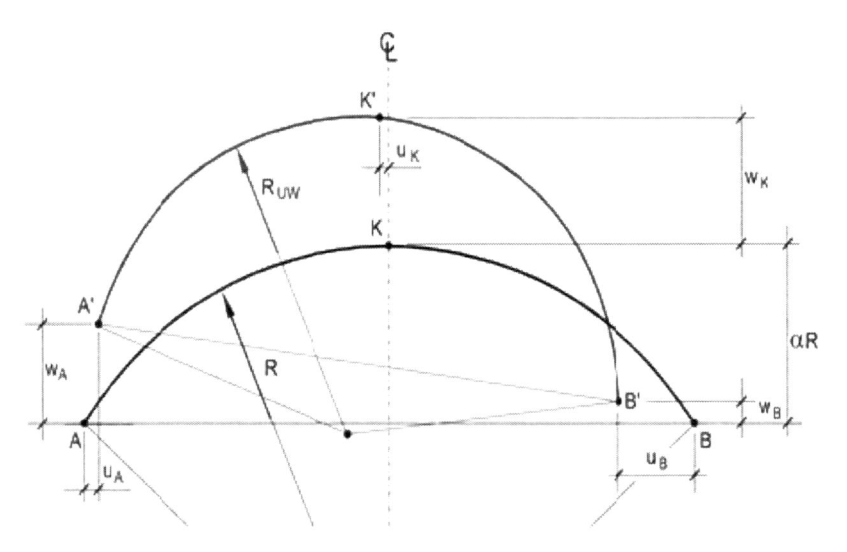

Figure 6. Change of shape of the top part of the arch.

Narrowing 2u at the same level is set as

$$2u = u_A + u_B \tag{3}$$

Deformations given at Figure 5 are not the same in general. Figure 6 presents deformations u_A and u_B constituing 2u in accordance to equation (3). Vertical inclination of the shell is specified by u_K.

The initial radius of the shell R, is specified as for circle drawn on an equilateral triangle, (Cowherd & Corda 2010), as per Figure 6, with basis AB and length C and height of $F = \alpha R$. Then:

$$R = \frac{4F^2 + C^2}{8F} \tag{4}$$

The change of such a radius related to the deformation of the shell, given by characteristic deformations w and u is given by an equation:

$$R_{uw} = \frac{4(F+w)^2 + (C-2u)^2}{8(F+w)} \tag{5}$$

To determine Ruw one can also use the formula of the circle drawn over equilateral triangle with radius R. Change of R from (4) is specified based on (5) and (4) as per equation:

$$dR = R - R_{uw} = \frac{w \cdot C^2 + 4Fu(C-u)}{8F(F+w)} - \frac{w}{2} \tag{6}$$

By using a parametric approach with α we can modify it to following equation:

$$R_{uw} = R_\alpha = \frac{R\left[\alpha(R+w) - u\sqrt{\alpha(2-\alpha)}\right]}{\alpha R + w} \tag{7}$$

And the equation (6) will be transferred to:

$$dR = R - R_\alpha = \frac{R\left[w(1-\alpha) + u\sqrt{\alpha(2-\alpha)}\right]}{\alpha R + w} \tag{8}$$

as

$$C = 2R\sqrt{\alpha(2-\alpha)} \tag{9}$$

Dimensionless indicator specifying the change of the curvature of the shell in the crown is given as:

$$\frac{dR}{R} = \frac{w(1-\alpha) + u\sqrt{\alpha(2-\alpha)}}{\alpha R + w} \tag{10}$$

The change of the radius of the curvature in the crown of the circle arch during backfilling is specified based on deformations w and u, which were specified at the level (A–B), as per Figure 7 ($\alpha = 0{,}2065$). Results of the calculations are given on Figure 8.

Assuming various leves of measurement points used for determination of narrowing u, one ontain various functions of dR(zg) as per (10). An important element of difference of dR/R is location of measurement level *u* which is a function of parameter α. For estimation of the maximum value of bedning moment in the crown when the backfill reaches the top of the crown one can use the following equation

$$M_{\max} = \frac{EI}{R}\frac{dR_{\max}}{R - dR_{\max}} \tag{11}$$

where *EI* bending stiffnes of the corrugated steel.

Maximum change of the curvature dRmax/R can be determined by specifying allowed normal stress σ, as a derivate of the bending moment *M*, based on flexure index for the corrugated steel plate

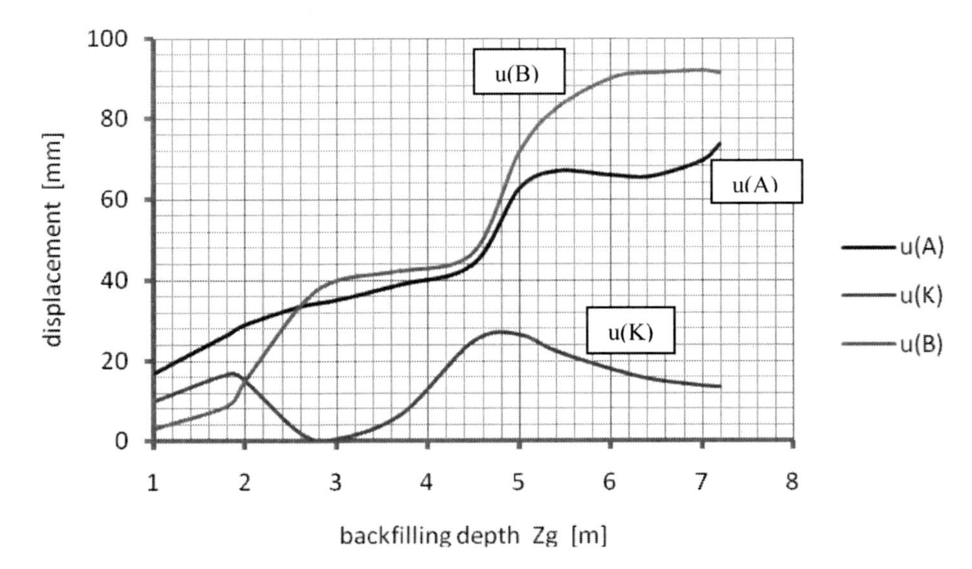

Figure 7. Horizontal deformations in the upper part of the steel shell.

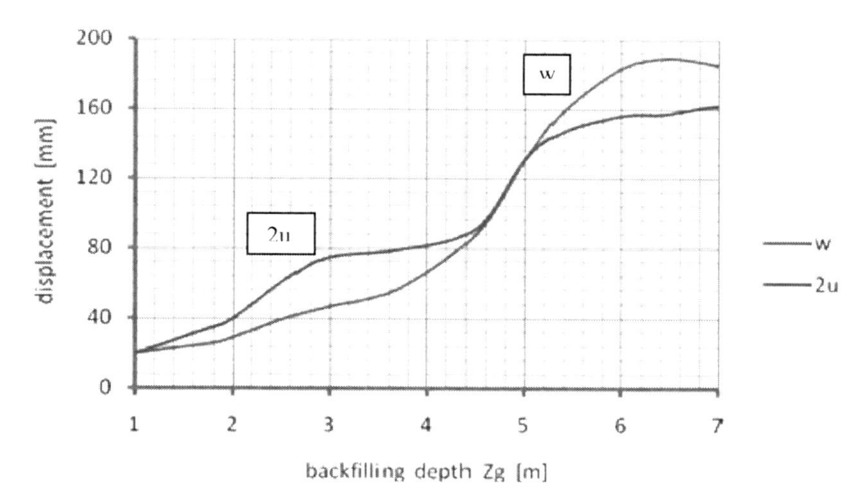

Figure 8. Deformations of upper part of the arch.

$$W = \frac{I}{(f+t)} \tag{12}$$

The value of bending moment can be specified as

$$M_{\max} = \sigma \cdot W = \frac{2\sigma \cdot I}{f+t} \tag{13}$$

Using the relations of bending moment and change of the curvature in the formulae

$$\frac{dR}{R} = \frac{RM}{EI + RM} \tag{14}$$

And by substituting M_{max} from (13) we obtain the maximum value

$$\frac{dR_{\max}}{R} = \frac{\sigma}{E\frac{f+t}{2R} + \sigma} \tag{15}$$

The fixed element in the equation (15) is a characteristic parameter for a given type of corrugation. In this case with corrugation MP 200 × 55 × 7 we get:

$$E\frac{f+t}{2R} = 205000\frac{55+7}{2 \cdot 7070} = 898{,}9 \approx 900\ MPa$$

By assuming maximum allowed stress e.g. σ = 180 MPa we will obtain from equation (15)

$$\frac{dR_{\max}}{R} = \frac{1800}{900+1800} = 0{,}1667$$

Thus values of change of curvature given on Figure 9 are 16,7% smaller then calculated as above.

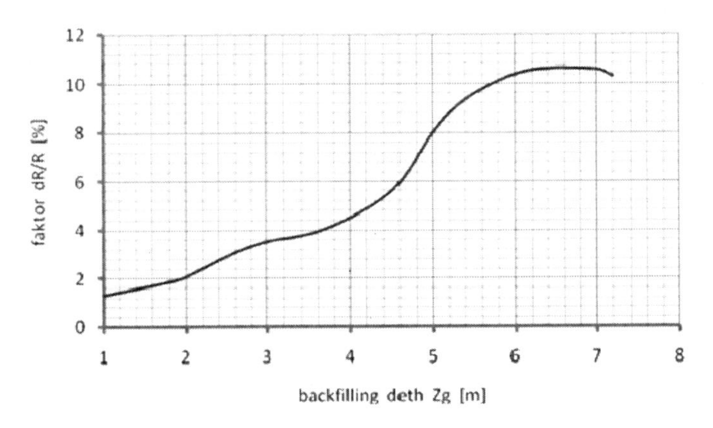

Figure 9. Change of the curvature.

4 SUMMARY

The corrugated steel structure arch embedded in the soil which is described in the paper behaves like a flexible structure during construction. It undergoes deformations related to soil-structure interaction. These deformations can be measured and based on them one can estimate bending moments and stresses in the steel. The paper proposes an algorithm to calculate these stresses and bending moments based on change of the curvature of upper part of the steel arch. Observations from various installations confirm that once structure is backfilled the deformations practically stops and therefore the construction stage is the most critical part for the stability of the soil-steel structures. This phenomenon allowed to apply fire-resistant payer of cement based spray with fibers which meets the requirements of class F4 for tunnels.

REFERENCES

Bakht B. 2007: Evoluation of the design for soil-metal structures in Canada. Archives of Institute of Civil Engineering. Poznań.

Cowherd D.C. & Corda I.J. 2010. Lessons Learned from Culvert Failures and Nonfailures. 89th Annual Meeting Transportation Research Board. Washington D.C.

Duncan J.M. 1977: Behaviour and Design of Long-Span Metal Culvert Structure. ASCE Convention, San Francisco.

Machelski C. & Janusz L. 2011: Estimation of bending moments in the crown of a soil-steel bridge structure during backfilling. 6th European Conference on Steel and Composite Structures. Eurosteel. Budapest.

Machelski C., Michalski J.B. & Janusz L. 2009: Deformation Factors of Buried Corrugated Structures. Journal of the Transportation Research Board. Solid Mechanics.Transportation Research Board of Nationalals Academies, Washington D.C. —literatura bez odniesienia.

Pettersson L. 2007: Full Scale Tests and Structural Evaluation of Soil Steel Flexible Culverts with low Height of Cover. Doctoral Thesis, KTH, Sweden.

Vaslestad J. 1990: Soil structure interaction of buried culverts, Institutt for Geoteknikk, Norges Tekniske Hogskole, Universitetet I Trondheim.

Underground Infrastructure of Urban Areas 3 – Madryas et al. (Eds)
© 2015 Taylor & Francis Group, London, ISBN 978-1-138-02652-0

Microtunneling as an alternative for crossing rivers during the execution of urban network infrastructure

C. Madryas & L. Wysocki
Wroclaw University of Technology, Wroclaw, Poland

J. Ośmiałowski & M. Ośmiałowski
Inwest Terma, Wroclaw, Poland

ABSTRACT: The paper presents possibilities of using microtunneling technology for crossing rivers when executing technical infrastructure networks as an alternative to traditional solutions such as laying networks on bridges or in open trenches below the bottom of a river. An analysis of costs of various solutions of crossing a river with a heating network is presented in a selected example.

1 INTRODUCTION

The traditional and most commonly used way of crossing rivers when executing technical infrastructure networks is by placing them on bridges or walkways suspended above the water. In the case of smaller rivers, an open trench under the bottom of a river is also used and the trench is usually protected by sheet piling. Although microtunneling technology is already well known, its usage for laying underground networks beneath the bottom of rivers is not yet common.

In this article, capabilities of microtunneling technology for executing a heating network due to the necessity of removing it from a bridge are presented. This necessity was caused by the bad technical condition of the bridge and the urgent need for the implementation of renovation works. The authors have proposed variant solutions of this problem with the use of microtunneling technology and traditional technology of placing the network under the bottom of the river in a trench secured by sheet piling.

2 TECHNICAL CAPABILITIES OF CROSSING RIVERS DURING THE CONSTRUCTION OF TECHNICAL INFRASTRUCTURE NETWORKS

Crossing an obstacle such as a large river is possible with the use of the following technologies:

- microtunneling,
- Horizontal Directional Drilling (HDD),
- traditional technology—in a trench secured by sheet piling,
- suspension on a supporting structure (e.g. bridge).

Microtunneling technology enables the implementation of networks of diameters up to 4500 mm and in nearly any soil and water conditions. In this technology it is also possible to execute a multi-utility tunnel in which can be placed multiple installations.

In the case of directional drilling technology difficulties may occur regarding the execution of the joining chambers and compensation chamber. Crossing a river using traditional technology in a trench secured by sheet piling is usually very difficult. The difficulties are greater in the case of navigable rivers and nearby hydropower stations as it can be seen in the presented case.

3 MATERIAL SOLUTIONS FOR CONDUITS MADE USING MICROTUNNELING TECHNOLOGY

The most important element in determining the type of material solutions for pipes to be used in this technology is the use of very large horizontal forces during pressing. Therefore, pipes must be adapted to transfer such forces. This criterion is not fulfilled by thermoplastic pipes (HDPE, PP, PVC), but is fulfilled by pipes made of reinforced concrete, polymer concrete, GRP (Glass Reinforced Plastic) and steel. Steel pipes are most commonly used as casing pipes in which are placed appropriate conduit pipes. Past experience resulting from the use of microtunneling technology in Poland indicates that GRP pipes are particularly useful. They can be used as technological pipes (sewer pipes) and also as casing of utility tunnels in nearly any soil and water conditions. The advantage of GRP pipes, apart from their great durability, is their relatively small wall thickness which makes their transportation significantly easier and reduces the volume of earthworks when compared to reinforced concrete pipes.

4 TECHNICAL AND ECONOMIC ANALYSIS OF ALTERNATIVE SOLUTIONS OF CROSSING THE ODRA RIVER

The objective of the project was to investigate the possibilities of transferring heating pipes under the Odra River in Wroclaw. These pipes are suspended to the historic bridge, which will be renovated in the near future. This situation made the Municipal Roads and Streets Department of Wroclaw decide that due to technical (reduction of loadings) and aesthetic (historic bridge in the city center) reasons, the pipes should be taken off the bridge and placed under the river. The view of the heating pipes suspended to the historic bridge is presented in Figure 1.

There were three options of solving the problem considered. In the first option the heating network (two conduits of 600 mm diameter each) was meant to be placed in a utility tunnel made of GRP pipes. The scheme of the tunnel construction is presented in Figure 2 and its location in Figure 3.

In the second option the execution of two independent tunnels (casing pipes) made of GRP pipes was considered which would enable the heating network to be transferred across the river without the possibility of its servicing. The scheme of this solution is presented in Figure 4, and its location in Figure 5.

In both options the length of the tunnels is equal to approximately 150 m. In the case of the utility tunnel it is necessary to build two chambers and in the case of the independent microtunnels three chambers are required. An additional third chamber is necessary in order to place compensators, which in the first option can be executed inside the tunnel.

There was also a third option considered—traditional technology for crossing the river in a trench secured by sheet piling. In the presented case, this technology was considered in relation to the periodic significant reduction of the water level in the river which was associated with the ongoing repair of levees and embankments near the bridge. A view of the building site in the vicinity of the bridge is presented in Figure 6.

Figure 1. The view of heating pipes suspended to the historic Pomorski Bridge *(fot. C. Madryas)*.

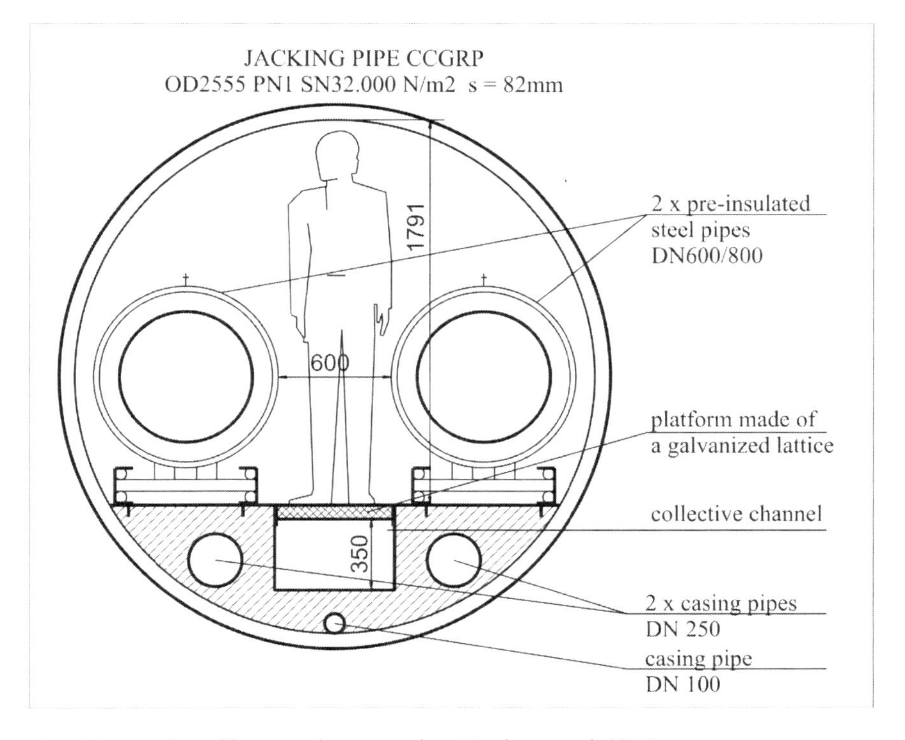

Figure 2. Scheme of a utility tunnel construction (Madryas et al. 2014).

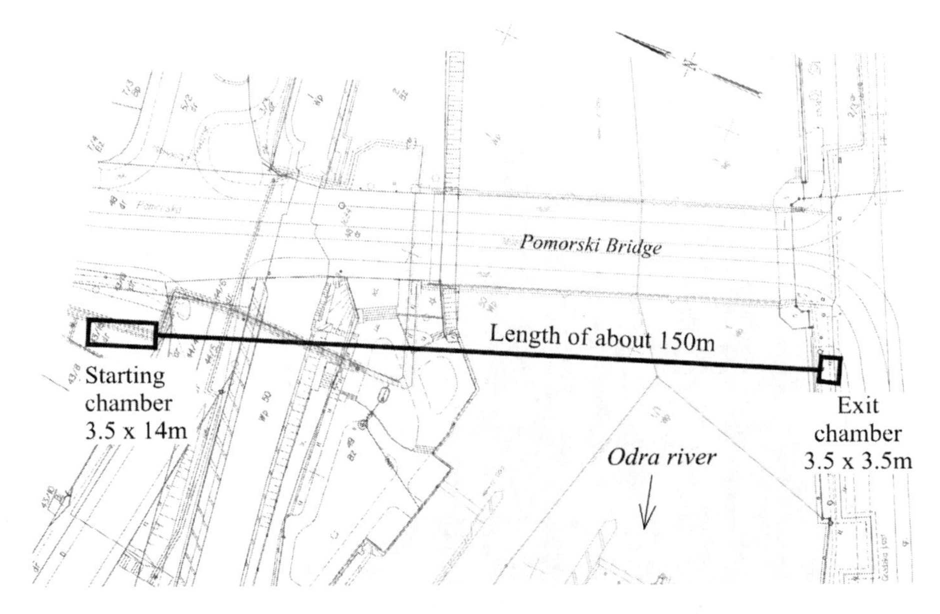

Figure 3. Route of utility tunnel (Madryas et al. 2014).

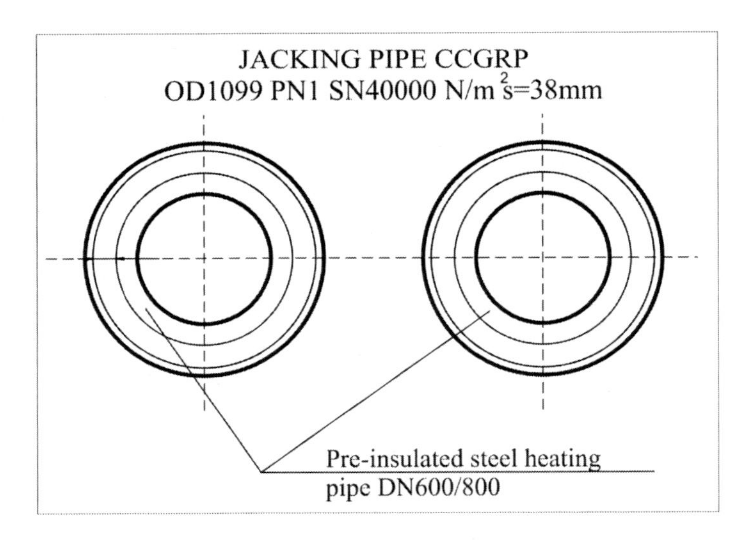

Figure 4. Scheme of two microtunnels (Madryas et al. 2014).

Comparative analysis of the solutions indicated that from a technical point of view, the most advantageous option is placing the network in a utility tunnel (first option). Such a solution would enable all the technical infrastructure networks to be removed from the bridge and placed inside the utility tunnel. Placing all networks in the utility tunnel, despite the obvious benefits associated with their operation, would also contribute to an increase in the security of the historic bridge structure (reduction of loading) and improve its aesthetics. The utility tunnel would also enable, if required in the future, the implementation of additional technical networks.

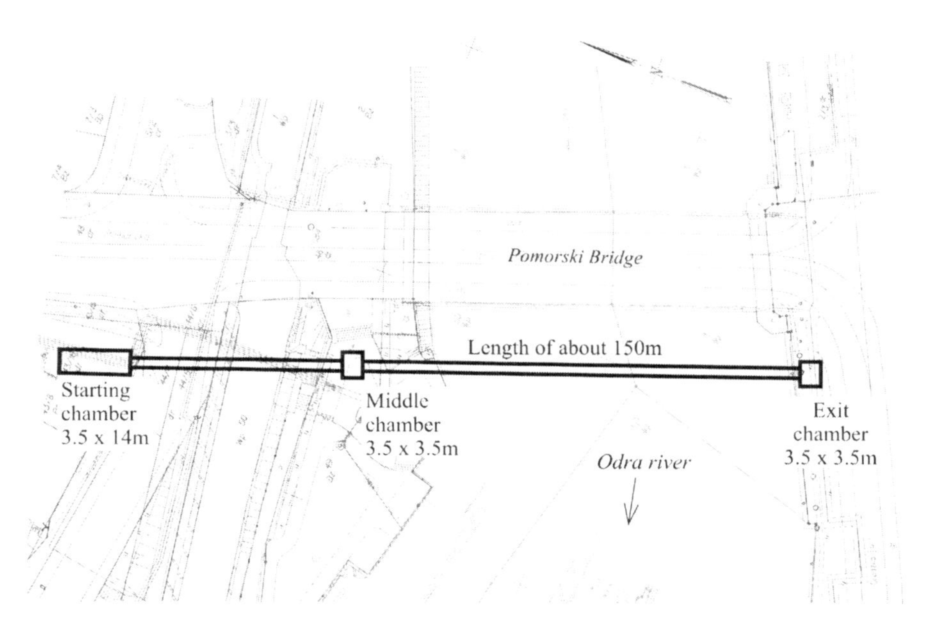

Figure 5. Route of microtunnels (Madryas et al. 2014).

Figure 6. The view of embankments under renovation and modernization works (*fot. C. Madryas*).

The economic analysis of different options was carried out and its results are summarized in Table 1.

Costs of crossing the river when executing works in the trench secured by sheet piling are summarized in Table 2.

When adding to these costs the cost of limiting the work of a power station which is equal to 1592000 PLN, it gives a total cost of about 5057000 PLN.

Table 1. Indicative costs of the transfer of two pipes DN600/800 outside the Pomorski Bridge construction using microtunneling technology (Madryas et al. 2014).

	Description	Microtunnel 2 × OD1099 [net price in thousands of PLN]	Microtunnel 1 × OD2555 [net price in thousands of PLN]
1	Execution of a tunnel with a length of about 150 m	1500	1500
2	Execution of a starting chamber	350	350
3	Execution of a middle chamber	200	0
4	Execution of an exit chamber	200	200
5	Casing pipes for microtunneling	1000	1500
6	Pre-insulated material DN600/800 of about 2 × 190 m	1300	1300
7	Installation of the heating network	400	400
8	Load-bearing structure in the tunnel with an assembly	50	425
9	Steel construction in the starting chamber with an assembly	180	180
10	Steel construction in the middle chamber with an assembly	120	0
11	Steel construction in the exit chamber with an assembly	120	120
12	Other technical installations in the tunnel	50	100
	Total	5470	6075

Table 2. Indicative costs of traditional technology (Madryas et al. 2014).

Description	Units	Amount	Unit price [net] in PLN	Value [net] in PLN
Implementation of sheet piling in the water 70 m × 2 = 140 m × 10 = 1400 m², on the land 20 m × 2 = 40 m × 14 = 560 m²	m²	1960	150	294000
Value of sheet piles	t	304	4000	1216000
Cutting sheet piling under the water	m	180	2260	406800
Digging excavated material from under the water	m³	1300	36	46800
Levelling the ground for the pipeline by a team of divers	m²	315	292	91980
Placement of pipelines under the water	t	36	5230	188280
Backfilling of pipelines	m³	1000	40	40000
Placement of panels with dimensions of 3,5 × 4.0 under the water on the pipeline	items	18	2760	49680
Cost of panels	items	18	3800	68400
Dismantling of walls	m³	50	6100	305000
Earthworks behind the walls—dismantling	m³	200	28	5600
Heating infrastructure 2 x DN700, building and installation works, chamber, modernization of K-1/8Z and K-1/8P, additional materials	set	1	700000	700000
Building Permit and Construction Design	set	1	35000	35000
Water Quality Impact Assessment	set	1	3000	3000
Report about the impact on the environment	set	1	15000	15000
Total				3465540

5 CONCLUSIONS

In many cases rebuilding or building of new underground infrastructure networks in cities causes serious technical problems which are related to the limited space for carrying out building works, the necessity of closing streets temporally, hazards to nearby buildings or the need of working in an environment under monumental protection. Microtunneling technology is an alternative to traditional methods of carrying out works in open trenches and it allows the elimination or a significant reduction of such problems. Economic analyzes indicate that this technology can also be competitive due to the cost of execution. The presented analysis did not include hard quantifiable social costs associated with difficulties in the functioning of cities. In many situations the social costs can be included as an important argument when choosing a particular technology. The rapid development of microtunneling technology causes a continuous reduction of the execution costs which should contribute to an increase in its use.

REFERENCES

Madryas C., Ośmiałowski J., Ośmiałowski M. & Wysocki L. Analiza techniczno-ekonomiczna zdjęcia sieci ciepłowniczej z Mostu Pomorskiego i ułożenia jej w nowej lokalizacji. Raport serii U nr 3/2014. Politechnika Wrocławska. Instytut Inżynierii Lądowej. Wrocław, 2014.

Underground Infrastructure of Urban Areas 3 – Madryas et al. (Eds)
© 2015 Taylor & Francis Group, London, ISBN 978-1-138-02652-0

The FEM back-analysis of earth pressure coefficient at rest in Brno clay K_0 with the homogenization of steel/shotcrete lining

J. Rott & D. Mašín
Faculty of Natural Sciences, Charles University in Prague, Czech Republic

ABSTRACT: In the paper we estimate the earth pressure coefficient at rest by means of back-analysis of deformation measurements in underground circular unsupported cavity (constructed for the purposes of detailed geotechnical survey) in highly plastic overconsolidated clay. Because of an inability of the adopted FEM code to assume 2 different components of primary lining, the homogenization of steel/sprayed concrete ("shotcrete") lining was considered. The main primary boundary conditions were conservation of static momentum and conservation of the axial and bending stiffness. In addition, the dependence of Young modulus of shotcrete on time was also taken into the account. Our research focuses on Královo Pole tunnel in Brno city, which is one of the key parts of Brno city road ring. The backanalyses of circular cavity indicate $K_0 = 0{,}81$. This value was subsequently entered in a numerical 3D model of the whole tunnel and verified by confrontation of measured and calculated displacements.

1 INTRODUCTION

1.1 *Role of K_0 in geotechnics*

The in-situ effective stresses represent an important initial condition for geotechnical analyses, however, especially for oveconsolidated soils, the value of K_0 can estimated only with approximation. The horizontal stress is computed from the vertical stress using the coefficient of earth pressure at rest

$$K_0 = \frac{\sigma_h}{\sigma_v} \tag{1}$$

where σ_h and σ_v are effective horizontal and vertical stresses, respectively. In the case of deep foundations (friction piles), retaining structures and tunnels, K_0 influences the mechanical behaviour dramatically.

1.2 *Estimation of K_0*

Lateral earth pressure coefficient at rest K_0 is very difficult to estimate while it significantly affects the predictions of the behaviour of geotechnical structures. The strategy adopted here was to back-analyze coefficient of earth pressure at rest K_0 in overconsolidated Brno clay using as much realistic numerical model of Královo Pole tunnel as possible. The computed values of ratio of horizontal and vertical displacement u_y/u_z in unsupported adit R2 were paired with initially entered values of K_0 and the relation $K_0 - u_y/u_z$ was constructed. For given value of measured ratio an appropriate value of K_0 was deducted. Because of uncertainties in some material characteristics, a number of parametric studies and verifications using the numerical model of tunnel tube were also carried out.

2 DESCRIPTION OF LOCALITY AND BRNO CLAY

2.1 *The Carpathian fore-trough*

From the stratigraphical point of view, the area is formed by Miocene marine deposits of the Carpathian fore-trough whose depth reaches several km. The top part of the overburden consists of anthropogenic materials. The natural quaternary cover consists of loess loam and clayey loam with the thickness of 0.5 to 15 m. The base of the quaternary cover is formed by a discontinuous layer of fluvial sandy gravel, often with a loamy admixture. The majority of the tunnel is driven through the tertiary calcareous silty clay. Brno Clay ("Tegel") of lower Badenian age reaches the depth of several hundred meters (300 m). The process of sedimentation took place between 16.4 to 14.8 Ma.

2.2 *Brno clay properties*

Sound Tegel has a greenish-grey color, which changes yellow-brown and reddish-brown within the zone of weathering closer to surface. Tegel exhibits stiff to very stiff consistency. The clay is overconsolidated but the height of eroded overburden is not known. The clay is tectonically faulted, the main fault planes are slicken-sided and uneven. The groundwater is mostly bound to quaternary fluvial sediments, and the collectors are typically not continuous. The clay is, however, fully water saturated. In Tegel there is about 50% of clay fraction, W_L is about 75%, I_P about 43%, the soil plots just above the A-line at the plasticity chart and its index of colloid activity I_A is about 0.9.

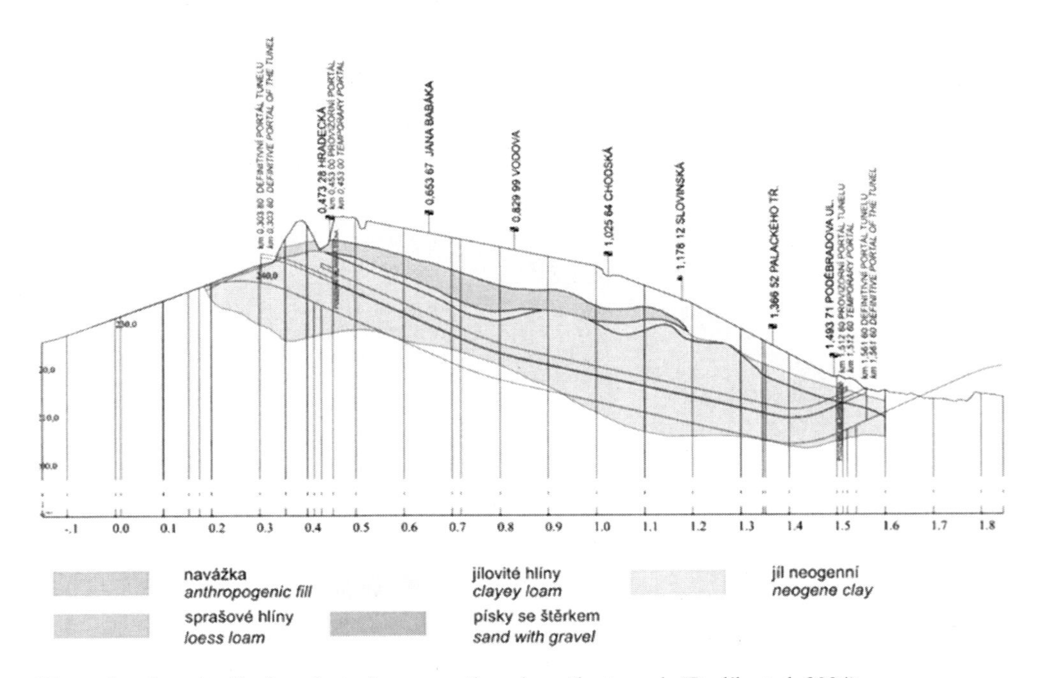

Figure 1. Longitudinal geological cross-section along the tunnels (Pavlík et al. 2004).

3 DESCRIPTION OF STRUCTURES AND NUMERICAL MODEL

3.1 *The Královo Pole tunnels (tunnel tubes TT1 and TT2)*

The Královo Pole tunnels (often referred to as Dobrovského tunnels) form an important part of the northern section (oriented SW-NE) of the ring road of Brno town in the Czech Republic. The tunnels consist of two mostly parallel tunnel tubes TT1 and TT2 with a separation distance of about 70 m and lengths of approximately 1250 m. Each tube conveys 2-lane road. The tunnel cross-section height and width are about 11.5 m and 14 m, respectively, and the overburden thickness varies from 6 m to 21 m. The tunnels are driven in developed urban environment. The excavation of the tunnels commenced in January 2008. The tunnels were driven by the New Austrian Tunneling Method (NATM), with sub-division of the face into six separate headings. As proposed in the project the excavation was performed in steps 1 to 6 with an unsupported span of 1.2 m. A constant distance of 8 m was kept between the individual faces, except the distance between the top heading and the bottom, which was 16 m. The face of main tunnel tubes was subdivided and the relatively complicated excavation sequences were adopted in order to minimize the surface settlements imposed by the tunnel.

3D numerical model of tunnel tube TT1 was created in PLAXIS 3D 2012 Software. Model was composed of 31,464 quadrilateral elements. The chosen constitutive material model of clay strata is represented by hypoplasticity with inherent stiffness anisotropy in the very small strain range. Simulated portion of the TT1 has 56.4 meters and corresponds to the tunnel chainage 0.6504–0.708 km. Results from numerical analysis of TT1 in chainage between km 0.6504 and 0.7068 were compared with monitoring data from inclinometer

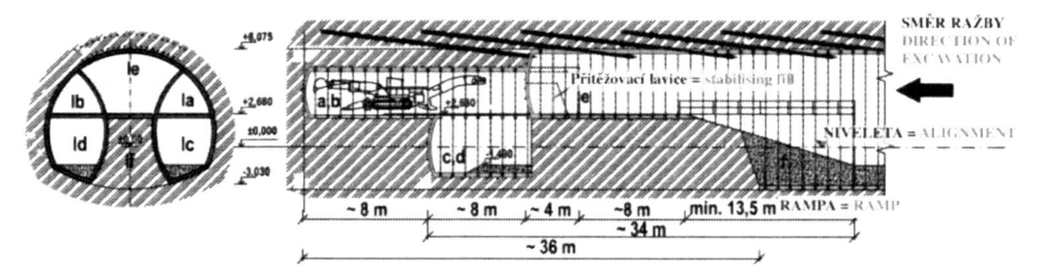

Figure 2. Sketch of the excavation sequence of the tunnel (Horák, 2009).

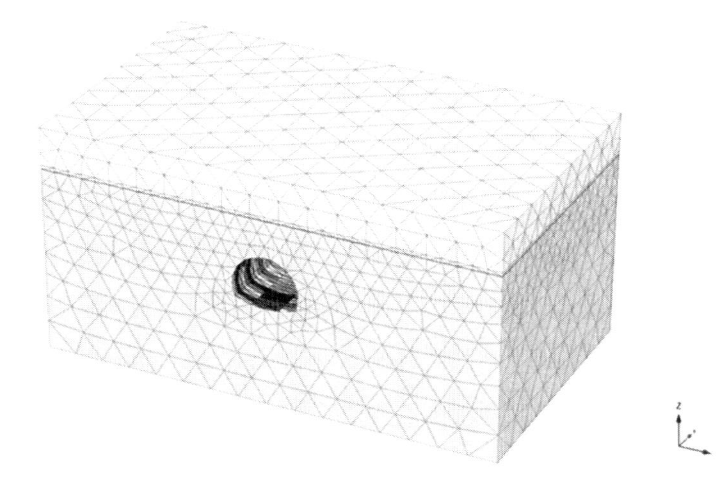

Figure 3. Model geometry of the region in the vicinity of TT1. Height of clay overburden is ca. 8 m.

in km 0.675 and from geodetically measured surface trough in km 0,740. The modeled field has dimensions 98 (width) × 50 (height) × 56.4 (length) meters. The thickness of clay overburden above the excavation is 7.9 m. The quaternary sediments in the model are represented by sand-gravel layer, sand-loess-loam and finally anthropogenic landfill which is only about 0.4 m thick. Due to its low thickness this layer was neglected. As a countermeasure, the unit weight of sand-loess-loam layer was magnified by 1 kN/m^3. The overall height of overburden is therefore 17.2 m as visible from overall geometry of the numerical model (figure 3). The whole numerical analysis is composed of 76 phases and each phase represents progress of excavation by 1.2 m. The last phase is relevant for evaluation of the surface trough and the deformations were evaluated 3 m from the front side of the model. Primary lining hardening was assumed using the principles of homogenization (see section 5). One excavation step 1.2 m long takes 8 hours. Always the first 1.2 m of excavation remains unsupported. The 6 partial tunnel headings (phases) are assigned as a,b, c, d, e, f as seen from the figure 1. Real excavation order according to scheme is b-a-d-c-e-f. The construction of TT2 was about 2 months in advance.

3.2 *Exploratory gallery IIB and unsupported adit R2*

Three exploratory galleries with triangular cross section and arched sides have width of 4.75 m and height of 4.5 m. With respect to TT1 and TT2 they are placed in the upper half of the profile. In routing of TT2 two equidistant exploratory galleries IIA and IIB have been excavated, where gallery IIB is one of the key structures for the back-analysis of K_0. The galery IIB was the first one, followed by the second one with the offset of several tens of meters. The length of structures is 831 m. In routing of TT1 only one exploratory gallery IA (the third overall) was excavated in the upper part of profile. The key for K_0 estimation was geotechnical monitoring in unsupported adit R2, which was chosen primarily for sufficient clay overburden and low urbanization. Four unsupported adits R1, R2, R3 and R4 were built for purposes of more detailed local exploration and all belong to exploratory gallery IIB. The unsupported adit R2 (as well as other three) is L-shaped. In each section there is one convergence profile, but only convergence profile in the part perpendicular to the gallery was taken into consideration. The part of adit R2 perpendicular to exploratory gallery IIB of Královo Pole tunnel is 5.38 m long. The diameter of circular cross-section of unsupported adit is 1.90 m. Support is made of mine steel arches in combination with steel wire meshes and acts only as a backup, with the offset of 50 mm from the excavation surface. It means that this is not in contact with soil so there is no interaction and stress redistribution. The height level difference between bottom of the adit R2 and bottom of the exploratory gallery IIB is 0.90 m. The depth of the center of the adit R2 is 22.15 m below the surface, the overburden is 16 m, out of which 6 m is quaternary cover. Figures 4a and 4b are photos of excavation faces in exploratory gallery IIB and unsupported adit R2, respectively. Chosen constitutive material model of clay strata is represented by hypoplasticity with inherent stiffness anisotropy in the very small strain range.

PLAXIS 3D Software was used. The chosen constitutive material model of clay strata is hypoplasticity with inherent stiffness anisotropy in the very small strain range. A Mohr-Coulomb model was assigned to the overburden.

Results from numerical analysis of exploratory gallery IIB and unsupported adit R2 were compared with geotechnical monitoring in the corresponding chainage (2.55 m from the junction of both structures) and time. Considering the rate of tunnel face progress (excavation of modelled portion of the exploratory gallery IIB and unsupported adit R2 took about 6 days) the analyses were modelled as undrained. The modeled section of the exploratory gallery IIB was 18 m long (figure 5b). The complete numerical analysis is composed of 28 phases and each of the phases represents progress of excavation by 1.2 m except the portion containing junction. The modeled field has dimensions 55 (width) × 37 (height) × 36 m (length) meters (figure 5a) to assure that the massif nearby the model boundaries remains intact. The quaternary sediments in model are represented by fluvial clay mould layer, then mixture of sand-loess-mould and finally anthropogenic landfill in overall height of strata of

Figure 4. a) View at the excavation face of exploratory gallery IIB and b) unsupported adit R2, Pavlík et al. 2004.

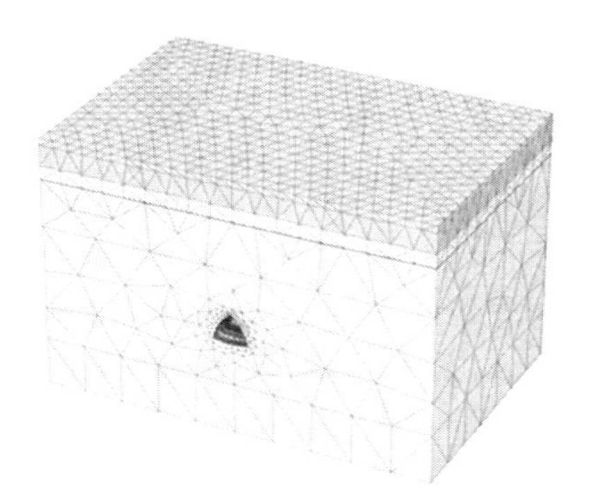
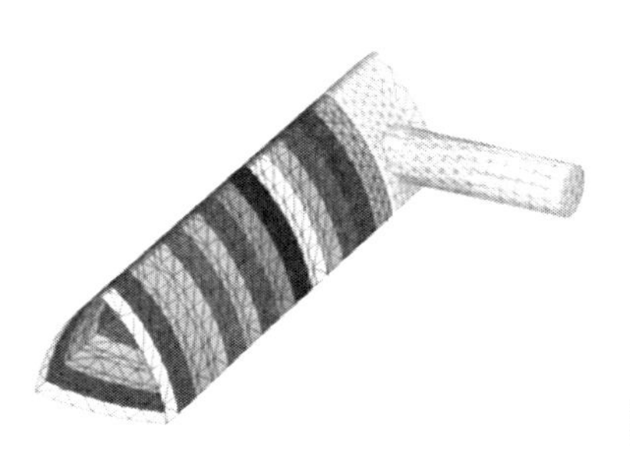

Figure 5. a) Detail of the geometry of exploratory gallery IIB and junction with adit R2 and b) model geometry in the vicinity of exploratory gallery IIB and unsupported adit R2.

6 m. The overall height of overburden is 22.1 m above the unsupported adit R2 and 20.4 m above the crown of exploratory gallery IIB so the thicknesses of clay overburden is 16.15 m and 14.4 m, respectively.

4 HYPOPLASTIC MODEL OF BRNO CLAY

4.1 *Model description*

The behavior of Brno clay has been described using hypoplastic model for clays incorporating stiffness anisotropy, developed by Mašín (2014). The model is based on the theory of hypoplasticity, which is governed by the following primary equation:

$$\overset{\circ}{\boldsymbol{\sigma}} = f_s(\boldsymbol{\mathcal{L}} : \dot{\boldsymbol{\epsilon}} + f_d \mathbf{N} \|\dot{\boldsymbol{\epsilon}}\|) \tag{2}$$

where $\overset{\circ}{\boldsymbol{\sigma}}$ and $\dot{\boldsymbol{\epsilon}}$ represent the objective (Zaremba-Jaumann) stress rate and the Euler stretching tensor respectively, $\boldsymbol{\mathcal{L}}$ and $\mathbf{N}$ are fourth- and second-order constitutive tensors, and f_s and f_d are two scalar factors.

The model parameters correspond to the parameters of the earlier model by Mašín (2005), which, in turn, correspond to parameters of the Modified Cam-clay model. The model thus requires parameters ϕ_c equal to 22° (critical state friction angle), N equal to 1.51 (position of the isotropic normal compression line in the space of ln p vs. ln $(1 + e)$), λ^* equal to 0.128 (slope of the isotropic normal compression line in the space of ln p vs. ln $(1 + e)$), κ^* equal to 0.015 (parameter controlling volumetric response in isotropic or oedometric unloading), and ν equal to 0.33 (parameter controlling shear stiffness).

In order to predict small strain stiffness, the model has been enhanced by the intergranular strain concept by Niemunis & Herle (1997). This concept requires additional parameters. In particular, the very small strain shear modulus G_{tp0} is governed by

$$G_{tp0} = p_r A_g \left(\frac{p}{p_r} \right)^{n_g} \tag{3}$$

With parameters A_g and n_g and a reference pressure p_r of 1 kPa. In addition, there are parameters controlling the rate of stiffness degradation R, m_{rat}, β_r, χ equal to 0.0001, 0.5, 0.2 and 0.8, respectively.

Finally, the adopted model allows for predicting the effects of stiffness anisotropy. An approach by Mašín & Rott (2014) has been followed. The most important parameter in this respect is the ratio of horizontal and vertical shear moduli α_G, defined as

$$\alpha_G = \frac{G_{ppo}}{G_{tpo}} \tag{4}$$

Two additional parameters characterize the dependencies of Poisson ratios and Young moduli:

$$\alpha_E = \alpha_G^{(1/x_{GE})} \tag{5}$$

$$\alpha_\nu = \alpha_G^{(1/x_{G\nu})} \tag{6}$$

Based on evaluation of an extensive experimental database, Mašín & Rott (2014) suggested default values of $x_{GE} = 0.8$ and $x_{G\nu} = 1.0$. In our analyses, these default values have been used primarily, however we also investigated sensitivity of the results on the selected parameter values.

4.2 *Model calibration*

Ratio of shear moduli α_G was measured by means of bender elements. For the purposes of the model, we considered the following value of α_G:

$$\alpha_G = 1.35 \tag{7}$$

During the calculation, the empirically estimated coefficients x_{GE} and $x_{G\nu}$ were used, however, the parametric study was also carried out. Model was further modified to predict the non-linear relation between shear modulus and mean effective stress in the small strain range:

$$G_{vh0} = p_r A \left(\frac{p}{p_r} \right)^{n} \tag{8}$$

where p_r is the reference stress equal to 1 kPa, p is mean effective stress, A, n are parameters equal to 5300 and 0.5, respectively. These parameters are the result of interpolation

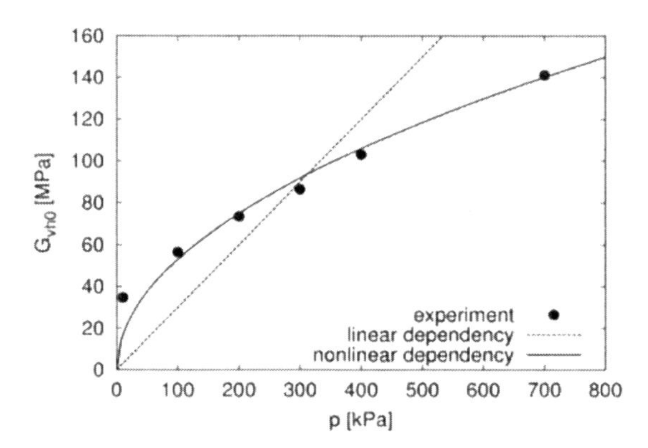

Figure 6. Calibration of vertical shear modulus in the very small strain range (Svoboda et al. 2010).

of experimental data (Figure 6). The related bender element measurements of vertical small strain shear modulus were carried out by Svoboda et al. (2010).

5 HOMOGENIZATION OF PRIMARY LINING

5.1 *Introduction*

The main tunnel tubes and exploratory galleries are supported by primary lining. This brings another uncertainty into the numerical model, since we do not exactly know time proportions and time-dependent material characteristics. To increase the model accuracy and because of inability of the adopted FEM numerical code to assume two different components of the primary lining, the homogenization of steel/sprayed concrete (shotcrete, further denoted as 'SC') lining was considered (Rott, 2014). The main primary condition was conservation of static momentum, conservation of the axial and bending stiffness and modification due to given constant value of the length of excavation step. In addition, the dependence of Young modulus of SC on time is taken into account. Královo Pole tunnels and exploratory galleries have primary lining composed of massive steel profile and SC applied in two layers. Then, in fact, we have three-material composite—older SC, younger SC and mine steel profile (exploratory gallery IIB) and "HEBREX" profile respectively (TT1 and TT2).

5.2 *Description*

In the numerical model, linear elastic material with time dependent SC stiffness was used, with parameters calculated using an empirical relationship where $E_f = 18$ GPa is the final Young modulus, α is a parameter equal to 0.14 and $t_r = 1$ is the reference time.

$$E_B = E_f\left(1 - e^{-\frac{\alpha t}{t_r}}\right) \tag{9}$$

Young modulus of homogenized and modified lining $E_m(t_1, t_2)$ is determined for bending moment in the direction of tunnel axis and can be expressed as

$$E_m(t_1, t_2) = \sqrt{\frac{E_f^2\left(1 - e^{-\frac{\alpha t_1}{t_r}}\right)^2 A_{NP}^2(t_1, t_2)}{12 I_{NP}(t_1, t_2) b_Z^2}} \tag{10}$$

where b_z is the length of tunnel step. If one considers rectangular cross section of the homogenized lining, the height of this shape $h_m(t_1, t_2)$ may be estimated from

$$h_m(t_1,t_2)=\frac{2\sqrt{3I_{NP}(t_1,t_2)}}{\sqrt{A_{NP}(t_1,t_2)}} \tag{11}$$

Substitute moment of inertia $I_{NP}(t_1, t_2)$ and substitute cross section area $A_{NP}(t_1, t_2)$ is calculated with the usage of conversion ratios. The conversion of steel to older SC is governed by equation (12) for conversion ratio $n(t_1)$:

$$n(t_1)=\frac{E_o}{E_f\left(1-e^{-\frac{\alpha t_1}{t_r}}\right)}>1 \tag{12}$$

and, analogically, for the conversion of younger SC to older SC the following equation (13) is valid:

$$m(t_1,t_2)=\frac{E_f\left(1-e^{-\frac{\alpha t_2}{t_r}}\right)}{E_f\left(1-e^{-\frac{\alpha t_1}{t_r}}\right)}=\frac{1-e^{-\frac{\alpha t_2}{t_r}}}{1-e^{-\frac{\alpha t_1}{t_r}}}<1 \tag{13}$$

These equations are valid with the assumption of the following main simplifications:

a. The overall stiffness (bending, axial and shear) does not involve SC cover of steel profile.
b. We assume full static interaction of 2 layers of SC.

For the primary lining of exploratory gallery IIB (consisting of 1 rolled mine steel beam, 2 layers of SC, overall static thickness 100 mm, $E_f = 18$ GPa, time difference $t_1 - t_2 = 1$ day, $\alpha = 0.14$ per one excavation step with the length $b_z = 1.2$ m) the resulting values for various time t_1 are summarized in following Table 1:

Figures 7a and 7b show the relation $t_1 - h_m(t_1, t_2)$ for primary lining of the exploratory gallery IIB and main tunnel tubes, respectively. Due to coupled effect of homogenization (conversion of Steel/SC and SC/SC) and modification (to satisfy the condition of constant width of tunnel step) the progress of the static thickness $h_m(t_1, t_2)$ seems to be surprising. Since the steel profile is in contact with the massif, the initial value of the above-mentioned static characteristics is non-zero.

Table 1. Resulting stiffness characteristics of homogenized and modified primary lining.

t_1 days	$E_m(t_1, t_2)$ MPa	$h_m(t_1, t_2)$ m	$E_m h_m(t_1, t_2)$ MNm[2]
1	5153	0.123	0.796
2	7114	0.117	0.941
5	11733	0.109	1.268
10	16277	0.105	1.581
15	18551	0.104	1.736
20	19683	0.103	1.813
25	20318	0.103	1.851

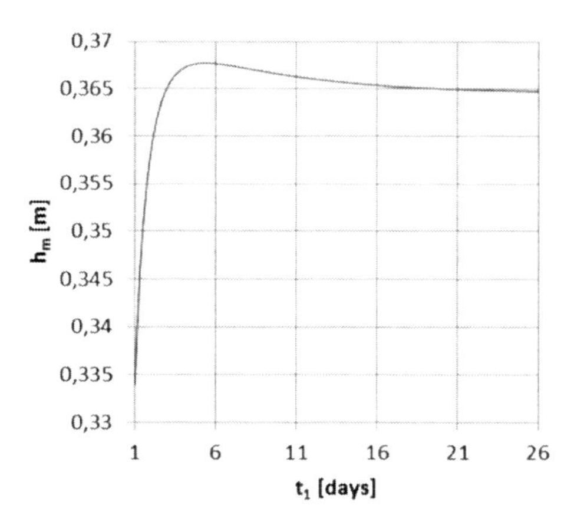

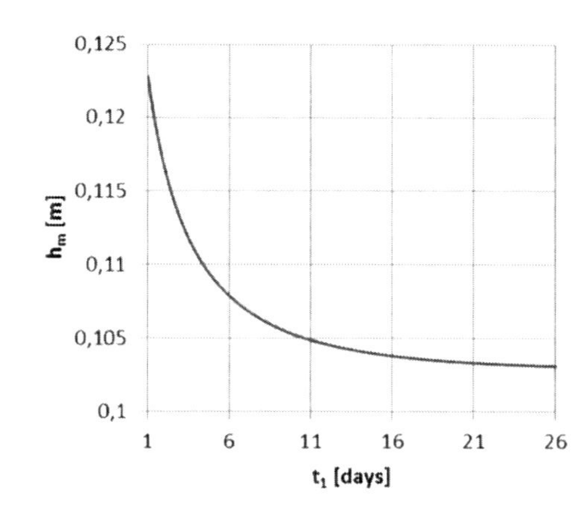

Figure 7. a) Progress of $h_m(t_1, t_2)$ in time for primary lining of TT1 and b) progress of $h_m(t_1, t_2)$ in time for primary lining of exploratory gallery IIB.

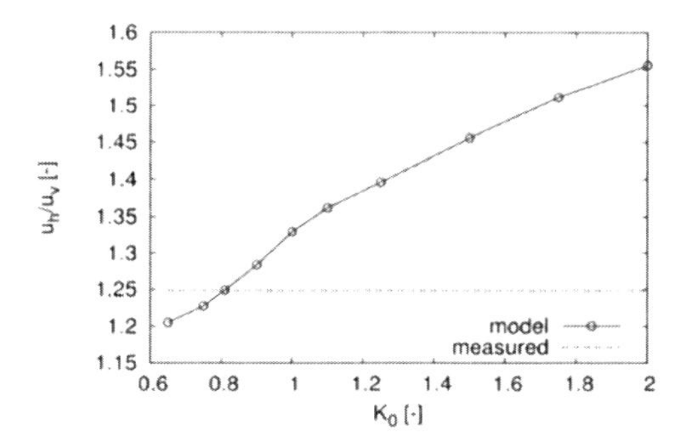

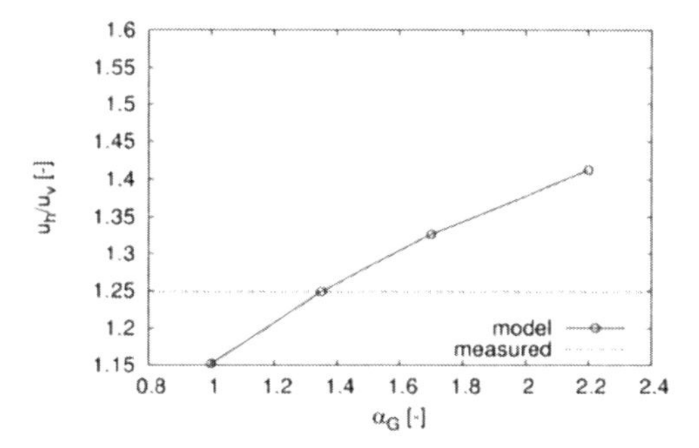

Figure 8. a) Resulting relation between K_0 and u_h/u_v (u_y/u_z in model) and b) relation between α_G and u_h/u_v (u_y/u_z in model).

6 RESULTS OF BACK-ANALYSIS OF K_0

The first step in numerical modeling was evaluation of the likely value of K_0 on the basis of back-analysis of exploratory gallery IIB and unsupported adit R2 according to the key convergence ratio $P = u_y/u_z = 1{,}25$. In the next phase the influence of each of the parameters α_G, x_{GE}, x_{Gv} was studied. The second step was to verify K_0 by means of numerical analysis of the complete Královo Pole tunnel tube TT1.

The dependence of convergence ratio on K_0 is obvious from Figure 8a and is very significant. The measured value $u_y/u_z = 1{,}25$ is exactly matched for $K_0 = 0{,}81$, which is considerably lower than value obtained by Mašín & Novák (2013) in the analyses with the hypoplastic model without anisotropy and without presence of exploratory gallery IIB.

It is apparent that the predicted ratio u_y/u_z is significantly affected by α_G as seen from Figure 8b. Fortunately, this value is easiest to be found out in laboratory. Increasing of α_G (increasing of degree of anisotropy) leads to increase of ratio u_y/u_z. Apart from the shear stiffness, anisotropy affects also the stress path and therefore change of effective stress around excavation under undrained conditions, which leads to the observed dependency of K_0 and α_G.

Since we have relations $u_y/u_z - K_0$ and $u_y/u_z - \alpha_G$, it is necessary to carry out an additional numerical calculation to obtain the dependence $K_0 - \alpha_G$. Therefore, for given boundary values of $\alpha_G = 1.00$ and $\alpha_G = 1.70$ (this is assumed as a maximum value for Brno clay) the corresponding values of K_0 are obtained. The search for K_0 corresponding to $\alpha_G = 1.00$ and $\alpha_G = 1.70$ and condition $u_y/u_z = 1.25$ was done by interpolation between 2 values of u_y/u_z. The resulting values are $K_0 = 1.01$ and $K_0 = 0.60$, respectively.

7 VERIFICATION OF K_0 USING SIMULATIONS OF THE COMPLETE TUNNEL

The real construction of TT2 was about 2 months in advance so the surface trough of TT1 is believed to be affected. Nevertheless, the affection is relatively small due to the distance between tunnel tubes axes. One of the main advantages of selected portion of TT1 is the absence of exploratory gallery IA. Numerical model does not fully respect the real time distances since they vary from date to date. However, the difference is relatively small, so the result is believed to be the same as through entering the mentioned average value. First of all, an analysis with $\alpha_G = 1.35$ and $K_0 = 0.81$ was performed, then an alternative pairs of α_G and K_0 consistent with the condition $u_y/u_z = 1.25$ (found on the basis of circular adit simulations) were entered. The purpose is to demonstrate that the "default" back-analysed combination of $\alpha_G = 1{,}35$ and $K_0 = 0{,}81$ is the most probable. The results of all analysed cases are summarized in following Table 2 and shown in Figures 9a and 9b.

The equidistant position of graphs is the verification that geotechnical structure is seated in correct depth. The surface settlement trough depth (Figure 9b) is almost unaffected and vary between 50–54 mm. The crucial reason is probably the influence of very stiff primary lining on the results of TT1 simulations which does not allow for significant bending due to the change of loading. All values are acceptable and realistic in comparison with the monitoring results in chainage 0.740, which in maximum yields 44 mm. We can state that variation of α_G and K_0 has only negligible influence on the surface trough. The surface trough for the combination $\alpha_G = 1.70$; $K_0 = 0{,}60$ is narrower than for other cases which is typical for anisotropic stiffness. Here the effect is further strengthened by the low value of K_0. Surface settlement trough is slightly non-symmetric. The asymmetry is caused partly by the tunnel headings offset and also by the fact that the tunnel tube TT2 was bored 2 months later. Even though the distance between the tunnel tubes is around 70 m, it is believed that terrain surface troughs of both tubes certainly influence each other. In contrast with the surface settlement trough, horizontal displacements in the tunnel vicinity vary relatively significantly (Figure 9a). The reason comes from the assumption of shear moduli of anisotropic clay. Theoretical shear modulus in the horizontal direction at zero strain is calculated as $G_{hh0} = G_{tp0}\alpha_G$ which means the change of stiffness in the horizontal direction only. For the "default" back-analyzed combination $\alpha_G = 1.35$ and $K_0 = 0.81$ the maximum obtained value is 14.5 mm. The monitoring data show 8.5 mm. The other results 19.5 mm and 10.5 mm belong to combination $\alpha_G = 1.00$ and $K_0 = 1.01$ and $\alpha_G = 1.70$ and $K_0 = 0.60$, respectively. It is now possible to say that the model was working properly and the obtained numbers are logical because the increase of α_G leads to a decrease of displacements and contrary, the increase of K_0 on the contrary leads to the increase of displacements.

Table 2. Resulting horizontal deformations from the verification analysis.

Case	α_G –	K_0 –	$u_{x,max}$ (mm)
1	1.35	0,81	14,5
2	1.00	1,01	19,5
3	1.70	0,60	10,5

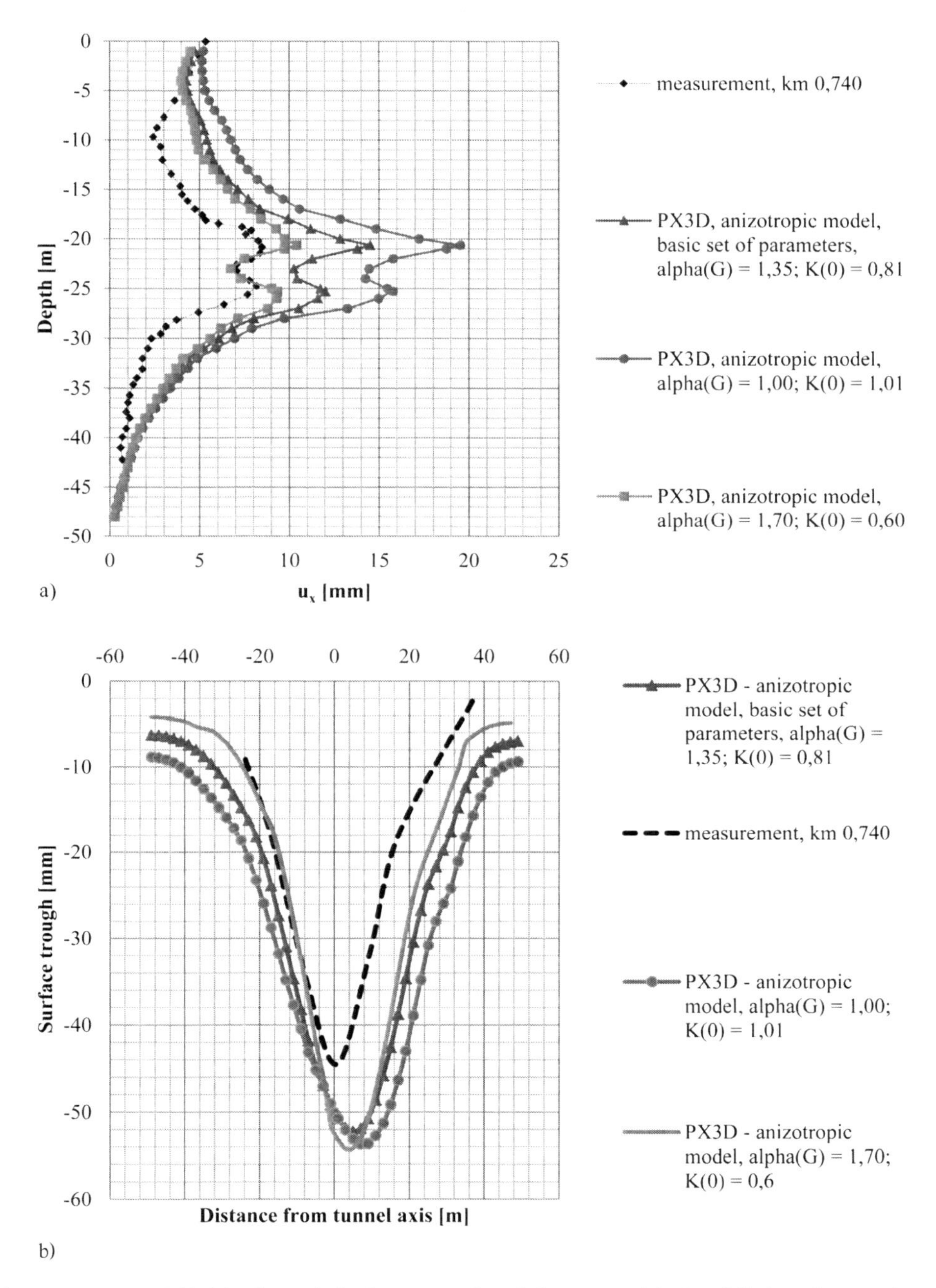

Figure 9. a) Modelled horizontal displacements for all 3 cases and b) Modelled surface settlement trough for all 3 cases.

8 SUMMARY AND CONCLUSIONS

The paper deals with an estimation of lateral earth pressure at rest in over-consolidated Brno clay. An indirect method of numerical simulation of boring process of Královo Pole tunnel was used, where the obtained values are fitted to geotechnical monitoring data through parametric study. We simulated Brno clay using a hypoplastic model with implemented inherent small strain stiffness anisotropy. The homogenization of primary lining was adopted to represent SC lining. The obtained values agree well with measurements. It was also proved that the variability of the degree of anisotropy of shear moduli α_G affects the results significantly, whereas the effect of parameters x_{GE} and x_{Gv} is small. The resulting lateral earth pressure coefficient at rest was $K_0 = 0.81$. Note that this value is valid for circular adit depth only.

ACKNOWLEDGEMENT

Scientific activity was financially supported by the research grant P105/12/1705 of the Czech Science Foundation and research grant 243-253370 of the Charles University Grant Agency.

REFERENCES

Horák, V. 2009. Královopolský tunel v Brně z pohledu projektanta. *Tunel*, 18(1): 67–72.

Mašín, D. 2005. A hypoplastic constitutive model for clays. *International Journal for Numerical and Analytical Methods in Geomechanics* 29, No. 4, 311–336.

Mašín, D. 2014. Clay hypoplasticity model including stiffness anisotropy. *Géotechnique* 64, No. 3, 232–238.

Mašín, D. and Novák, V. (2013). Evaluation of the earth pressure coefficient at rest by backanalysis of circular exploratory adit in Brno clay. In Proc. *12th International Conference Underground Construction Prague 2013*, CD proceedings, ISBN 978-80-260-3868-9.

Mašín, D. & Rott, J. 2014. Small strain stiffness anisotropy of natural sedimentary clays: review and a model. *Acta Geotechnica* 9, No. 2, 299–312.

Niemunis, A. & Herle, I. 1997. Hypoplastic model for cohesionless soils with elastic strain range. *Mechanics of Cohesive-Frictional Materials 2*, 279–299.

Pavlík, J., Klímek, L. & Rupp, D. 2004. Geotechnický průzkum pro nejvýznačnější stavbu na velkém městském okruhu v Brně—tunel Dobrovského. *Tunel*, 13(2), 2–12.

Rott, J. 2014. Homogenization and modification of composite primary lining with increase of elasticity modulus of shotcrete in time (In Czech), *Tunel*, 3/2014, (in print).

Svoboda, T., Mašín, D. & Boháč, J. 2010. Class A predictions of a NATM tunnel in stiff clay. *Computers and Geotechnics* 37, No. 6, 817–825.

Underground Infrastructure of Urban Areas 3 – Madryas et al. (Eds)
© 2015 Taylor & Francis Group, London, ISBN 978-1-138-02652-0

Risk assessment for infrastructures of urban areas

P.M. Mayer, A.S. Corriols & P. Hartkorn
ITC Engineering GmbH & Co., KG, Stuttgart, Germany

M. Messing
VMT GmbH, Bruchsal, Germany

ABSTRACT: The development of infrastructure on urban areas such tunnelling works requires continuous monitoring of project risks, reporting and evaluation of the project progress and the service cost assessment as part of management duties. Web-based, modular and thereby project-related configurable technology platforms offer real-time connection of so far autonomous systems such as document management, design and cost estimation data, and machine and site data. The monitoring and analysis system based on application software allows to a complete control of all tunnelling works as well as in their area of influence. The main objective is to achieve the highest level of safety throughout the construction process both in TBM driving or traditional tunnelling methods and secondary processes required, such as compensation grouting stabilization or earth stabilization, shaft drilling. Controlling must be supported by the data analysis of measurements taken by the extensive network of sensors available. When any sensor measurement overpasses the alarm levels then a planned action protocol will immediately followed to prevent any incident or inconvenience during construction works. The timely availability of current process data monitored on the TBM in the tunnel, at the surface and in the buildings is the crucial prerequisite to control and optimize the complete operating processes, thus contributing to risk minimization and ensuring safety.

1 INTRODUCTION

Risks during construction are always present. Underground works must be carried out by assuring the maximum level of safety. The image on Figure 1 depicts the collapse during the digging works for Cologne's new metro line, on March 2009.

Standard quality levels and risk minimization of accidents due to the changes introduced in the ground by infrastructure projects require the control of the construction processes, especially for tunnels and other underground works. This is performed by placing a sensor network that continuously records measurement data of a number of geotechnical parameters: settlement and deformations, groundwater level and tensional state of the terrain, among others. Alarm trigger levels are set according to the standards. If exceeded, it could pose a risk and therefore the implementation of preventive or corrective measures should be performed.

Traditionally monitoring measurements were collected manually and analyzed sometimes a few hours later. This delay entailed an increased risk and the incapability of detecting problems in tunneling works. Technological advances such as the introduction of automation of Tunnel Boring Machines (TBMs) and data registers from the entire sensors place on them

Figure 1. Cologne's ground collapse during the construction of the new line of Metro (March, 2009).

allowed a better control of the drilling process. However, analysis and interpretation of such amount of data meant serious difficulties in making relatively quickly estimations and taking decisions accordingly.

Data acquisition should also be performed effectively by a sensor network placed on the surface to control works performed underground as well as monitoring of buildings and other structures affected by the tunnel progression. The integration of all readings from the TBM sensors as well as measurements provided by total stations, inclinometers, piezometers, strain gauges and other devices that conform the instrumentation sensor network in the same data pool further justify the use of global data management systems.

2 PECULIARITIES OF INFRASTRUCTURE WORKS IN URBAN AREAS

The construction or renovation of infrastructures in urban areas has certain peculiarities, as they require stricter safety conditions than other works performed outside of the urban environment. The most remarkable aspects are presented below:

- Works performed in a densely populated urban area. The objective of the planned new infrastructure is to improve communication networks, especially in areas of the city with high population density. Also new service network lines are needed to supply the growing demand in these areas, most of which are planned underground.
- High degree of technical difficulty. Pre-existing infrastructures significantly influence the proposed solution and the constructive process. Basic city services such as power lines, water or gas pipes passing underground should not be disrupted to guarantee normal supply. An action plan must be defined in case of accidental interruption.
- Effect of tunnelling excavation on buildings and other structures on the surface (settlements, vibrations, noise). The performance of works alters ground structure. Settlement and deformations must be strictly controlled because of their negative impact on the stability of buildings and other structures on the existing urban area, such as bridges, water tanks or power line towers. Moreover, the use of construction equipment could generate

some vibrations that might cause structural damage to buildings and affect the daily lives of citizens.
- Dense and highly active transport network. The most important infrastructure works carried out in urban areas are transportation lines such as the Metro network, railways or city highways. These lines can be completely new, in which case the project should include not only the design of the new layout, but also the connection to the existing network through the construction of a new metro station or a transport interchange. In other cases the main goal is to bury an existing line to minimize the traffic density on the surface with a positive impact on city liveability, leaving more space for pedestrians and green spaces for leisure and enjoyment of citizens.
- Connivance with other contemporary construction projects. Plans to remodel the communications network or services may contemplate the simultaneous execution of several projects at the same time, in order to concentrate all the impacts generated by the works on the same time period. An overall plan regarding collateral effects between different works perform simultaneously is then required, especially if they are located on the same area, with a proper coordination that minimizes negative impacts on the residents.
- Social acceptance and sensitivity of the population to the inconvenience and risks to be assumed during the construction works. The implementation of projects to improve existing infrastructure entails a sacrifice from people for the sake of further benefit. However, some citizens might not be willed to take the hassle and risks involved in such improvements. The use of data management systems would ensure full control of the construction processes and a better risk assessment, in turn facilitating social acceptance.

3 ADVANTAGES OF INTEGRATED DATA MANAGEMENT

3.1 *Web-based software platforms*

The development of infrastructure on urban areas such tunnelling works requires continuous monitoring of project risks, reporting and evaluation of the project progress and the service cost assessment as part of management duties. Web-based, modular and thereby project-related configurable technology platforms offer real-time connection of so far autonomous systems such as document management, design and cost estimation data and machine and instrumentation data.

The need to overcome these obstacles and avoid the lack of information accompanying the decision making during the underground works in urban areas motivated to civil engineers, geologists, surveyors and software developers working together on the design and creation of integrated data management monitoring systems.

Their main features are:

- Full integration of general project information, on-going process data, construction schedule, geological information and geotechnical/environmental measurements into a single data base system.
- Ongoing data evaluation during the construction process.
- Transparency by presenting complex data in standardized reports and charts.
- User-defined reporting and analysis tools. The user can add their own analysis modules to IRIS in order to obtain tailored solutions.
- Integrated alarm systems for all measurements.
- Worldwide access to data of on-going or completed project, stored on a web-based database system.
- Standardized reporting reduces the site team's daily paperwork and provides a single report standard over the entire project.
- As shown in Figure 2, data acquisition can be performed manual or automatic from different data sources: TBM, suppliers, or jobsite documents. Data are sent through automatic upload, FTP server or manual input to a gateway or server where they are stored. Then, every software user can access to all the information through the Internet.

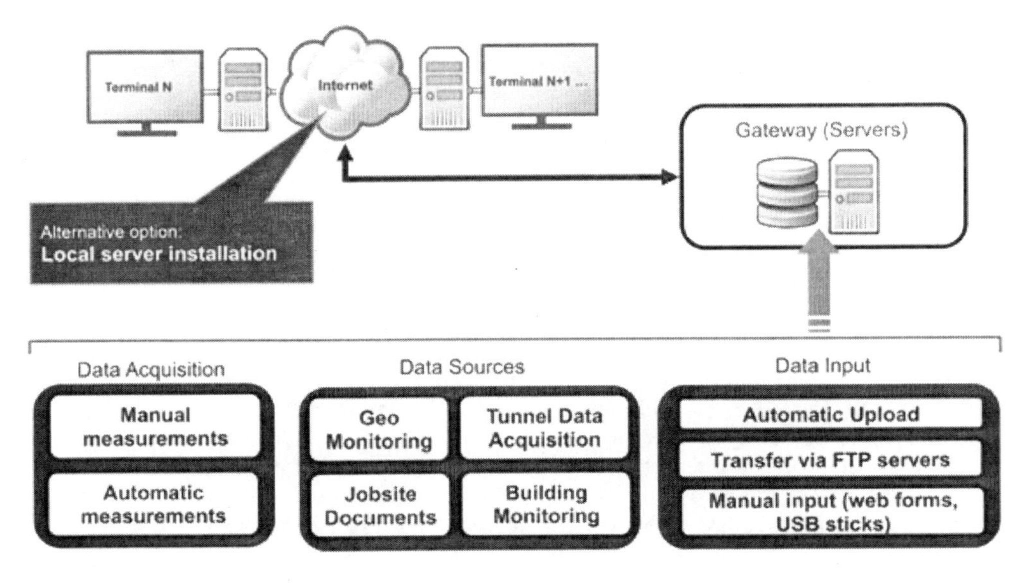

Figure 2. Data acquisition and upload on the data management structure.

Who are the users of the so-called web-based data management systems? Contractors, tunnelling crew—which manages the TBM guidance system and the other machine sensors—and survey teams provide all data to the universal data pool, which is accessible to the external project team—contractors themselves or the owner/client—and also for authorized guests who take care about the evolution of works as the client representative at the jobsite.

Data management systems consist on several modules that focus on the Key Process Indicators (KPI), which are the basis for the quantitative and qualitative evaluation of tunnel driving and also for the entire sensor network as part of the surface instrumentation. Some of these modules are described below:

- Visualization tools for instant access to navigation machine, geotechnical and monitoring data. A map viewer presents the TBM position in plan-view with high accuracy provided by the GPS tool, as shown in Figure 3, top.
- Alarm notifications when a measured value exceeds a predefined threshold configured by the user. The software communicates alarm messages through three different channels: pop-ups (website), e-mails and SMS. Sensors are shown in the visualization maps with a colour code: green when trigger levels are not exceeded, yellow for warning levels exceeded and red when a sensor reading overpasses the alarm level). Figure 3, middle, shows two trigger levels on a chart that display settlement readings from a levelling point.
- Integration of TBM sensor data with geotechnical sensors data from surface monitoring. Figure 3, bottom, shows an example of a time history chart that includes data registers from the pressure excavating chamber together with settlement readings of a sensor placed on the surface. Also the influence of the TBM on settlements registered along a surface line is displayed in Figure 4. Every settlement curve corresponds to a certain time step with a fixed distance between the TBM and the monitored surface line. It can be observed that the settlement readings are stabilized as the TBM moves away from the surface line.
- Detailed view of building monitoring with interactive pictures that shows sensor location and data charts when clicking over the sensor symbol. Figure 5 shows an historic building with a series of sensors placed on the main wall. The green colour indicates that readings didn't exceed the trigger levels, unlike sensors shown in Figure 3, top, are coloured in yellow and red.

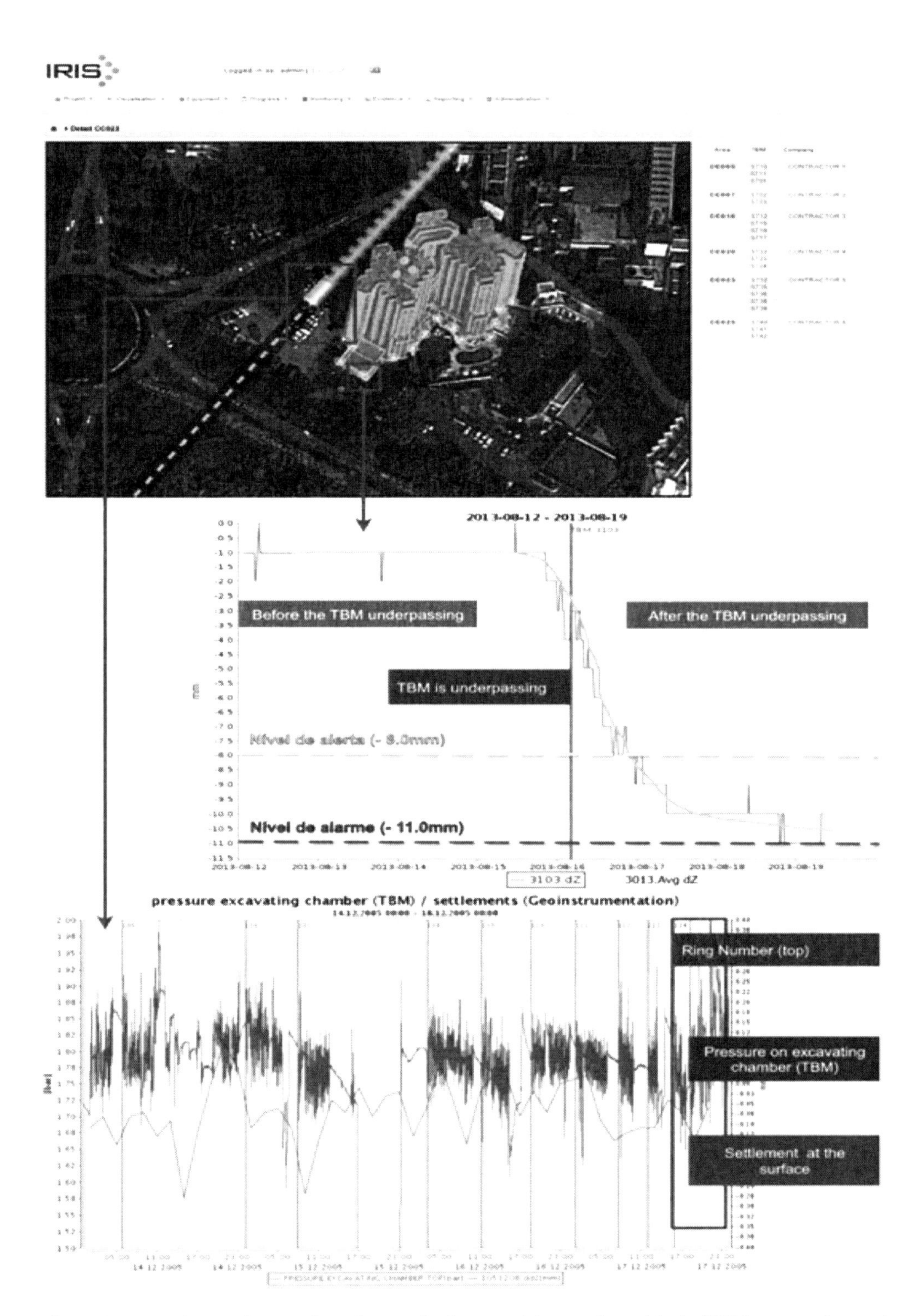

Figure 3. Data integration on the software platform: aerial map view with the TBM progression and building monitoring (top), settlement time history with trigger levels (middle) and combination of a TBM sensor with settlement at the surface (bottom).

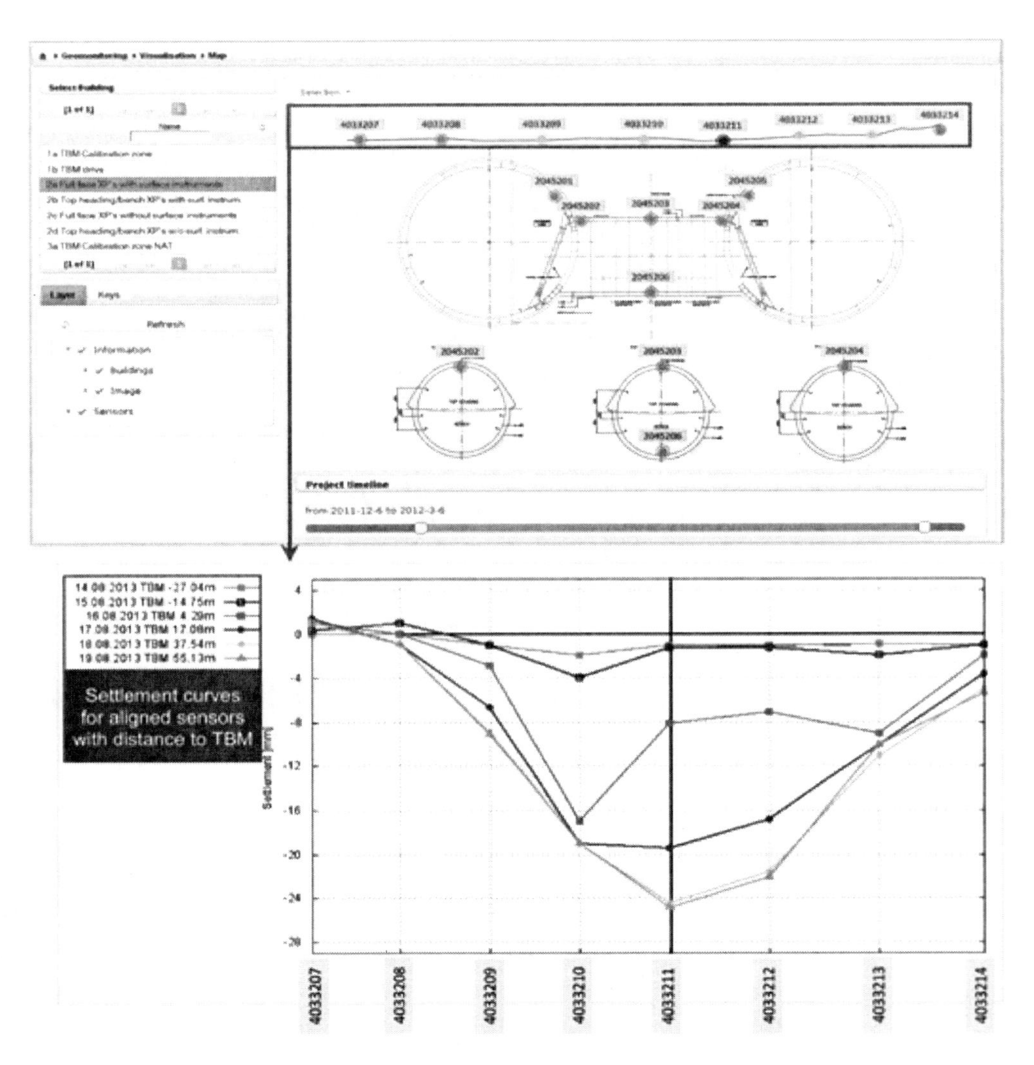

Figure 4. Top: General cross section of the works, which shows sensors placed at the surface with sensors located inside the tunnel track for controlling ring deformation. Bottom: Chart that shows the influence of TBM on registered settlements at the surface.

- The data management and monitoring of several TBMs on the same project facilitates the control of the overall process. Figure 6 shows the visualization map of a multi-project with the current status of every TBM identified by a colour code. Measurements of sensors located on the tunnel and also distributed on the surface are stored on the database.
- Machine data like cutting wheel torque and advance rate are plotted. Comprehensive target-performance analysis tools allow the user to update TBM advance predictions according to the encountered geology. Ring processing and installation, stand still phases and some other sub-processes are registered on the work shift reports, as shown in Figure 7, top. Visualization of sensors placed on the cutting head (Figure 7, middle) allows controlling the functioning of the excavation process and detecting any problem during operation such as pressures, torque, rotation speed, volume of air, volume of grouting and bentonite, transported mass and conveyor belt speed, etc.
- Automatic generation of summaries and reports available for daily, weekly and monthly periods with a quick and detailed overview concerning advance rates, duration and causes of

Figure 5. Monitoring of an historic building for control of settlements and inclinations (Building Monitoring Module).

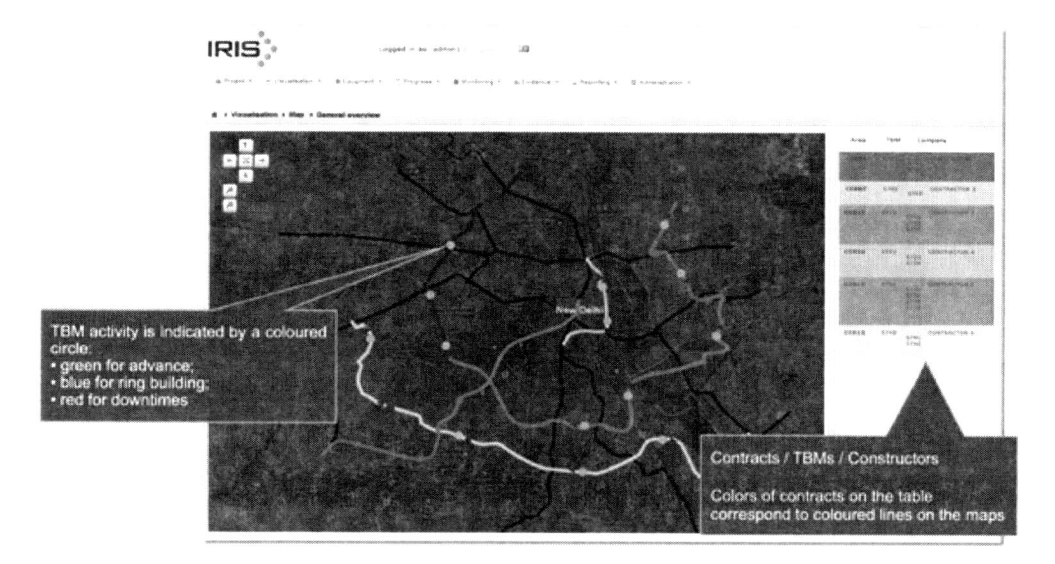

Figure 6. Map visualization of a global project with a number of TBMs running simultaneously. Their current state identified by a color code.

downtimes, etc. The machine data reports include full information on the TBM from its starting date (advance speed, downtime) as well as complete information about incidences. These reports can be downloaded as a PDF format. An example is shown in Figure 7, bottom.

- Segment management module provides overview of all segments used on ring construction. It is therefore possible to track the history of each segment from delivery, storage to installation in the tunnel and afterwards. Segment damage and the associated repair works can be reported as well. This tool is being successfully used in the Koralm Tunnel Project, a 32.9 km-long railway tunnel in the Austrian Alps which will by finished by 2022.

Some manufacturers of sensors and measuring instruments have created a platform use for the data management of their own devices. In standard underground works it is normal to use devices from different brands and therefore it becomes necessary to configure the

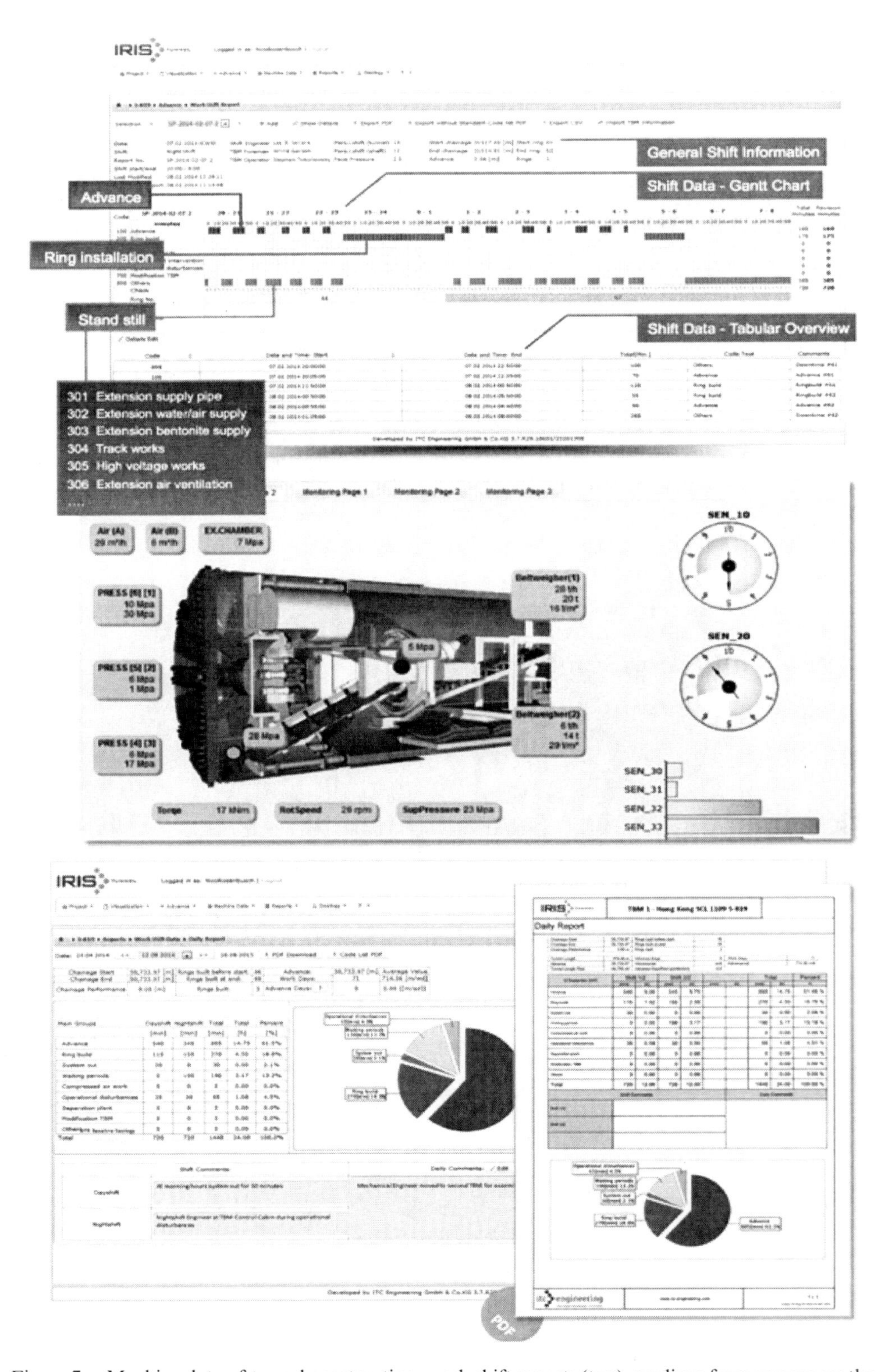

Figure 7. Machine data of tunnel construction: work shift reports (top), readings from sensors on the TBM on real time (middle) and generation automatic reports in PDF (bottom).

required interfaces for including all them in a common platform. In recent years there have been some companies exclusively specialized on this matter. However, only very few have gone a step further and data management software system allow the complete integration of data from TBM sensors together with data from the sensor network on the surface: total stations, tachymeters, inclinometers, piezometers, strain gauges and load cells, among others.

This cross-linking of elements from underground works and surface influence makes a progress-related interpretation of the monitoring data possible. For example, deformation of retaining walls measured with inclinometers can be visualized together with the TBM driving at the location, or the pressure excavating chamber on the TBM can be compared with the vertical settlement on one point right on the surface, as shown in Figure 3, bottom.

The need of integrating all data under the same common platform for infrastructure projects that consist on the simultaneous construction of several lines demanded global solutions for data management able to deal with a huge amount of measurement and registers from the tunnel progression and surface monitoring.

What about projects that don't use TBM for performing tunnel excavation? In this case, basic concepts of data management systems remain valid. Both sensor data for controlling the evolution of works can be integrated in the universal data pool: ones coming from the tunnel track and the others from the sensor network set on the surface. The New Austrian Tunnel Method (NATM) is also very extended on tunnel construction. It demands additional information gathered from on site measurement to perform structural models and design the system support and tunnel reinforcement. The integrated data management systems are again a reliable and useful tool for this purpose.

The data management systems can be considered in the near future as a useful tool of risk analysis and assessment of underground infrastructure projects performed by insurance companies. Their use would reduce insurance costs given its positive effect on risk minimization.

The benefits of such systems can be summarized in the following points:

- Integration of Information Technologies (IT) in global construction works: tunnel + affected area.
- Relative low costs versus remarkable benefits for the overall process.
- Global real-time web-based platform usable for all parties involved.
- Identify and reduce risks.
- Surface monitoring (total stations, piezometers, inclinometers, etc.) and TBM data.
- Manage growing flood of data from automatic monitoring systems.
- Achieve transparency.

3.2 *Process automation*

Automatic sensors allow meter readings automatically and send them directly via a network connection to the server where the database is stored where they are. Once the meter is installed and begins the process requires minimum equipment topography: Any revision and correction in case of malfunction. In the case of the hand sensors, measurements must be performed by a survey team that is responsible for inserting into the database, either through manual typing or by uploading data files generated by the device every few time. Automatic sensors are more expensive, but they represent a time and handwork saving, especially for bigger projects.

TBM sensors are all automated control requirements for operation and reliability when reading data. Manuals sensors are still widely used, although the use of automatic sensors has grown rapidly in recent years. The clear trend in infrastructure projects is the progressive replacement of manual sensors by automatic devices due to the numerous advantages that compensate their price difference: faster decision and available data for analysis, less probability of reading errors and human mistakes, better control of risk and therefore greater safety. Data management systems are optimized for the use of automatic sensors, since their advantages clearly overcome their higher costs.

The use of integrated data management systems allows performing partial automation of manual sensors for data acquisition, as shown in Figure 8. Multiple sensor data files can be

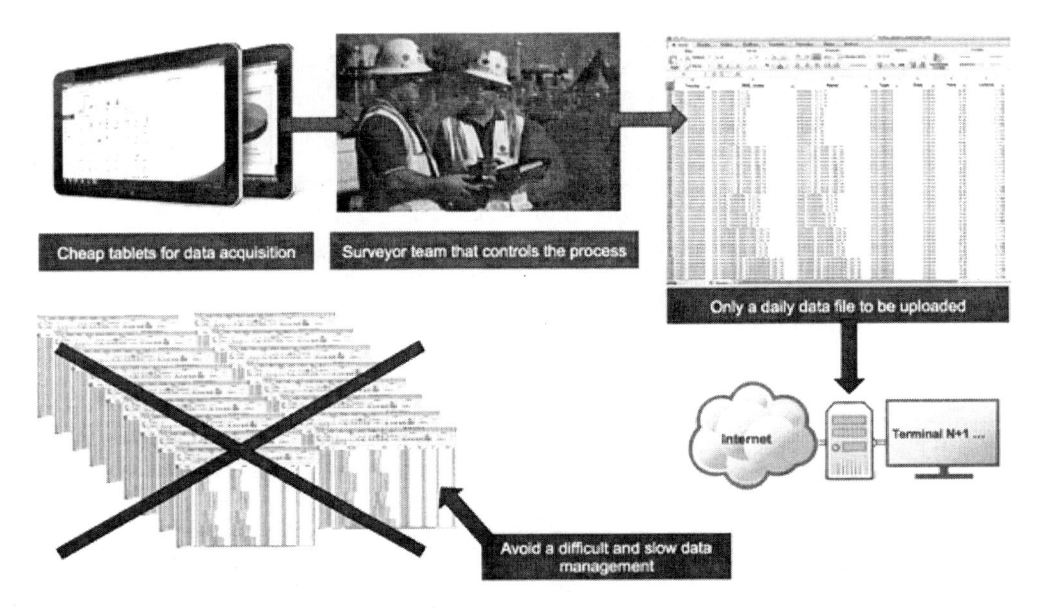

Figure 8. Scheme of a proposed solution for partial automation of data acquisition with manual sensors.

replaced by a single file that includes all sensor readings, every of them identified by a sensor index. This unique file is then sent to the gateway server are added to the database.

Subsequent data analysis can be performed through the web platform data management system. Alarm notifications, reporting and other tasks are performed automatically, so surveyors' work becomes easier and simpler. Digital devices such as tablets can be used to optimize data acquisition instead of entered later from hand-written field sheets thus shortening the working time.

3.3 *Use of other sensor types and new data base structure*

Data volume of sensors measuring dynamic parameters is well above those read by standard sensors used in geotechnical instrumentation. The growing use of dynamic sensors requires a consistent database to store high-frequency registers, such as and vibration amplitudes and accelerations. Manual readings would become very difficult and expensive. On the other hand, fiber optic sensor technology is becoming of very common use in cases where a high densely spatial instrumentation is needed—e.g. measuring points every 2 cm—since it enables a better characterization of settlement lines. Database structures based on NoSQL schemes-Not only Structured Query Language—allows to store a large amount of information while maintaining the web-based platform operability for data analysis and risk assessment.

These new data management systems could even be used as a visualization and analysis tool by some Transport Authorities for training. Engineers and surveyor teams could access to the data base of the first phase of a metro line performed 10 years ago if the data base structure is easily adaptable to the management system's. They would visualize all data and learnt from this to use this knowledge in the second phase of the metro line being managed with the same software data management system.

4 OUTSTANDING EXAMPLES

Integrated data management systems are successfully being used on several mega-projects around the world, such us London, Copenhagen, San Francisco, Auckland, Sydney, Istanbul, New Delhi, Hong-Kong and Singapore. The complexity of their different work phases, together with the inconveniences of being performed mostly in urban areas led to the establishment of an universal data pool that deliver all agents involved of a common platform for data analysis, decision making faster and safer.

The main characteristics of two of these examples are shortly presented below.

4.1 *San Francisco Central Subway*

The Central Subway in San Francisco (USA) is an extension of the Muni Metro light rail system, which will cross the City Downtown connecting Union Square and Chinatown—the most densely populated neighbourhood—with the East district. The scope of works, with a total budget of $1.6 billion includes the structural and geotechnical instrumentation and monitoring of project structures and adjacent existing utilities, buildings and infrastructure, installation of compensation grout pipes and subsequent compensation grouting for protection of adjacent buildings, as well as handover of temporary tunnel utilities and sump dewatering to follow on contracts for construction of the subway stations, track way, traction power and control systems. The extreme difficulty of controlling the TBM underpassing the existent tunnel below Market Street and the control of the building foundations is one of the remarkable aspects that need an accurate monitoring system backed by the experience of use in other similar construction works like in Copenhagen.

4.2 *Doha Metro Green Line*

The Doha Metro Green Line is the biggest monitoring project where the IRIS software will be used. This includes the construction of a twin bore tunnel with a length of 16.6 km and the construction of 8 underground metro stations at an average depth of 20 meters in a time period of 5 years and the use of 6 TBMs, with a total budget of $2.2 billion. The new metro Green Line will feature ten stations that extend from the coastline to the interior of Doha from Msheireb (Central Station) through to Education City and on to Al Rayyan (East-West direction). There will be a modal transport interchanger—Switch Box—that will provide the connection to other means of transportation medium and long distance, so it will be an important communication hub in the heart of the city. All areas around the stations will feature a vast monitoring sensor network to control the progress of works to foresee risks, especially those related to the settlements of the main buildings within the area of influence and the water level. Also the driving of the 6 TBMs will be integrated in the software thus allowing a centralized control of the monitoring process on each tunnel boring. This project is essential in the delivery of the 2022 FIFA World Cup and is a key contributor to Qatar's 2030 vision.

5 CONCLUSIONS

Mobility needs in the cities demand an increase of underground infrastructures that requires the construction of new tunnels but also controlling affected areas. Other infrastructure projects for city services such as power lines, water pipes or gas conductions are also underground works to be considered.

The construction or renovation of infrastructures in urban areas has certain peculiarities, as they require stricter safety conditions than other works performed outside of the urban environment.

Monitoring data management systems allow to a complete control of all tunnelling works and their entire area of influence in urban areas, in turn offering to all users a standardized platform for data management by automating purchasing, reducing analysis time, minimizing risk and enabling faster decision-making to ensure higher safety levels in the construction process. A high level of safety can be achieved throughout the construction process both in TBM driving or traditional tunnelling methods and secondary processes required (compensation grouting stabilization, earth stabilization, shaft drilling). Controlling must be supported by the data analysis of measurements taken by the extensive network of sensors available.

The benefits of such systems can be summarized in the following points:

- Integration of Information Technologies (IT) in global construction works: tunnel + affected areas.
- Relative low costs versus remarkable benefits for the overall process.
- Global real-time web-based platform usable for all parties involved.
- Identify and reduce risks.
- Surface monitoring (total stations, piezometers, inclinometers, etc.) and TBM data.
- Manage growing flood of data from automatic monitoring systems.
- Achieve transparency.

Two example projects using an integrated monitoring data management system are shortly presented: San Francisco Central Subway and the Doha Metro Green Line.

REFERENCES

Brinkgreve R.B.J.; Schad, H.; Schweiger H.F.; Willand, E (editors). 2004. *Geotechnical Innovations. Studies in Honour of Professor Pieter A. Vermeer on Occasion of his 60th Birthday and Proceedings of the Symposium held at Stuttgart on June 25th 2004*. Stuttgart.

Chmelina, K. 2009. Integrated tunnel monitoring and surveying supported by an information system. In: *Proceedings of the 35th ITA-AITES World Tunnel Congress*. Budapest: Pál Kocsonya.

Ehrbar, H. and Lieb, R.H. 2011. Gotthard Base Tunnel—Risk Management for the world's longest railway tunnel—lessons learnt. In: *Proceedings of the 37th ITA-AITES World Tunnel Congress*. Helsinki.

Hamaide, G.; Humbert, E.; Robert, A. 2012. Basics for efficient risk management process in tunnelling: need for an evolution in practice in France. In: *Proceedings of the 38th ITA-AITES World Tunnel Congress*. Bangkok.

Hu, M.; Zhou W.B.; Zhu, B., Pu, J.L. 2011. Risk Monitoring based on the State Transaction Graph. In: *Proceedings of the 37th ITA-AITES World Tunnel Congress*. Helsinki.

Loganathan, N.; Turner, D.; Pan, J. 2011. Tunnelling Induced Risks to Buildings—A Case Study. In: *Proceedings of the 37th ITA-AITES World Tunnel Congress*. Helsinki.

Matsui, H.; Iriji, Y.; Kamemura, K. 2012. Risk management methodology for construction of underground structures. . In: *Proceedings of the 38th ITA-AITES World Tunnel Congress*. Bangkok.

Mayer, P.M. 2000. *Verformungen und Spannungsänderungen im Boden durch Schlitzwandherstellung und Baugrubenaushub*. Thesis (PhD in Civil Engineering)—Institutes für Bodenmechanik und Felsmechanik der Universität Fridericiana in Karlsruhe, 178p. (in German).

Mayer, P.M. et al. 2014. Risk mitigation in urban infrastructure projects. In: *Proceedings of the 40th ITA-AITES World Tunnel Congress*. Iguassu Falls (Brazil).

Mayer, P.M. et al. 2014. Data management and risk analysis for tunnelling projects. *Tunnel* (2/2014): 18–24. Gütersloh (Germany): Bauverlag.

Pöllath, K; Mayer, P.M.; von Lübtow, B. 2014. Construction and monitoring of infrastructure works as illustrated by the example of the Nordhavnsvej Tunnel in Copenhagen. In: *Proceedings of Darmstädter Geotechnik-Kolloquium*. Darmstadt (Germany).

Schmuck, C. 2012. Data management in tunnelling projects—Challenges and Possibilities. In: *Proceedings of the 38th ITA-AITES World Tunnel Congress*. Bangkok.

Vähäaho, I.T.; Van Alboom, G.; Korpi, J.M.; Satola I.S.; Vergauwen, I. 2011. Geotechnical and geological data management in urban underground areas. In: *Proceedings of the 37th ITA-AITES World Tunnel Congress*. Helsinki.

Underground Infrastructure of Urban Areas 3 – Madryas et al. (Eds)
 ISBN 978-1-138-02652-0

Influence of experience acquired in the construction of the Warsaw Metro on current designs of underground structures

H. Michalak
Warsaw University of Technology, Warsaw, Poland

ABSTRACT: The article presents the issues related to designing and technologies for the construction of facilities with multi-storey underground parts in high-density municipal development. Due to the pioneering nature of the works related to the construction of the Warsaw Metro in this respect, experiences of the 25-year long construction project were used in investments of the first Polish buildings with multi-storey underground sections and followed by the development of individual procedures.

1 INTRODUCTION

The concept of improving the municipal transport of Warsaw with a metro dates back to 1925, when the authorities passed a resolution to develop a design for the "Metropolitan" underground train. Due to the economic crisis in the 1930's, work on the project ceased. In 1938, Stefan Starzyński, the Mayor of Warsaw, established the Bureau of Underground Railway Research which was responsible for updating and continuing the earlier engineering operations. After a break resulting from the 2nd World War, the idea was brought back and ended with the government resolution of December 1950 on the construction of the underground train system. Construction works were started in 1951 on the Praga District side of the Vistula, however as early as in 1953 they stopped due to problems which existed relating particularly to difficult hydrogeological conditions.

After almost 30 years, in 1982 the government passed another resolution on the construction of the 1st underground train line, based on which a team of designers representing different disciplines was formed (B.P.B.K.iS. "Metroprojekt). The 1st Warsaw Metro line with its route arranged almost parallel to Vistula connects the southern and northern districts of Warsaw along the Kabaty–Młociny section. Construction of the 1st Metro line which includes 21 track tunnels and 21stations took 25 years, i.e. from 1983 to 2008 (Fig. 1).

Construction of the track tunnels was performed by mining methodology with the use of a semi-mechanical tunneling shield and the stations were constructed according to an opencast method.

Due to the pioneering nature of such an investment in Poland, construction works started in the southern part of the city, in less developed areas. The first tunnels and stations, as well as specially assigned test fields, were used to acquire experience, improve and verify calculation methods based on geodetic measurements in the natural scale of displacement of the support and adjacent buildings.

The end of the 1990's was a period of continuation of the underground structures in highly urbanized areas of the city center and also the start of many projects for facilities with a multi-storey underground and a foundation depth similar to the underground stations. In many cases, such investments incurred major construction problems whilst they were being carried out and work was temporarily stopped due to emergency or pre-emergency situations.

Figure 1. The scheme of the 1st line of Warsaw Metro (B.P.B.K.iS. "Metroprojekt").

A few projects involved construction disasters covering a large area adjacent to the edge of the pit and buildings located within it.

As a result of repeated cases of safety hazards in new facilities with multi-storey underground areas, as well as adjacent buildings, it became necessary to extend the scope of works related to preparing the investment, engineering and realization with the issues of the interaction of a deeply founded facility and the adjacent building structures.

Due to the similarity of buildings with multi-storey underground sections to the cut-and-cover station projects, the engineering principles applied in the Warsaw Metro were preliminarily adopted and later enhanced with the acquired experience and additional procedures.

2 GENERAL CHARACTERISTICS OF THE CONSTRUCTION METHODS USED FOR THE WARSAW METRO

The soil conditions at the foundation level of approx. 20 m may be usually classified as complex, in the Warsaw area. There are heterogeneous, discontinuous layers with varying lithology and genetic properties, at the ground water table (usually two) found above the foundation level. The water table is usually identified as an unconfined aquifer or the upper surface of a confined aquifer.

Metro stations in the southern part were constructed with open pits with a support made as a Berliner wall which was usually supported at three levels with struts or grout anchors.

Uniform support with struts or grout anchors or both methods at the same time were used. The Berliner wall support included steel double-tee bars with timber lining behind the double-tee shelves. Approximate pit dimensions were usually: length 300 m, width 23 m and depth 13 m. Due to the ground water level being above the level of the bottom of the pit, the installation of drains was usually necessary. Solutions to lower the water table with dewatering wells located along the outer line of the pit and local drainage of the bottom of the pit were installed.

The middle section of the 1st Metro line included a diaphragm wall as the support for the pit, with struts or grout anchors at specific levels. In the case of the immediate proximity of buildings, diaphragm walls were provided with supports prior to pit shafting—floor method.

3 THE PROGRAM OF ASSESSING THE IMPACT OF THE WARSAW METRO STATION CONSTRUCTION ON FACILITIES IN ITS IMMEDIATE PROXIMITY

For the development of the program of observation of the Metro facilities, a special research office was established (Julian Gajda, Wojciech Kożon & Hanna Michalak) in the B.P.B.K.iS. "Metroprojekt". Based on the results of research and analyses performed by the office, a routine observation program was established for the pit support, buildings and underground reinforcements.

The scope of the impact of the pit on soil deformation depending on pit support technology at the elevation was determined.

In the case of strutting, four subsequent zones of the impact of the pit on adjacent facilities marked with letter R were introduced:

- zone R_0—located directly behind the support of the pit with the largest movements of soil base, determined approximately with a failure plane,
- zone R_1—another zone with the risk of exceeding the limits of usability and the load capacity of the structures of the existing facilities and the necessity of their protection or reinforcement,
- zone R_2—area with soil deformations that should not have an impact on the safety of the existing building structures,
- zone R_3—area with minor, diminishing dislocations of soil base caused mainly by lowered ground water level.

Depending on the type of support of the pit, the following ranges of the zones were defined (h—pit depth, h_f—building foundation depth), particularly in the case of:

- Berliner wall (1 ÷ 4),

$$R_0 = (h - h_f)\,\mathrm{ctg}(45 + \varphi/2) = 0.577(h - h_f) \quad (1)$$
$$R_1 = 2(h - h_f)\,\mathrm{ctg}(45 + \varphi/2) = 1.15(h - h_f) \quad (2)$$
$$R_2 = 2(h - h_f) \quad (3)$$
$$R_3 = 3 \div 4\mathrm{h} \quad (4)$$

- bored pile wall (5 ÷ 8),

$$R_0 = 0.577(h - h_f) \quad (5)$$
$$R_1 = 1.0(h - h_f) \quad (6)$$
$$R_2 = 2.0(h - h_f) \quad (7)$$
$$R_3 = 3.0(h - h_f) \quad (8)$$

- diaphragm wall (R_2 zone not identified; 9 ÷ 11)

$$R_0 = 0.577(h - h_f) \quad (9)$$
$$R_1 = 1.0(h - h_f) \quad (10)$$
$$R_3 = 3.0(h - h_f) \quad (11)$$

- floor method for construction of the underground part—R_1 and R_2 zones not identified (12 ÷ 13):

$$R_0 = 0.577(h + t - h_f) \tag{12}$$

where t is a theoretical average depth of the wall support below the pit bottom,

$$R_3 = 3.0\text{h} \tag{13}$$

For the anchored support of the pit, 4 zones marked as K symbol were identified (14 ÷ 17):

- zone K_0—represented the range of R_0 zone, limited by the soil wedge plane,

$$K_0 = R_0 = 0.577(h - h_f) \tag{14}$$

- zone K_2—located behind the K_0 over anchor strands, the range of soil deformations in this zone depends on the quality of anchors. With the correct construction process, only minor soil deformations are expected,

$$K_2 = l\cos\alpha \tag{15}$$

where l is the anchor length, and α is anchor angle to the horizontal axis,

- zone K_1—the area with the largest soil deformations (apart from K_0), located at the end of the anchor bulb,

$$K_1 = l\cos\alpha + 0.577(h - h_f) \tag{16}$$

- zone K_3—the area with diminishing soil deformations

$$K_3 = 3.0h \tag{17}$$

Within the range of impact of the pit on soil deformations, all building structures and underground reinforcements in an individual zone were combined. The scope of tests and works required to perform the project safely was defined.

For facilities located in zones R_0, K_0, R_1 and K_1 the following requirements for supports installed prior to the works were introduced, including:

- developing an opinion on the technical condition; for poor technical conditions of the facilities, a reinforcement or temporary protection scheme was established for the time of construction and implemented before the works were started,
- cataloguing all internal and external damage of the structure and equipment,
- installation of benchmarks on structural elements in the facades of buildings for measuring vertical dislocations,
- installation of benchmarks on the supports of underground reinforcement lines with a large diameter for measuring vertical dislocations,
- significant defects (e.g. cracks) were sealed to observe the advance or stabilization of deformations.

During the construction project, land surveys and visual inspections of buildings, underground reinforcements and pit support horizontal dislocations with the frequency based on the work phase were performed.

For facilities located in zones R_2 and K_2, it was necessary to prepare a general technical opinion on the condition of the structure, install benchmarks for measuring vertical dislocations and catalogue defects in the façades. For facilities with a poor technical condition, the scope of examinations was identical as for K_0, R_0 zones. During the project, land surveys and visual inspections were performed with a frequency much lower than for R_0 and K_0 zones.

For zones R_3 and K_3 benchmarks were installed on building structures to measure vertical dislocations at 1 month intervals.

For metro stations constructed in highly urbanized areas (from Wierzbno station), a "Program of surveys and observation of buildings located in the area impacted by the underground train construction" was developed and during underground station construction the results of tests and measurements were reviewed on a current basis and conclusions were drawn (Gajda & Michalak, 1990). Where dislocations close to the limit values occurred, recommendations were provided for further work methodology. The limit values for the horizontal dislocation of the support of the pit were defined depending on the type of applied walling and support method at the elevation, based on earlier experiences on the test field, while for the buildings—based on the literature studies, particularly related to facilities built in areas used for mining activities.

When the construction of the station was completed, graphical and tabular comparisons of results were prepared and general conclusions and design recommendations were provided, particularly for the following: observations and measurements of lateral dislocations of the support of the pit in specific work phases, vertical dislocations of the terrain, compression of the struts and grout anchors, as well as the observation of buildings and underground reinforcements.

4 EXPERIENCE IN THE CONSTRUCTION OF THE 1ST LINE OF THE WARSAW METRO

Conclusions prepared in the 1990's covered the experience acquired in the 1st stage of the construction of the 1st underground train line, i.e. Kabaty–Politechnika section. Based on test and observation results, collected through many years, the following general design recommendations were introduced connected to the engineering and construction of underground railway stations:

- the pit support made as a Berliner wall with horizontal timber lining can be used if buildings are located outside zones R_1 and K_1,
- steel piles made of narrow-footed double-tees should be inserted into drilled holes filled completely with self-hardening slurry; wide-foot steel double-tee piles were recommended,
- lateral timber lining was recommended only with "isolation" of the ground water flow to the pit by introducing effective drainage or a grout curtain behind the pit support (vertical grout barrier),
- in order to limit the lateral dislocations of the support of the pit, strut pre-stressing was introduced,
- for pit support made as a bored pile wall without pile overlapping and expected seepage of ground water into the pit, a grout curtain seal was required,
- grout curtains were recommended to correct the depth to ensure stability of the bottom of the pit during the works,
- for ground drainage cases, the water table level was controlled on a current basis, e.g. in piezometers.

5 GENERAL CHARACTERISTICS OF CONSTRUCTION METHODS USED FOR BUILDINGS WITH MULTI-STOREY UNDERGROUND PARTS IN A HIGH-DENSITY MUNICIPAL DEVELOPMENT

The depth of pits for multi-storey underground sections of buildings is, depending on the designed number of storeys, approx. 7 m (2-storey), 11 m (3-storey) up to 20 m (5-storey). They are often constructed in immediate, or close proximity, of existing facilities, built at a small depth below the terrain level. The support of such pits are usually diaphragm walls, which play a dual role: pit support and walls of the underground building section, transferring—apart from the lateral loads (earth pressure, hydrostatic force, surcharge loading)—also vertical loads exerted by the ground-based building part.

For very extensive excavations, an underground structure is often constructed to support the struts. In such a case, an approx. 4 m deep initial excavation is made on the entire surface whilst keeping the support-based static stabilization of pit support. Next, in the central section, the pit is shafted down to the bottom level and the wall is supported with a buttress wall of earth. When the central section of the underground part is constructed in the shaft, struts are installed. During the construction symmetrical load distribution should be maintained to eliminate the risk of generating a damaging load on the constructed structure.

6 FAILURES AND CONSTRUCTION DISASTERS RELATED TO THE CONSTRUCTION OF BUILDINGS WITH MULTI-STOREY UNDERGROUND PARTS

Many of the first buildings with multi-storey underground parts in dense municipal developments incurred various problems, in the worst cases—with local or more extensive construction failures and disasters.

The problems occurred due to various causes, including those resulting from negligence or errors made at different phases of the investment project. Generally speaking, the most frequent problems arising from the preparation phase of the investment and the project documentation were:

- incorrectly adopted technology for construction and support of the pit,
- insufficient identification of the ground and water conditions of the substrate. This identification has an impact on the adopted statistical calculation value representing the earth pressure on the support of the pit and, as a consequence, on the endurance parameters or geometry of such a support, requiring the ground anchoring of the support below the

Figure 2. The failure during the construction of multi-storey underground sections.

bottom of the excavation, as well as on the selected construction technology for the pit and underground part of the building,
- insufficient identification of the arrangement of the water-bearing layer causing the risk of localized failures of the substrate at the bottom of the pit or a loss of the stability of the bottom and walls of the pit,
- incomplete cataloguing of the arrangement of underground reinforcements, particularly sewage and water pipes that had been shut down. Such problems are difficult to avoid, particularly in cities, which have suffered a war and significant damage to buildings, as well as their documentation. Insufficient knowledge about the network of underground pipes may be the cause of the lost stability of gaps when the slurry walls are installed, by uncontrolled leaks of thixotropic slurry.

During the construction of buildings with multi-storey underground sections problems occurred which were related to the failure to follow the technological regimes and deviations from the design recommendations for the works. Such modifications were usually applied to make the works faster or reduce the costs (Fig. 2).

7 CURRENT PROCEDURES APPLIED IN THE DESIGNING AND CONSTRUCTION OF BUILDINGS WITH MULTI-STOREY UNDERGROUND PARTS IN A HIGH-DENSITY BUILDING DEVELOPMENT

Failures and disasters occurring during the construction of multi-storey underground sections of buildings always lead to a halt in the work and the necessity to perform tests and analyses to determine the causes of problems and also to indicate procedures for repairs and the continuation of the work. For such works, teams of experts, often including surveyors with experience and documented participation in the design of the Warsaw Metro, were usually assigned. Due to these reasons, the procedures implemented in the Warsaw Metro project were initially directly transferred to analogical projects of facilities with multi-storey underground sections. With the increased demand for buildings with extended underground parts, used mainly for parking space, multi-aspect studies were also performed and finally brought about the determination of procedures for such investments.

Common procedures covering preparation, designing and construction stages for facilities with multi-storey underground parts in a high-density municipal development were supplemented with the necessity to define the impact of such facilities on adjacent objects (Michalak, Pęski, Pyrak & Szulborski, 1998; Michalak, 2006 & 2009). These include in particular:

a. Determination of the range of impact of the pit on vertical shifts of the terrain, depending on the soil substrate:
The study (Kotlicki & Wysokiński, 2002) defined pit impact zones as ground areas around the pit, in which excavations result in vertical and horizontal shifts of soil. It was estimated that the highest deformations occur in the soil in the immediate proximity of the pit, in an area with the following length (h stands for pit depth): for sand $0.5h$, for clay $0.75h$, for loam $1.0h$. The range of impact of the pit, depending on soil conditions, was specified as $2.0h$ in sand, $2.5h$ in clay and $3 \div 4h$ in loam.

The studies (Michalak, Pęski, Pyrak & Szulborski, 1998; Michalak, 2006) show that the largest vertical shifts of the terrain are found in the area with a width of 0.5 up to $0.75h$ from the edge of the pit. The shifts cease at a distance of $2h$ from the edge of the pit, or, $3 \div 4h$ with a lowered ground water table (dewatering wells located outside the pit outline).

A comparative analysis of lateral shifts v_x of the support of the pit and vertical shifts of soil located immediately behind that support v_y confirmed the following relationship $v_y = 0.5 \div 0.75v_x$. This applies to Berliner walls with supports, anchors, as well as pile walls and slurry walls with supports or anchors.

It should be noted that the evaluations mentioned concern the influence of the construction of the pit, while they do not cover deformations that occur during the construction of the underground building structure, and subsequently, the ground-based structure.

b. Definition of the impact of a lowered ground water table on the range of the impact of the pit.
Lowering the ground water table causes the settling of the soil in adjacent areas due to the increased bulk density of the drained soil. The value of the settling depends mainly on the type of soil and the value of the table lowering. It can be estimated that the settling in moraine soils is approx. 1 mm per 1 m of lowered ground water table. Respective test results were presented in the study (Kotlicki & Wysokiński, 2002).

c. Definition of requirements for limiting dislocations of soil and the support of the pit.
Due to the fact that lateral shifts of the support of the pit are one of the major factors generating lateral soil shifts in adjacent areas, for projects executed close to existing buildings it is necessary to provide measures to limit such dislocations. Due to the requirement to ensure the safety of use and the bearing capacity of buildings and the technical infrastructure within the range of impact of the pit, for dense development it is preferable to use a "rigid" support such as diaphragm walls or secant piled walls, supported with pre-stressed struts or floors of the underground levels in the constructed building—floor method.

The studies (Michalak, Pęski, Pyrak & Szulborski, 1998; Michalak, 2006) show that the limitation of soil shifts can be ensured by constructing the underground section of buildings with the floor method or diaphragm walls supported with pre-stressed struts (Fig. 3).

Due to the necessity to limit the lateral shifts of the diaphragm walls, they should be sufficiently rigid, anchored to the required depth below the bottom plate and sufficiently strutted. Depending on the depth of the pit, strutting should be used at several levels, while the limitation of the support of the pit lateral dislocations is the most effective if strutting is installed along with the progress of earth works, starting with level "0" (floor on the top underground storey).

d. Definition of the total range of impact of a deeply founded building in a dense municipal development on vertical soil movements.
Vertical soil movements in the area adjacent to the new building result from their superposition in individual work phases, covering: execution of the support, pit shafting and successive supporting of its walling, execution of the underground part of the building, then the entire structure and its use.

The examinations and calculation studies (Gajda & Michalak, 1990; Michalak, Pęski, Pyrak & Szulborski, 1998; Michalak, 2006) show that the value of vertical soil movements and the shape of the subsiding trough depend mainly on the type of soil substrate, the support of the pit and its static operation parameters, the technology used in building the underground part and the construction phase.

In buildings with their underground sections executed according to the floor method (starting with floor "0") or with excavation shoring (the locking of supports in the capping beam or pre-stressing from floor "0"), lateral movements of the pit are minor when compared to the vertical shifts of the support of the pit.

For the technologies mentioned for underground sections, the zones of the impact of the new building on the dislocation of the terrain and adjacent buildings in such areas were specified (Michalak, 2006). Within the range of impact exerted by the new building, four zones were specified ($S_{0.75}$, $S_{0.50}$, $S_{0.25}$ and S_0) with their range depending on the absolute value of vertical movements at the edge of the pit v_0 not exceeding $0.75v_0$, $0.50v_0$, $0.25v_0$ and 0 (no dislocations), respectively.

The studies relate to the following construction phases: phase II—construction of the underground structure and phase III—construction of the building and applying the full service load.

The studies (Michalak, 2006) indicated the following conclusions, with reference to the range of impact of new buildings on soil movements (Table 1):

a)

b)

Figure 3. Construction of building basement: a) roof method, b) propped diaphragm walls.

Table 1. Impact zones of the construction of new buildings upon the vertical displacements of the surrounding ground surface (Michalak, 2006).

Construction phases	Zone range			
	$S_{0.75}$	$S_{0.50}$	$S_{0.25}$	S_0
II	$0.5h$	$0.7h$	$1.1h$	$1.7h$—sands, $5.4h$—silts
III	$0.5h$	$0.8h$	$1.3h$	$2.8h$—sands, $5.4h$—silts

- the largest vertical movements of soil with an absolute value up to $0.75v_0$ occur at a distance of up to $0.5h$ in phases II and III,
- vertical movements of soil with an absolute value of $0.75 \div 0.50v_0$ occur at a distance of up to $0.7h$ in phase II and $0.8h$ in phase III,
- vertical movements of soil with an absolute value of $0.50 \div 0.25v_0$ occur at a distance of up to $1.1h$ in phase II and $1.3h$ in phase III,
- vertical soil movements disappear depending on the type of soil substrate and as follows: for sand in phase II at a distance of $1.7h$ from the edge of the pit, and in phase III at a distance of $2.8h$ from the edge of the pit, while for loam-based substrate at $5.4h$ from the edge of the pit for both phases,
- the range of $S_{0.75}$ does not depend on the phase of building construction,
- beyond zone $S_{0.75}$, phase III shows an extended range of individual areas as compared to phase II.

The process of soil movements does not disappear when the execution of the pit is completed. It is estimated that for sand substrates it practically stops when the construction is completed, while in cohesive soil it can continue for up to three years after that moment. On average, it can be estimated that in heterogeneous substrates the process lasts approx. one year after the completion of construction and with the full service load on the structure.

For buildings with underground sections constructed according to the floor method in diaphragm walls supported with pre-stressed struts, for heterogeneous soil substrates below a sand or loam bottom, functions are defined to allow the forecasting of the total values (covering soil substrate unloading and subsequent loading with the weight of the new building) of vertical soil movements in the proximity of such buildings (Michalak, 2006).

e. Diagnostic survey of the existing buildings located in the area which are impacted upon by the new deep-founded construction.
Examinations of the existing buildings and reviews of their technical conditions should be performed in the project preparation phase and constitute an element of its design documentation. The results should enable the determination of the possibility or lack of possibility to carry additional loads resulting from the expected irregular vertical soil dislocations in the area of foundations of such buildings, through their structures. If the technical condition of a building—as a result of diagnostic operations—is deemed as insufficient to carry additional loads, it is necessary to design appropriate reinforcement for the structural elements of the building, soil substrate, or both.

When surveying the building, the following items should be verified in particular: bearing structure system, foundation type, the materials used for structural components, designed occupancy mode and the compliance of service conditions with the design occupancy, random factors which have an impact on the changes in the technical conditions of the structure elements and existing damage of the structure and equipment.

The existing damage of the building should be catalogued and the results should be reported with photos, photogrammetric printouts or graphics.

Based on the results of the structural survey and after: a review of the existing technical documentation, endurance tests of materials and verification calculations of structure static parameters, an assessment of the technical condition of the building, including its natural and technical wear, should be made.

f. Requirements related to the observation of the condition of existing buildings.
The scope of observations of the existing buildings should be adjusted to their locations and distance from the outer edge of the constructed building. Based on internal examinations, it can be estimated that that ¾ of vertical soil shifts (for support with slurry wall and strutting or floor method installation) occurs at a distance of up to $1.3h$ from the pit and then decreases gradually. Due to this fact, for this area the structures of existing buildings should be covered by a thorough land survey and visual inspections of the following:

- vertical movements of the support structure elements (columns, main walls). The number and locations of benchmarks depends on the length of the existing building and should allow for the determination of deformation on every solid block including the movement of joint separation,
- lateral shifts of the wall panels of buildings adjacent to the designed building,
- lateral shifts of the pit support cap and on the levels with intermediate supports,
- vertical shifts of the bed of the new building.

A database for the results of the measurements mentioned (stabilization of benchmarks and initial measurement) should be made before the launch of construction works related to the new building. The preparation stage of the project should also include the following: photogrammetric or photographic documentation of the existing damage found on external and internal structural elements and general construction damage on the adjacent buildings, a catalog of the identified damage, with specifications of the width of scratches and cracks in the building structures, or, installation of seals (e.g. gypsum type) in the major cracks of building structural elements.

With regard to buildings located in the area impacted by the diminishing soil movements, i.e. more than $1.3h$ from the edge of the designed building, their surveys can be limited to the observation of vertical shifts of the load-bearing structural elements (columns, main walls), however, the number and arrangement of benchmarks should allow for the determination of deformation on every solid block including movement joint separation. It is also helpful to prepare photogrammetric or photographic documentation of the existing external and internal damage of structural and general construction damage in the existing buildings, as well as to catalogue the existing damage, covering the width of scratches and cracks, before the project is started.

The frequency of observations should be adjusted to the progress of construction works and can be reduced when the construction of the underground section of a new building is completed. However, observations should be carried out until the moment when soil deformations are stabilized, which is approx. one year after the commissioning of a new building on heterogeneous soils.

The results of land surveys should be immediately analyzed and reviewed for compliance with the predicted values, as specified in the design documentation. If alarming effects occur, e.g. excessive displacements of the support of the pit or existing buildings, damage, etc., all construction works should be immediately stopped. Participants in the construction process should then make a decision about the rules and possibilities for work continuation by themselves or with the help of a surveyor.

8 FINAL CONCLUSIONS

The methods of designing and constructing buildings with multi-storey underground sections were largely influenced by the engineering experiences acquired on the 1st Line of the Warsaw Metro on the Kabaty–Politechnika section.

Selection of engineering technologies for the construction of underground building sections, as well as the protection and support of deep foundation walls must follow many analyzes aimed mainly at the estimation of the impact of the designed project on soil stability.

Designing buildings, particularly their ground-based parts, in a high-density urban development should be handed over to professional designers and respective works should be performed by contractors who can guarantee adherence to the technological regimes and high quality of work. The supervision of a professional investor, as well as geotechnical and geodetic supervision, is also very important.

The safety requirements for designed structures and adjacent buildings were specified in applicable regulations, as follows: (Technical Specifications, to which the buildings and their locations ought to conform, 2002):

§204.1. *The building structure should meet conditions that ensure compliance with the load-bearing and service limits by all the elements and the entire structure.*

5. The construction of a building in the immediate proximity of a building structure must not cause hazards to the users of the facility or a reduction in its service capabilities.

§206.1. *In cases mentioned in §204 para. 5, the construction should be preceded by a technical evaluation of the existing facility, in which safety and fitness for use is verified, with consideration of the impact exerted by the new building.*

Such legal regulations oblige the investor to prepare technical documentation with an evaluation of the existing adjacent buildings, as well as to design and select execution technology while considering the interactions between the new building and existing buildings.

REFERENCES

Domurad J. & Miros G. 2009. Rozwiązania konstrukcyjne i realizacja stacji „Wawrzyszew" i „Młociny" I linii metra w Warszawie. *„Inżynieria i Budownictwo", nr 8/2009.*

Gajda J. & Michalak H. 1990. Praca konstrukcji i tuneli metra w ośrodku sprężystym. *Informator Generalnego Projektanta nr 28. Biuro Projektów Budownictwa Komunalnego i Specjalnego „Metroprojekt".*

Gajda J. & Pokora M. 1990. Wpływ realizacji stacji „Rakowiecka" (A-10) na obiekty znajdujące się w strefie oddziaływania budowy metra. *Informator Generalnego Projektanta nr 28. Biuro Projektów Budownictwa Komunalnego i Specjalnego „Metroprojekt".*

Kotlicki W. & Wysokiński L. 2002. Ochrona zabudowy w sąsiedztwie głębokich wykopów. *ITB, praca nr 376/2002.* Warszawa.

Madryas C., Kolonko A., Szot A. & Wysocki L. 2006. Mikrotunelowanie. *Dolnośląskie Wydawnictwo Edukacyjne,* Wrocław.

Michalak H., Pęski S., Pyrak S. & Szulborski K. 1998. O diagnostyce zabudowy usytuowanej w sąsiedztwie wykopów głębokich. *„Inżynieria i Budownictwo", nr 6/1998.*

Michalak H., Pęski S., Pyrak S. & Szulborski K. 1998. O wpływie wykonywania wykopów głębokich na zabudowę sąsiednią. *„Inżynieria i Budownictwo", nr 1/1998.*

Michalak H. 2006. Kształtowanie konstrukcyjno-przestrzenne garaży podziemnych na terenach silnie zurbanizowanych. *Oficyna Wydawnicza Politechniki Warszawskiej,* Warszawa.

Michalak H. 2008. Selected problems of designing and constructing underground garages in intensively urbanised areas. *X International Conference on Underground Infrastructure of Urban Areas,* Wrocław 22 ÷ 24.10.2008.

Michalak H. 2009. Garaże wielostanowiskowe. Projektowanie i realizacja. Arkady, Warszawa.

Siemińska-Lewandowska A. 2001. Przemieszczenia kotwionych ścian szczelinowych. *Oficyna Wydawnicza Politechniki Warszawskiej,* Warszawa.

Szulborski K. & Pyrak S. 1998. O katastrofie obudowy wykopu głębokiego pod budynek przy ul. Puławskiej w Warszawie. *„Inżynieria i Budownictwo", nr 12/1998.*

Szulborski K., Michalak H., Pęski S. & Pyrak S. 2001. Awarie i katastrofy ścian szczelinowych. *XVI Ogólnopolska Konferencja „Warsztat Pracy Projektanta Konstrukcji". PZITB, Oddział w Krakowie,* Ustroń, 21–24 February 2001.

Szulborski K. & Michalak H. 2003. Uwarunkowania realizacji budynków z kilkukondygnacyjną częścią podziemną w strefach zabudowy zwartej. *XLIX Konferencja Naukowa KILiW PAN i KN PZITB „Krynica 2003".* Warszawa-Krynica, 14–19.09.2003.

Technical Specifications, to which the buildings and their locations ought to conform (Warunki techniczne jakim powinny odpowiadać budynki i ich usytuowanie*). Rozporządzenie ministra infrastruktury z 12 kwietnia 2002 r. Dz. U. z 2002 r. nr 75 poz. 690.*

Wysokiński L., Kotlicki W. & Godlewski T. 2011. Projektowanie geotechniczne według Eurokodu 7. *Instytut Techniki Budowlanej.* Warszawa.

Underground Infrastructure of Urban Areas 3 – Madryas et al. (Eds)
© 2015 Taylor & Francis Group, London, ISBN 978-1-138-02652-0

Modern devices for detecting leakages in water supply networks

K. Miszta-Kruk, M. Kwietniewski, A. Malesińska & J. Chudzicki
Environmental Engineering, Department of Water Supply and Wastewater Disposal, Warsaw Technical University, Warsaw, Poland

ABSTRACT: Leakages in water supply networks constitute a problem that has always had to be faced by waterworks all over the world. Water losses resulting from leakages range from 3 to 70% of water introduced into a network. In some cities, they are even greater. Events that originate water losses are not only of economic importance. They also constitute a threat that affects water safety, as locations where leakages occur are potential sources of contamination of water distributed through a water supply network. Traditional methods of water loss assessment and control—although they are well-tested and commonly implemented—do not always provide a basis for loss reduction. Moreover, proposed methods and devices do not fully meet the needs of all water suppliers. Therefore, they cannot be implemented in all water supply systems on an equal basis. This situation triggers new ideas and patents for new devices that facilitate identification of leakage location and thus allow for water losses in distribution systems to be reduced. This article presents examples of such technical solutions that make it possible to identify quickly both preliminary and precise location of water leakages.

1 INTRODUCTION

Water leakages or escapes are natural phenomena to be faced during the operation of a water supply network. It is impossible to predict precisely the time or place, where they can occur. In many cases, it is not even known that there is a leakage, until it produces noticeable effects or even serious disasters. An effective manner of fighting against water leakages consists in detecting them as soon as possible and remedying them, before they generate serious losses.

In Poland, before 1990, if water losses did not exert a negative impact on other elements of infrastructure, adjacent to a water supply conduit, they were disregarded for a long time and no remedial actions aimed at repairing such conduit were taken. Oftentimes, no attempts at locating the source of small water losses were made, as their presence was not considered to constitute a threat for network users and adjacent facilities.

Water losses constitute a serious problem faced by water supply systems all over the world. Amounts of water escaping from water supply systems may occasionally reach amazing levels. According to a World Bank research, water losses—i.e. amounts of Non Revenue Water (NRW), to be precise—in the world are estimated to reach annually about 48 billion cubic meters, which translates to 14 billion USD of financial losses annually (Kingdom et al. 2006). The same report shows that about 55% of the NRW volume in the world fall to developing countries. At the same time, according to WHO/UNICEF (2010), 884 million people all over the world do not have access to healthy and clean water, while almost all the examined cases were identified in developing countries.

Actions aimed at minimizing water losses are propagated all over the world within the framework of programs for environment protection and rational management of environmental resources. In Poland, such losses continue to constitute a considerable problem, yet one can notice clear progress in the reduction of water losses. On the other hand, many operators do not disclose the actual amounts of water losses affecting their networks. An interesting assessment of this situation is presented in the most recent report of the Polish Supreme Audit Office (Najwyższa Izba Kontroli, 2011). This document states that at least some irregularities in water loss monitoring and reduction were identified in the case of not less than 33% of inspected water supply operators. In this group of enterprises:

- 12.5% do not keep any record of water losses that would give consideration to their reasons, while
- at least 50% do not analyse the reasons for the appearance of water losses.

Moreover, none of these entities undertakes actions aimed at limiting water losses.

The above statistical data show that the water supply enterprises under examination did not pay proper attention to the issue of minimizing water losses. This can be explained by the fact that the major part of the financial losses inflicted on the enterprises as a result of water leakages is covered by the tariffs paid by water consumers.

Many water distribution systems in Poland, however, take measures that result in reducing water losses. One of the measures consists in systematic modernization and improvement of the technical condition of water supply networks and in the development of consumers' awareness. However, there is still a need to undertake further actions for this purpose, as leakage removal means reduction of financial losses of a water supply enterprise, as well as the possibility of supplying water to a greater number of consumers.

Growing amount of water pumped into a water supply network generates an increase in the operating costs of water supply pipelines and arouses concern of water supply enterprises. Hence, measures aimed at preventing water losses (Hotloś H., 2003) are implemented, which includes, for instance, regular inspections and modernization of water supply pipelines, as well as intensive search for efficient methods and technological solutions that would allow water leakages to be effectively detected.

There are two main methods of fighting against water leakages:

- detect existing leakages and remove even those that are imperceptible
- remove the leakages that generate noticeable effects.

The majority of water supply enterprises react to and remove leakages that generate noticeable effects. The problem of imperceptible leakages, however, is disregarded. This is so primarily for economic reasons. A water supply operator itself considers mainly the economic aspects of leakage level and cost-effective intervention in a water supply network. At present, the market offers numerous technical methods of effective detection and research programs aimed at identifying imperceptible leakages. This purpose is served by the creation and implementation of advanced technologies and corresponding specialist technical solutions.

Before setting to analyse the most interesting solutions proposed to detect leakages in water supply networks, one should consider the nature of defects affecting such networks. Among the basic reasons for pipeline breakdowns, one can count:

- excessive pressure and pressure variations
- defective pipeline construction (laying and assembly)
- water hammer
- implementation of inappropriate materials
- excessive wear and tear of pipelines.

Upon analysing the reasons for breakdowns of water supply networks, Berger and Ways (Berger M., Ways M., 2003) proposed to classify them in terms of leakage presence and characteristics (Fig. 1). This classification illustrates well the issue of leakage generation. Therefore, it was adopted in this article as the basis for further considerations.

The above diagram distinguishes between two general types of failures of water supply networks, namely failures that do not result in leakages (e.g. a jammed valve) and failures that result in leakages. In the second case, one faces water losses that are defined as "amounts of water that escaped from a water supply network as a result of lack of watertightness or

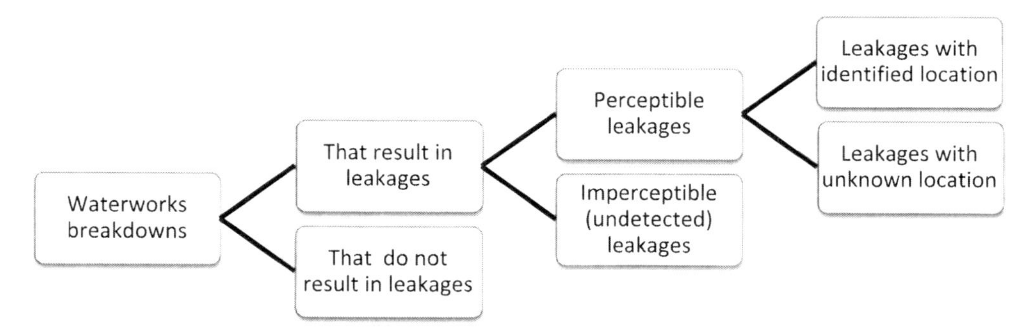

Figure 1. Classification of damages to water supply networks according to (Berger M., Ways M., 2003).

failure" [(Dohnalik P. przy współpracy Wojtoń K., 2000)]. Generally, however, two types of water losses can be distinguished:

- actual losses,
- apparent losses, i.e. losses that do not constitute actual water escapes from a leaking water supply system, but result primarily from imprecise measurements of water consumption.

The basic causes of actual losses include (Dohnalik et al., 2000; Speruda S., 2004, 2007; Yi Wu., 2011):

- water escapes through leakages in:
  - water tanks
  - pumping facilities
  - water conditioning facilities
  - pipeline networks with appurtenances
- water escapes from damaged pipelines and fittings
- water overflows in expansion tanks
- leakages in devices at the clients'
- leakages in buildings and within private premises

On the other hand, apparent losses result from:

- errors of measuring instruments
- non-concurrent readings of water meters and the master water meter.

It is debatable, how to classify thefts of water from water supply networks, yet it seems they should be counted among actual water losses.

Table 1 presents values of water losses depending on leakage size, as measured in laboratory conditions at the pressure of 5 bar. It follows from the data presented in the table that an opening that is 5 mm in diameter can generate a water leakage directly into the atmosphere that reaches about 32,000 litres per 24 hours.

The data published in the SAO report (Najwyższa Izba Kontroli, 2011) and the above-quoted research may be used to estimate the number of inhabitants that could be supplied with water saved by limiting water losses in the waterworks under examination. Assuming the average water consumption per one consumer per 24 hours amounts to 130 litres, the following correlation can be formulated:

$$\frac{\text{leakage-generated water losses/24 hours}}{\text{(average water consumption)?(consumer?24 hours)}} = \frac{32\,000 \text{ l/d}}{130 \text{ l/M d}} = 246M \quad (1)$$

The above formula shows that reduction of water escapes allows 246 consumers to be supplied with drinking water during 24 hours. This is a very substantial profit, especially in the case of smaller-scale water supply systems.

Table 1. Water losses at the pressure of 5 bar (0.5 MPa), depending on leakage size [(*Hermann Sewerin GmbH*)].

Hole diameter mm	Water outflow l/min	l/h	l/doba	m^3/doba	m^3/miesiąc	m^3/rok
0,5	0,50	20,00	480	0,48	14,40	173
1,0	0,97	58,00	1390	1,39	41,60	500
1,5	1,82	110,00	2640	2,64	79,00	948
2,0	3,16	190,00	4560	4,56	136,00	1632
2,5	5,09	305,00	7300	7,30	218,00	2616
3,0	8,15	490,00	11700	11,70	351,00	4212
3,5	11,30	680,00	16300	16,30	490,00	5880
4,0	14,80	890,00	21400	21,40	640,00	7680
4,5	18,20	1100,00	26400	26,40	790,00	9480
5,0	22,30	1340,00	32000	32,00	960,00	11520
5,5	26,00	1560,00	37400	37,40	1120,00	13440
6,0	30,00	1800,00	43200	43,20	1300,00	15600
6,5	34,00	2050,00	49100	49,10	1478,00	17736
7,0	39,30	2360,00	56800	56,80	1700,00	20400

2 COMPARISON OF WATER LOSSES IN POLAND AND ABROAD

The above-mentioned SAO inspection (Najwyższa Izba Kontroli, 2011) involved an analysis of the values of water production, sales and losses during the period 2008–2010 in a selected group of inspected waterworks in Poland. On the basis of collected data, it was found out that—out of the entire amount of water (71,724 thousand m^3) produced in the period under examination—only 62.2% (44,600 thousand m^3) of water was sold, while unsold and lost water amounted for not less than 37.8% (27,124 thousand m^3). As no source data were available, water losses were averaged at the level 18,948 thousand m3, i.e. 26.4%. Simultaneously, it was found out that water losses were highly diversified. In the least efficient water supply pipeline, the losses oscillated between 67% and 70% (and tended to grow), while the losses in the best managed water supply network ranged from 11% to 12% (and tended to grow as well) (Fig. 2).

Next, the research conducted by the MPWiK [Municipal Water Supply and Sewage Disposal Enterprise] in Bydgoszcz and covering water supply systems in the biggest cities of Poland shows that the averaging value of water losses in all the water supply systems under examination amounted to 14.3%, which was considered to be a satisfactory result. It proved the water supply networks in the regions under examinations to be in good condition. (Fig. 3)

Similar positive results were yielded by the research covering smaller water supply systems (Kwietniewski, 2013; Choma A. et al, 2013).

In Holland, leakages in water supply networks are reported to generate losses generally within the range of 3–7%. In the USA, it is estimated that about 22 million m3 of water is lost per 24 hours. The average level of water escapes in the USA amounts to 15%, yet it varies from 7.5% to 20%, depending on the state. In the USA, the annual financial losses caused primarily by discontinuation of water supply are estimated to be similar to the losses resulting from floods and amount to over 2 billion USD. In Great Britain, oftentimes seen as the leader in water leakages management, about 3 million m3 of water are lost daily, which allows one to estimate that the level of water leakages was relatively stable during the last decade and amounted to about 20–30% of supplied water. The majority of companies in Great Britain use the method of Economic Levels of Leakage (ELL) to assess water losses, based on current tools, techniques and technologies. In Italy, the NRW levels fall within the range from 15 to 60%, while the average equals to 42%. In Portugal, the level of leakages

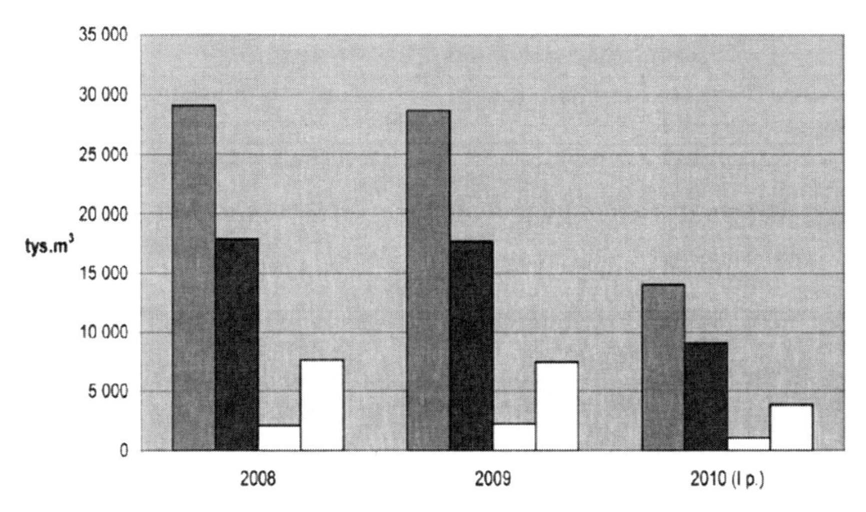

Figure 2. Production, sales and water losses in enterprises under inspection (*Najwyższa Izba Kontroli, 2011*)—water production; water sales; water consumption for in-plant consumption of a water supply and sewage disposal enterprise; water losses.

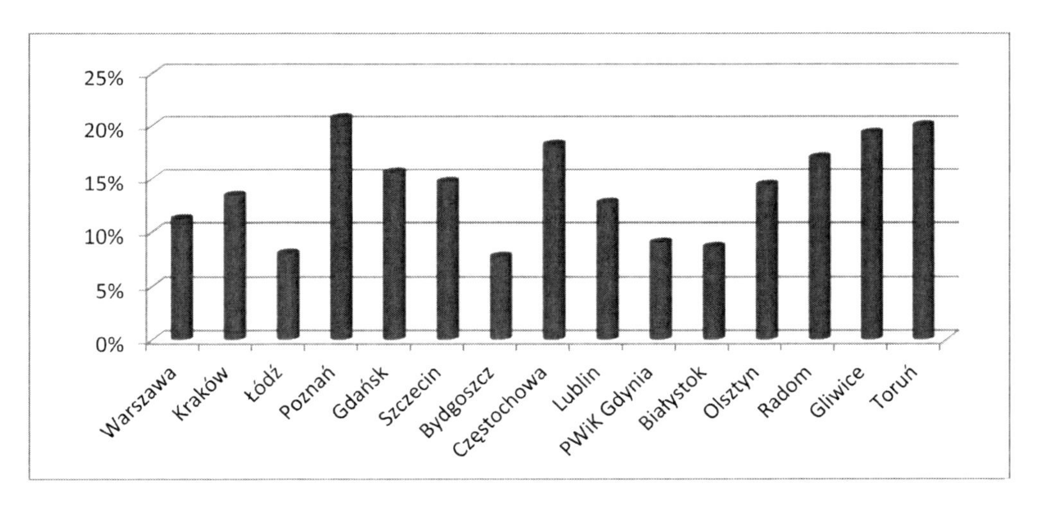

Figure 3. Water losses in selected Polish cities in 2009 (Report on the activities of the MPWiK Bydgoszcz in 2009, 2009).

varies from 30% to over 50%, while the average figure quoted in literature is 34.9%. In the town of Larisa (Greece), the NRW was estimated to reach 6,000,000 m3 annually (or 34% of the System Input Volume [SIV]). In Australia, NRW calculated for the set of 10 water supply systems varies from 9.5 to 22%, with the average of 13.8%. In the province of Ontario, Canada, drinking water worth 1 billion dollars sinks every year into the ground due to leakages in municipal water supply networks. It is estimated that the average amount of lost water is within the range from 7% to 34% of the system input volume (Mutikanga, H., et al., 2013).

Despite the above-quoted results presented in the SAO report, Poland comes out relatively well, compared to other countries. This thesis is substantiated by the schedule of water losses for selected places in the world presented in table 2.

Table 2. Water losses in the world [(McKenzie RS., Wegelin W., 2008)].

Country	Total water losses	actual losses
Afganistan/Kabul	60%	–
Australia/New South Wales	30%	–
Australia	–	9,5%
Canada/Montreal	40%	–
England/London	42%	–
England and Wales	22%	16%
Fiji	55%	–
France	17%	–
Francje/Marsylia	36%	–
Germany/Monachium	14%	–
Ghana	Ponad 50%	–
Ireland	50%	–
Italy/Rome	25%	–
Japan/Tokio	5,4%	–
Kenya	50%	–
Malta/Gozo	–	26%
The Netherlands	5%	–
Saudi Arabia	30%	–
Scotland	47%	–
Senegal/Dakar	30%	–
Singapore	–	1%
Spain	28%	–

3 LEAKAGE DETECTION METHODS

Leakage detection is a labour-consuming process and it requires waterworks personnel to have wide knowledge in the domains of hydraulics and materials engineering, as well as knowledge of the relevant methods and access to various tools. However, as in the case of any control activity, the basic condition for the success of leakage detection activities is their systematic performance. Among leakage detection methods, one usually distinguishes methods of observing network pipeline routes, pressure measurement and flow measurement (Berger and Ways, 2003) that can be generally termed as traditional methods. Water escapes, however, take different forms and magnitudes and, unfortunately, not all of them become manifest. Considerable water escapes may usually be visible in the form of water pools in the vicinity of a pipeline breakdown. In the case of less permeable soils, water tries to find other ways to flow and it can penetrate into adjacent buildings and flood their basements. Therefore, there are many methods of preliminary leakage detection.

The diagram below presents a general classification of methods of leakage detection and localization in water supply networks, broken down into classical methods and modern methods, dedicated to identification of leakages that are hard to detect (Fig. 4).

3.1 *Traditional leakage detection methods*

The first method—and the least precise, although the simplest one—consists in visual observation and appraisal of ground surface along and in the vicinity of a pipeline route. Such examination allows one to detect and preliminarily localize leakages. However, the results are far from specific and not always correct. The lack of precision of this method results from multiple factors, including primarily:

- the type of soil, mainly in terms of water permeability,
- the type of building development along a water supply pipeline,
- leakage intensity.

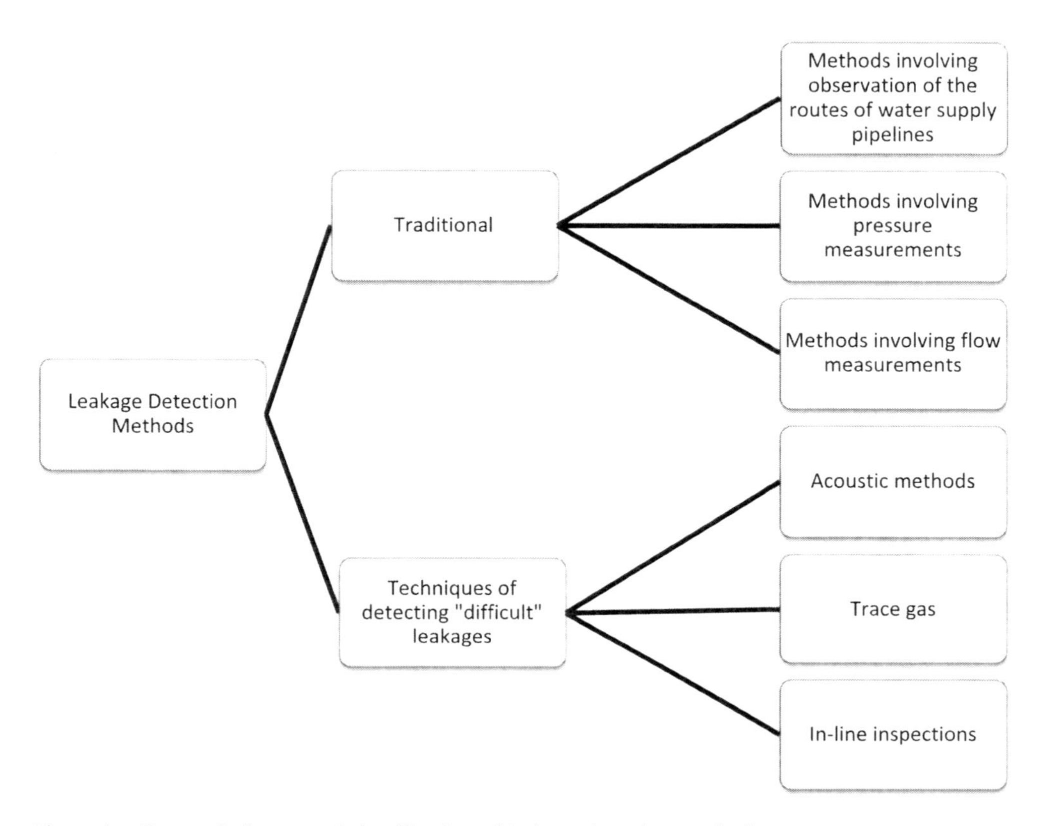

Figure 4. Proposal of a general classification of leakage detection methods.

It is common knowledge that it usually takes very long for a small leakage to become manifest. Occasionally, a leakage develops so slowly that water dissipates in the ground and never surfaces to reveal the defect. Therefore, this method—being very simple—is the least precise one. Generally, it just determines the fact that a leak occurred, unless it is accompanied by particular events, such as:

- caving-in of ground or pavement,
- appearance of icing in winter always in the same places,
- sudden appearance of springlets and pools that are not correlated with atmospheric precipitation in the vicinity of a water pipe route,
- regular surfacing of water during freezing periods in winter,
- rampant vegetation, especially around fitting boxes, compared to poor vegetation in their vicinity.

The next visual method of detection and preliminary localization of leakages consists in ground probing along water supply conduits. Such examination is performed with the so-called probing pin, i.e. a steel rod that is 2.25 m long and about 0.02 m in diameter, pointed at the bottom end and fitted with an insulated handle at the other end. Leak detection consists in driving the pin into the ground about every 1 m along a pipeline route and in assessing pin moistening once it is pulled out, which indicates ground moistening and the level of groundwater. This method yields very subjective results.

In order to reduce the duration of repairs and their costs, the method of probing excavations and bore-holes was used. This method consisted in preparing a small-sized bore-hole or excavation in the place of identified leakage, before repair works were commenced. Then, the damaged pipeline section was connected to the water supply network. If water surfaced through the performed bore-hole/excavation, it meant that the defect location was correctly identified.

Methods involving pressure and flow measurements were created as a result of growing knowledge on the part of water supply network operators, concerning the possibilities of identifying leakage locations. At present, the majority of water supply networks is monitored to a larger or smaller extent. First of all, it is the water pressure in the network that is recorded, but water flow is often recorded as well. Implemented equipment provides the possibility of on-the-run monitoring of network condition and transferring collected data to a control room. This provides the possibility of reacting quickly to potential undesirable events. Sets of continuously collected data allow one to assess the possibility of leakage by means of additional advanced calculation tools, such as, for instance, artificial neural networks, fuzzy inference systems, Bayesian systems, breakdown/crack detection based on recursive linear quadratic estimation and adaptive filtering of hydraulic measurements, or else by means of the ANN & GA hybrid model, as well as support vector machines and contextual classification, or even commonly known genetic algorithms.

The listed methods, however, are not flawless and—apart from accurate knowledge about a given water pipeline—they also require for the operators to have considerable theoretical knowledge concerning the implemented tools. Moreover, they are highly labour-consuming. Thus, waterworks have not been fully prepared to implement these modern methods, although one can observe actions aimed at preliminary preparation of collected data to function as an input for the listed models.

3.2 *Modern solutions in the domain of detecting leakages in water supply networks*

The main area, where new measurement techniques are developed covers detection of leakages that are hard to identify. Due to the need to reduce water losses for both ecological and economic reasons, this domain is very interesting for manufacturers of new technologies, who offer every time better solutions. Their complexity is often aimed at ensuring simple and effective diagnostics of water supply networks, or even identification of leakage locations. This holds true, among others, in the case of digital correlators that use low frequency hydrophone sensors, where sound intensity measurement (upon sound power averaging) is performed with at least two hydrophones. As a result, one acquires a spatial distribution of the acoustic field. This method consists in simultaneous recording of acoustic wave speed and pressure at two points of a conduit. Preliminary testing of this method in water environment are conducted by the navy. Devices of this type proved effective, when applied to large-diameter water supply conduits made of plastics and operating at low pressures. To be implemented, however, they need access to a conduit at several points.

Leakages from plastic conduits are also detected successfully by means of the trace gas method. This method proved particularly useful to detect small leakages occurring at water supply service line connections. The method also produces measurable results in detecting small leakages in new plastic conduits, where pressure-based methods failed.

Methods of direct in-line inspection inside conduits and the relevant tools have been quickly conquering the domain of leakage detection in water supply networks. In-line inspections inside conduits are usually performed by means of sensors introduced into a pipeline by means of a cable or a free-floating plastic ball. Two devices are used: SmartBall® and Sahara®. Both devices are designed to diagnose networks of raw and drinking water, or even sewage disposal networks made of any material. The only limitation of both technologies consist in the requirement for the pipeline diameter to be smaller than 150 mm (this is not a problem in the majority of currently operated water supply networks) (Pure Technologies Ltd.; Yi Wu Z. et al. 2011). Both methods are innovative systems for in-line control of active networks, aimed at localizing leakages.

In the case of the SmartBall® (an intelligent ball—the authors' translation) technology, examinations are performed by means of a foam ball that contains a core made of aluminium alloys, containing all the measuring instruments. This apparatus does not require any control, as it was designed to be carried by the flowing medium along the entire length of a pipeline under examination. Thanks to the implemented measuring instruments and the technology of in-line examination performed inside conduits, SmartBall® detects and localizes even very small leakages and gas pockets inside pipelines (Pure Technologies Ltd.). SmartBall®

is inserted into and extracted from a pipeline during normal pipeline operation (Fig. 5). The ball carried by water flow travels through a conduit for even 12 hours, while collecting data concerning leakages. This solution ensures that one can inspect a network that is many kilometres long during one session (Pure Technologies Ltd.).

The apparatus is available in different sizes. Generally, however, the selected SmartBall® diameter should be less than one third of the pipeline diameter.

The operation of this device is based specifically on a technology that resembles the one implemented in classical noise loggers and correlators. It means that SmartBall® moving inside a conduit examines its entire length by means of an integrated microphone that identifies acoustic symptoms of leakages. The device records in its memory the course of the measuring process, so that the data can be further processed and the results can be presented to identify the spots, where leakages are detected (Pure Technologies Ltd.).

The device calculates the exact location of leakages by means of additional receivers installed on conduit fittings (on the ground surface). When SmartBall® is at a place in a conduit, where it recorded a leakage, it starts to generate signals to the receivers installed on the sruface. This allows the device to establish the precise location of the leakage, on the basis of the difference in the time required by signals to reach particular receiving units. Thus, the device implements the same principle that is used in correlation-based methods (Pure Technologies Ltd.).

SAHARA® is another device that operates inside a normally operating conduit. Thanks to implemented solutions, the device has the same functionalities as the above-mentioned SmartBall®. Additionally, it is equipped with a camera that transmits (in real time) images to the operator's computer through a cable connected to the apparatus. This inspection system is currently one of the most precise systems of the type and it offers the greatest range of functionalities. Apart from detecting leakages, trapped gas pockets, structural defects in laid networks, it can also locate illegal service line connections (Pure Technology Ltd.; Yi Wu Z et al 2011; Hamilton S. 2013).

The Sahara® device sensor is introduced into a pipeline through surface-accessible fittings. It is subject to only one limitation, namely the diameter of a fitting element to be used must be at least 50 mm. Once the apparatus is introduced into a conduit, it opens a small drag chute (resembling a parachute) that takes advantage of water flow to transport the sensor along an examined pipeline section (Pure Technologies Ltd.).

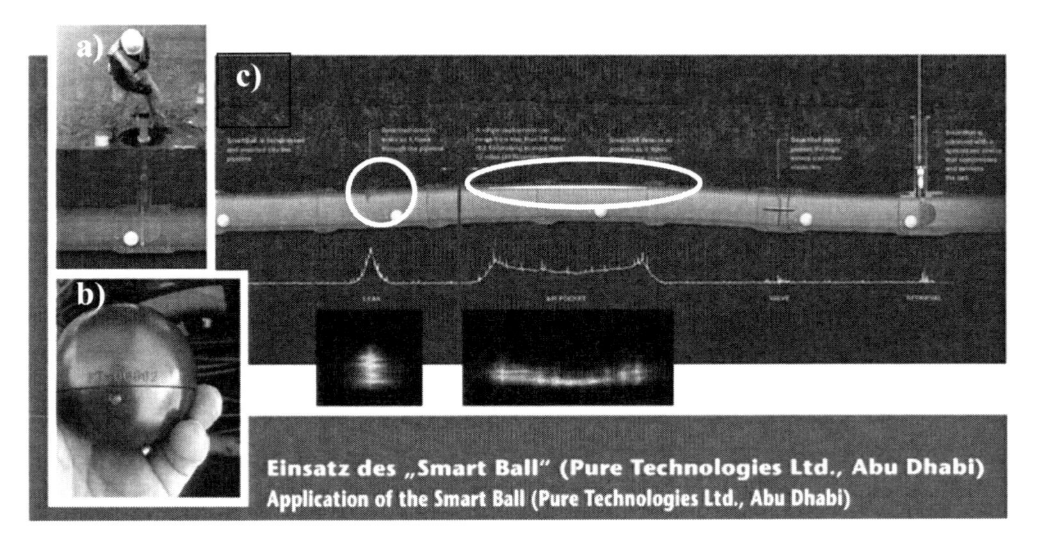

Figure 5. Example of detecting leaks and air pockets by means of the SmartBall device (Rieder et al. 2012) a) extraction of the "ball" from a conduit, b) the "ball", c) visualization of measurement results (the fields marked in white indicate leakages and gas pockets).

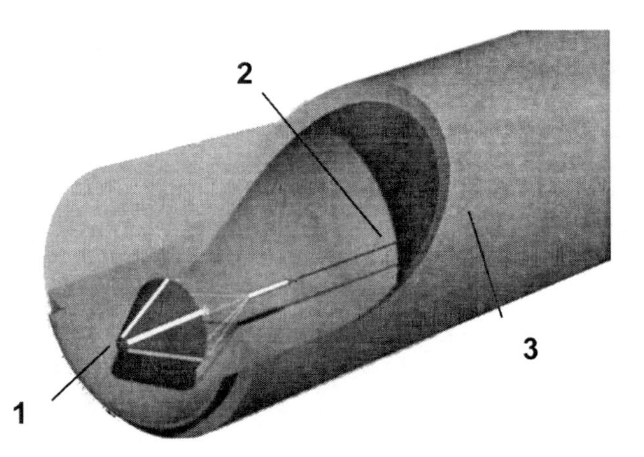

Figure 6. Illustration of the SAHARA® system ((Pure Technologies Ltd.)) 1—unfolded shell (parachute) with a camers, 2—cable connecting the device to a computer, 3—examined w środowisku (Pure Technologies Ltd.).

In the case of this method, the sensor is connected through a cable to a computer on the ground surface. This allows the operator not only to receive the video image in real time, but it also ensures a wide range of control over the apparatus. The Sahara® device is able to identify locations of leakages and elements of fittings with the precision of up to 500 mm (Pure Technologies Ltd.).

4 SUMMARY AND CONCLUSIONS

The analysis performed in this article was aimed at presenting leakage in water supply networks as an important problem that has to be faced by waterworks both in Poland and abroad. The importance of this problem is also underscored by the conferences periodically organized by the Water Loss Specialist Group of the International Water Association. These conferences usually assemble several hundred participants from all over the world.

The methods of leakage detection in waterworks and the corresponding devices constitute a very important element of a water loss reduction strategy. The range of issues related to measurement techniques is very wide. The article concentrates primarily on the group of most recent solutions in the domain of water leakage localization, as illustrated by several examples of the state-of-the-art devices. It is worth pointing out that some of them are still in the phase of implementation testing.

The described devices perform in-line measurements and recording of irregularities of conduit functioning (the Sahara® device is additionally fitted with an integrated camera for real-time image recording, which allows the conduit condition to be thoroughly inspected). Modern devices do not require any additional software, while allowing one to acquire very accurate diagnostic results. Moreover, these technologies eliminate totally the impact of the human factor on the acquired results. The operator's role boils down to introducing a device into and removing it from a pipeline and—in the case of the Sahara® device—to supervising the examination process through the monitoring of the transmitted image.

Research works financed from the resources of the National Centre for Research and Development within the development project No. PBS1/B9/15/2012, 2013–2014

REFERENCES

Berger M., Ways M., 2003: *Poszukiwania przecieków sieci wodociągowej*. Poradnik. Seidel-Przywecki Sp. z o.o. Warszawa 2003.

Choma A., Iwanek M, Kowalska B., Kowalski D.; 2013: *Analiza strat wody w sieci wodociągowej eksploatowanej przez zakład gospodarki komunalnej w Puchaczowie*. Aktualne zagadnienia w uzdatnianiu i dystrybucji wody. Rozdział III, Innowacyjne Technologie Ujmowania i Uzdatniania Wody 2013, vol.3 pp.313–322.

Dohnalik P. przy współpracy Wojtoń K., 2000: *Straty wody w miejskich sieciach wodociągowych*. Polska Fundacja Ochrony Zasobów Wodnych. Bydgoszcz 2000.

Hamilton S., Charalambou B., 2013: *Leak Detection*. IWA Publishing, London 2013.

Hermann Sewerin GmbH: Materiały wewnętrzne firmy. Gütersloh, Niemcy.

Hotloś H., 2003: *Analiza strat wody w systemach wodociągowych*. Ochrona Środowiska 2003.

Kingdom, B., Liemberger, R., and Marin, P., 2006: *The Challenge of Reducing Non-Revenue Water (NRW) in Developing Countries*, The World Bank, Washington, DC, USA.

Kwietniewski M.; 2013. *Zastosowanie wskaźników strat wody do oceny efektywności jej dystrybucji w systemach wodociągowych*. Ochrona Środowiska, vol. 35, nr 4/2013; s. 9–17.

Mutikanga, H., Sharma, S., and Vairavamoorthy, K. (2013). *Methods and Tools for Managing Losses in Water Distribution Systems*. J. Water Resour. Plann. Manage., 139(2), 166–174.

Najwyższa Izba Kontroli, 2011: *Informacja o wynikach kontroli. Prowadzenie przez gminy zbiorowego zaopatrzenia w wodę i odprowadzania ścieków*. Nr ewid. 128/2011/P/10/140/LKI.

Pure Technologies Ltd.; SmartBall®, Sahara®–Leak Detection for Water Mains; http://www.puretechltd.com/applications/pipelines/water_wastewater_pipelines.shtml

Rieder von Adrian, Hammer M, Buchler M., 20012: *Schach der korrosion in wasser & abwassersystemen*. aqua press international 3-2012.

Speruda S., 2004: *Ekonomiczny poziom wycieków. Modelowanie strat w sieciach wodociągowych*. Warszawa. Translator 2004.

Speruda S., 2007: *Straty wody w polskich sieciach wodociągowych*. Warszawa: WaterKEY 2007.

Yi Wu Z., Farley M, Turtle D., Kapelan Z., Boxall J. et al.; 2011: *Water Lost Reduction*. Bentley Institute Press, Exton, Pennsylvania 2011.

Underground Infrastructure of Urban Areas 3 – Madryas et al. (Eds)
© 2015 Taylor & Francis Group, London, ISBN 978-1-138-02652-0

Evaluation of deep excavation impact on surrounding structures—a case study

M. Mitew-Czajewska
Warsaw University of Technology, Warsaw, Poland

ABSTRACT: In the paper several methods of evaluation of the width of impact zone as well as values of the terrain displacements around deep excavations are presented. A case study, concerning the estimation of deep excavation influence on surrounding structures is described. Finite elements model of the excavation and surrounding soil body was built in order to estimate the range of displacements (impact zone) as well as to calculate values of vertical and horizontal displacements of soil body and adjacent structures. Comparison of the results of FEM analysis with the values estimated using several simplified methods is shown. At the end a summary and conclusions are presented on the suitability of using simplified methods of determining the excavation influence on surrounding structures in geotechnical conditions of Warsaw.

1 INTRODUCTION

In the last ten years, in large urban areas structures with multilevel underground part are often built. The main reason of this phenomenon is the lack of free space in dense urban areas, which could be used as a parking space of new investments. Theses structures usually require execution of excavations characterized by significant dimensions in plan and considerable depth up to 20–25 m. One of the criteria for the design of deep excavations is the assessment of horizontal and vertical displacements of surrounding soil, Siemińska-Lewandowska & Mitew-Czajewska (2008). Displacements of the ground around the excavation depend on several factors, such as: the type of casing, excavation dimensions, soil parameters, levels of groundwater, methods of construction, etc. There are different methods of evaluation of the displacements within the impact zone, theses can be classified as follows, Krajza (2014):

- Empirical methods—displacement is determined on the basis of data from similar sites executed in similar geotechnical conditions,
- Semi-empirical methods—the nomograms are developed for typical cases and are further applied for the estimation of displacements,
- Numerical methods—in most of the cases use the finite element method for the modeling of the excavation and adjacent soil body.

In the following sections different empirical and semi-empirical methods of calculation of displacements around the excavation are described.

1.1 *ITB Instruction n° 376/2002 "Protection of structures in the vicinity of deep excavations"*

The method of evaluation of the range of the impact zone around the excavation and estimation of soil displacements within this zone presented in the Instruction n° 376/2002 issued by Building Research Institute (ITB) is the most commonly used in Poland, Kotliński & Wysokiński (2002).

The impact zone is divided into two zones S_I—the direct impact zone and S—impact zone (Fig. 1); its ranges are expressed as multiples of the depth of the excavation H_w in dependence of the type of surrounding soil (Tab. 1).

The Instruction proposes semi-empirical method for the assessment of values of maximum displacement around the excavation basing on schematic distributions of displacements shown at Figure 2. Line 1 represents the maximum positive displacements distribution,

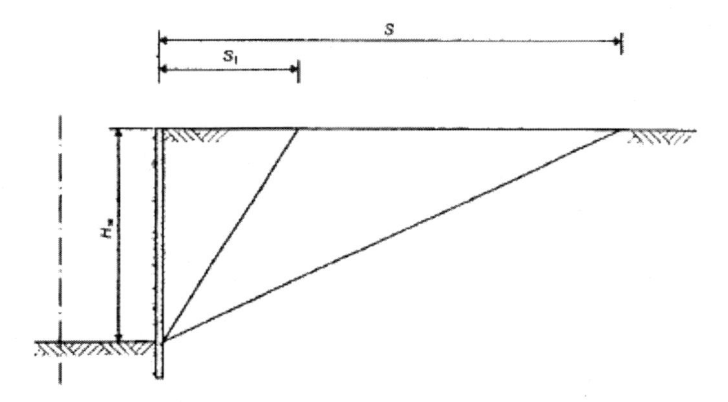

Figure 1. The range of excavation impact zones S_I and S.

Table 1. The range of excavation impact zones S_I and S.

Type of soil	SI	S
Sand	0,5 H_w	2,0 H_w
Silt	0,75 H_w	2,5 H_w
Clay	1,0 H_w	3 ÷ 4 H_w

*H_w – depth of the excavation.

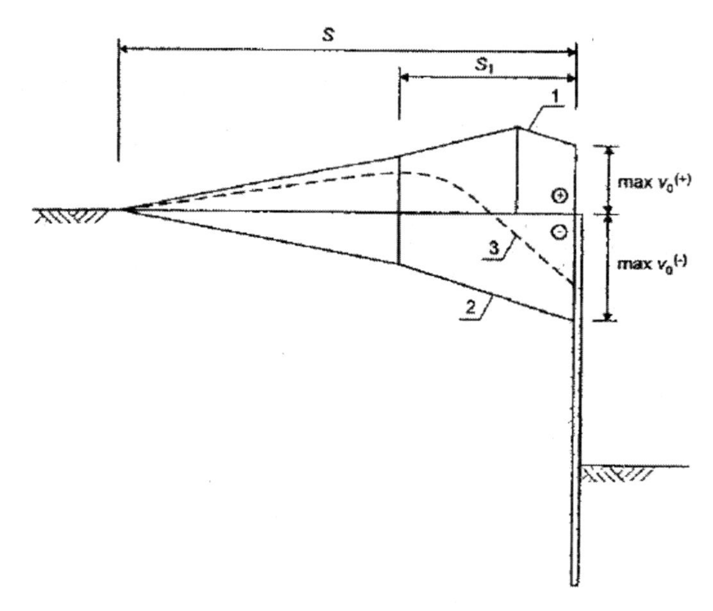

Figure 2. Schematic distributions of displacements in the vicinity of the excavation.

Line 2—the maximum negative displacements distribution, Line 3—the average distribution of displacements. Assessment of negative displacements is obligatory in all cases (line 2), while positive displacements are usually classified as favorable, reducing settlements around the excavation that is why its calculation is not always necessary and have been omitted from further consideration.

The distributions presented in Figure 2 can be characterized by three values of soil (terrain) displacements:

- maximum v_0—in the immediate vicinity of the excavation wall,
- medium v_{01}—at the border (end) of direct impact zone SI, equal to half of the value of the maximum displacement v_0,
- zero—at the end of the impact zone S.

The value of maximum negative displacement $v_0(-)$ must be determined as the sum of maximum displacements caused by:

- execution of the excavation wall—v_i,
- horizontal (lateral) displacement of the wall—v_u,
- lowering the water table around the excavation—v_w.

$$\max v_0(-) = v_i + v_u + v_w \tag{1}$$

The displacement caused by the execution of the wall (v_i):

$$v_i = \alpha \cdot \sqrt{H_w} \tag{2}$$

where:

v_i—vertical displacement of the soil (mm)
α—empirical coefficient
H_w—depth of the excavation (m)

The value of α may be considered as follows:

$$\alpha = 1{,}0 \div 1{,}3 \text{ for stiff moraine soils} \tag{3}$$

$$\alpha = 1{,}3 \div 5{,}0 \text{ for soft, highly permeable soils} \tag{4}$$

The displacement caused by the horizontal (lateral) deflection of the wall (v_u):

$$v_u = 0{,}75 \cdot \max(u_k) \tag{5}$$

where:

v_u—vertical displacement of the soil (mm)
$\max(u_k)$—maximum lateral displacement of the wall (Fig. 3).

The displacement caused by lowering the water table around the excavation (v_w):

$$v_w = \theta \cdot v_{(w,\max)} \tag{6}$$

where:

v_w—maximum vertical displacement of the soil caused by the lowering of water table (mm)
θ—empirical coefficient
$v_{(w,\max)}$—increment of terrain settlement caused by lowering of water table, for standard moraine soils it may be considered equal to 1 mm of soil settlement per 1 m of water table lowering.

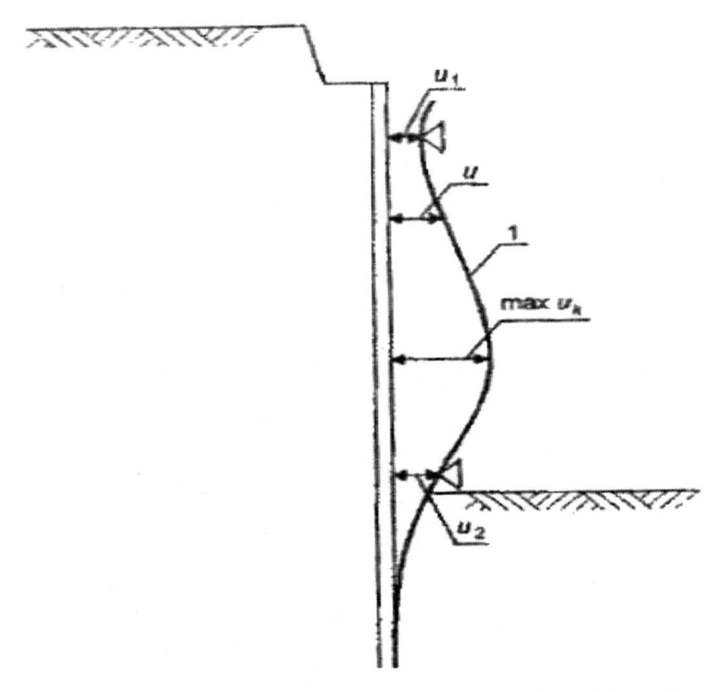

Figure 3. The displacement caused by the horizontal (lateral) deflection of the wall.

In addition:

$$\theta = \frac{L}{R} \tag{7}$$

where:
L—length or width of the excavation perpendicular to the excavation (m),
R—the range of depression trough (m).

1.2 *Bowles's method*

The basis of the analytical Bowles's method created in 1986, Bowles (1988), is the relationship of vertical displacements of the ground with horizontal displacements of the excavation wall described by a function of one variable (l_x—distance from the end zone of subsidence, as the origin). The principle of this method is shown in Figure 4 and the analysis steps are the following, Ou (2006):

- calculation of lateral wall displacement using FEM or other method, calculation of the area of the lateral wall deflection (a_d),
- estimation of the range of ground surface settlement, (D):

$$d = (H_e + H_d) \cdot \tan\left(45^\circ - \frac{\phi}{2}\right) \tag{8}$$

where:
H_e—excavation depth; $H_d = 0$ if $\phi = 0$ and $H_d = 0{,}5 \cdot B \cdot \tan(45^\circ + \frac{\phi}{2})$ if $\phi \geq 0$ where: B—excavation width and ϕ—soil strength parameter

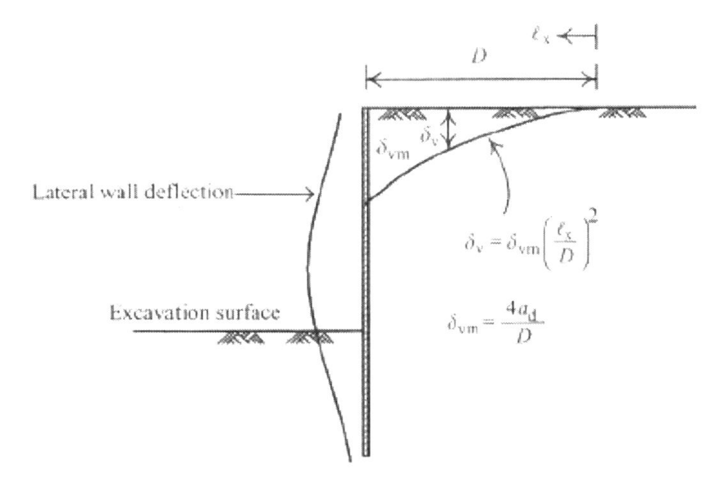

Figure 4. Bowles's method, estimation of vertical displacements around excavation.

- calculation of maximum ground surface settlement located immediately behind the wall:

$$\delta_{vm} = \frac{4 \cdot a_d}{D} \tag{9}$$

- calculation of settlement (δ_v) at l_x, assuming that the ground surface settlement has parabolic distribution:

$$\delta_v = \delta_{vm}\left(\frac{l_x}{D}\right)^2 \tag{10}$$

where: l_x—distance from a point at the distance D from the wall, δ_v—the settlement at that distance l_x.

1.3 *Clough and O'Rourke's method*

Semi empirical method of Clough & O'Rourke (1990) proposes functions representing changes in the value of vertical subsidence (δ_v) depending on the distance from the excavation wall (d) and the depth of the excavation (H_e). The maximum value of the settlement is found just behind the retaining wall and decreases with the distance from the wall depending on the soil type. The authors classified settlement envelopes into following groups (Fig. 5):

- Triangular shape—when the excavation is made in non cohesive soils (Fig. 5a) and the influence zone is equal to $2 \cdot H_e$, or in stiff clays (Fig. 5b) for which the influence zone is $3 \cdot H_e$; where H_e is the final excavation depth.
- Trapezoidal shape—in case of excavations executed in soft to medium clays. Maximum value of the settlement occurs up to a distance of $0{,}75 \cdot H_e$ from the wall then decreases within the transition zone, which ends at $2 \cdot H_e$.

1.4 *Ou and Hsieh's method*

Semi-empirical method developed in 2000 based on research conducted in Oslo, San Francisco, Chicago and Taiwan, Ou (2006). In this method the influence zone of the excavation and values of vertical displacements, which occur in this zone depend on the shape of the surface settlement curve. There are two types of curves proposed—spandrel and concave

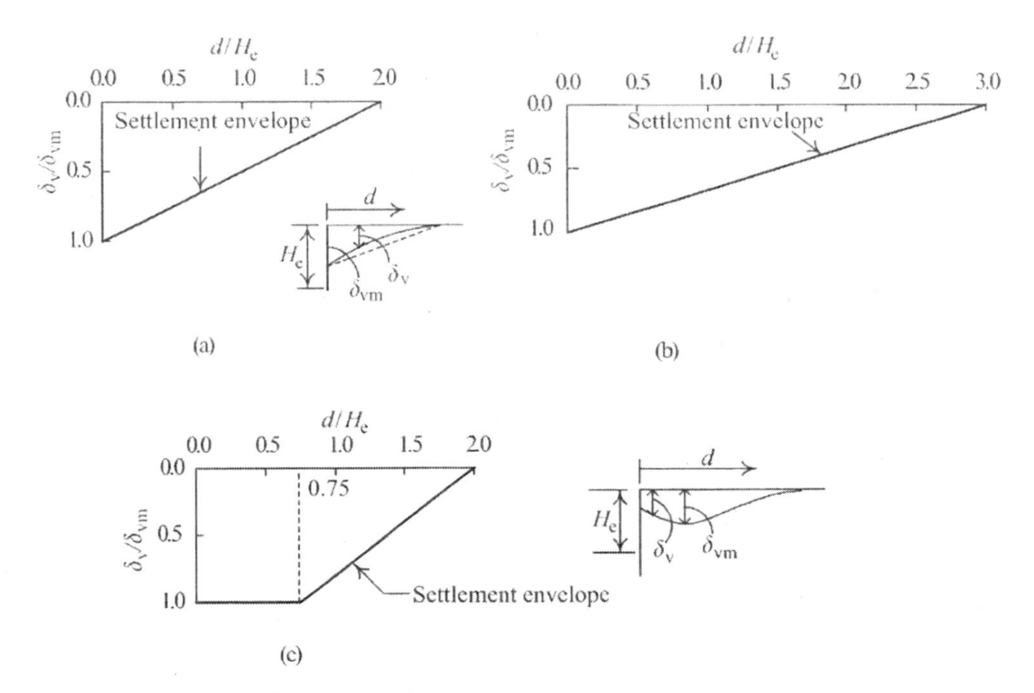

Figure 5. Clough and O'Rourke's method.

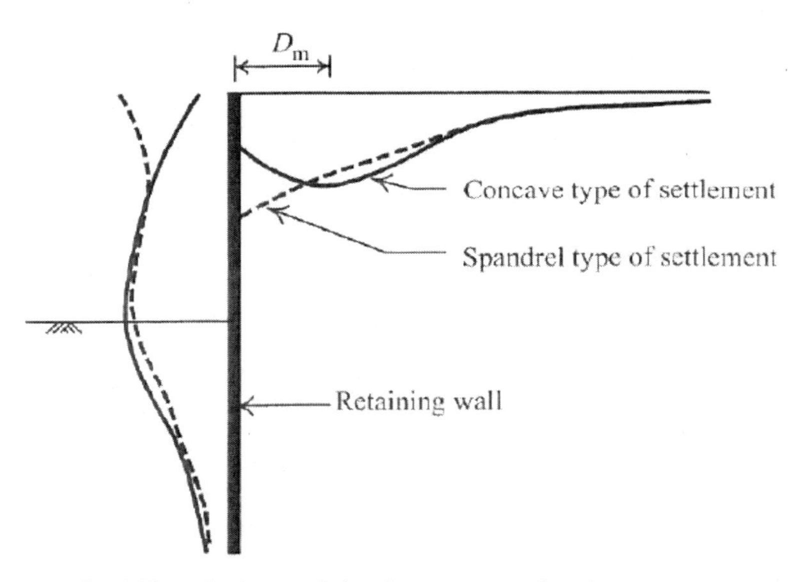

Figure 6. Ou and Hsieh's method—spandrel and concave type of settlement curve.

(Fig. 6), for which two types of graphs for the estimation of surface settlements are proposed respectively (Fig. 7). Ou and Hsiegh's method distinguishes between two zones of impact: PIZ (primary influence zone) and SIZ (secondary influence zone). The maximum vertical displacement of the ground δ_{vm} may be determined basing on the diagram proposed by (Ou et al. 1993) showed at Figure 8.

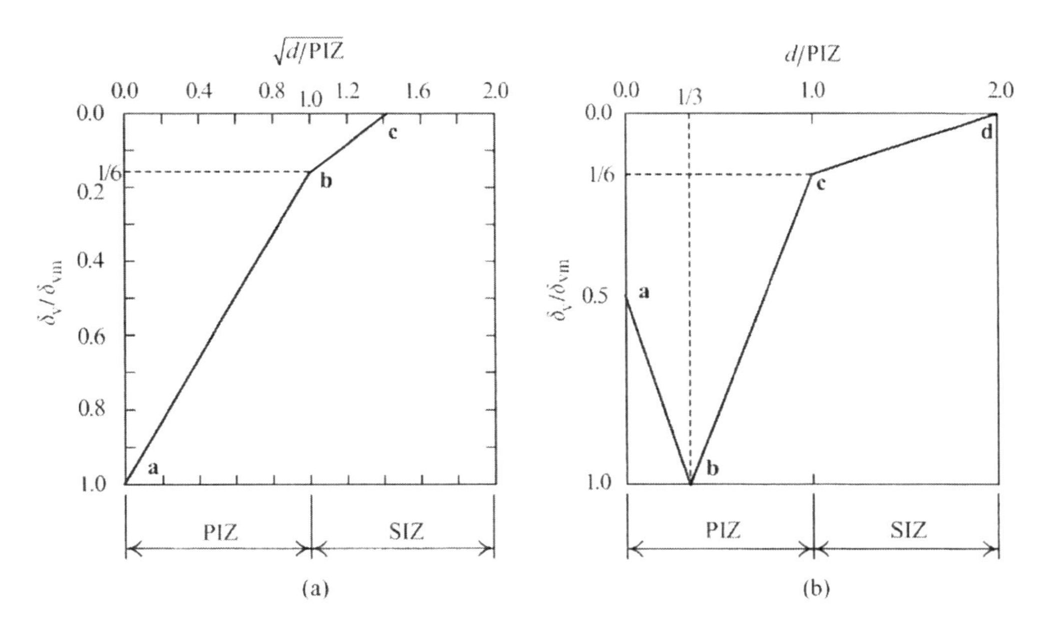

Figure 7. Ou and Hsieh's method—estimation of surface settlements, spandrel and concave type of settlement curve.

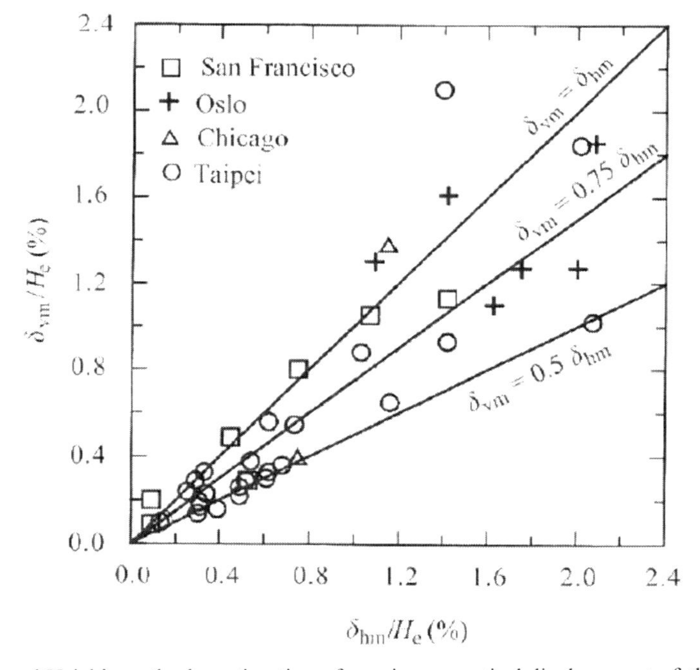

Figure 8. Ou and Hsieh's method—estimation of maximum vertical displacement of the ground δ_{vm}.

2 CASE STUDY

2.1 *Geotechnical context*

The case study analysed in the paper is a high-rise building with multilevel underground car park situated in Warsaw on the left bank of the Vistula river within the area of glacial plateau. The elevation of the surface of this area is approximately 33.4 ÷ 34.2 m above the "0" level of Vistula river. Geological conditions of the area are complex. According to the geological report, (Kuszyk et al. 2013), following soil layers may be distinguished (Fig. 9):

- directly under the surface, 2–4 m thick anthropogenic soils (so called fills) occur (I),
- below the layer of fills, 4–6 m thick moraine clays of Warta glaciation occur, mostly sandy clays or clayey sands (IIIa),
- below the cohesive layer up to great depths sandy formations occur in form of dense and very dense fine and medium sands (IVb, IVc).

The main groundwater level with free and locally slightly confined water table stabilizes at the level 28,00 ÷ 28,25 m above the Vistula river level.

2.2 *Geometry and construction stages*

Analyzed section of the building has three levels of underground car park, the final excavation depth is 13 m. The underground part will be constructed with the use of so called Milan method with deep external diaphragm walls and underground slabs serving as a support of diaphragm walls in the temporary and permanent stages. The geometry of the underground part of the building is shown at Figure 9. Following construction stages were designed and considered in further numerical analysis, Buro Happold (2013):

- stage 1—initial state (activation of initial, geostatic stresses in the soil body),
- stage 2—shallow, initial excavation and execution of diaphragm walls,

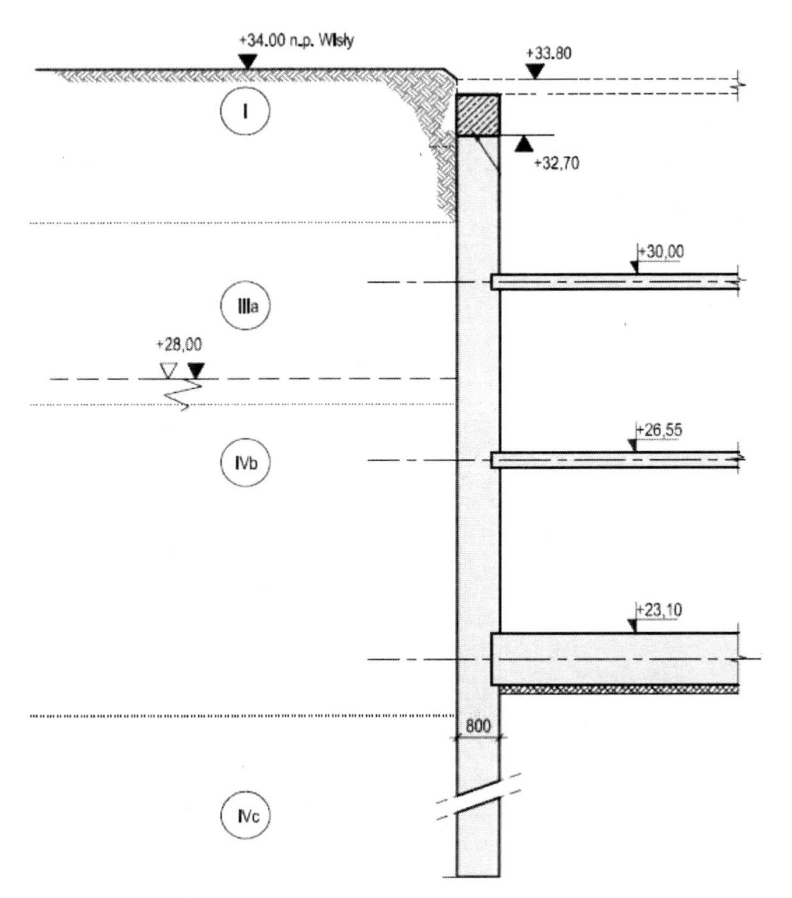

Figure 9. Geometry of the underground part of the building.

- stage 3—excavation below the –1 slab (at the level +30 m above Vistula river),
- stage 4—construction of the –1 slab (30 cm thick),
- stage 5—excavation below the –2 slab (at the level +26,55 m above Vistula river),
- stage 6—construction of the –2 slab (30 cm thick),
- stage 7—excavation till the final depth (+21,80 m above Vistula river),
- stage 8—construction of the foundation slab (1,2 m thick),
- stage 9—construction of "0" slab (30 cm thick) and final service loading.

2.3 *Numerical analysis*

2.3.1 *Numerical, FE model*

Finite element plain strain analysis were carried out using GEO5 FEM software, Fine Ltd (2014). Two phase elastic perfectly plastic constitutive material model—modified Coulomb-Mohr was chosen for modeling the soil body, diaphragm walls as well as foundation slab were modeled as 3-nodes, linear beam elements. Non-associated plastic flow law was considered. For modeling wall frictions at the soil—structure interface, contact elements with zero thickness and Coulomb-Mohr plasticity low were used. Model dimensions were: 62 m high and 144 m wide (Fig. 10). FEM model mesh, generated automatically, was built of 8219 nodes, 4873 (3121 15-nodes triangle surface elements, 438 beam elements and 1314 contact elements).

For the initial state, geostatic stresses in the soil body were assumed. Construction stages were modeled as listed in chapter 2.2.

There were 4 soil layers considered in the analysis, basic parameters of which are specified in Table 2.

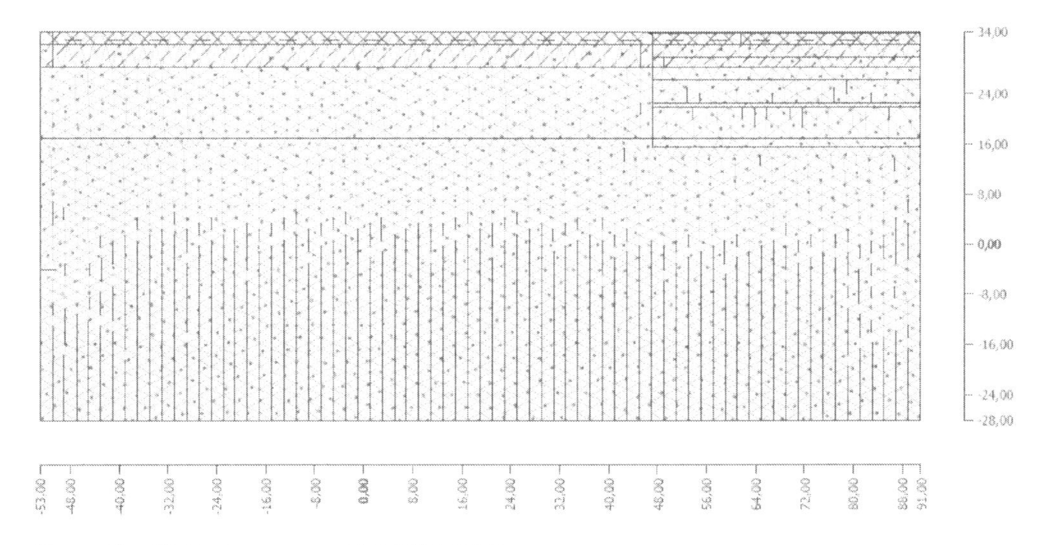

Figure 10. Finite elements model (GEO5 FEM).

Table 2. Material parameters of soil layers.

Type of soil	φ' °	c' kPa	E MPa
Fill (I)	20	0	25
Clay (IIIa)	28	5	37
Sand (IVb)	33	0	150
Sand (IVc)	33	0	200

2.3.2 *Results of FE analysis*

As a result of FE analysis the displacements and stresses in the soil body, displacements of the diaphragm wall and settlements of the surface were obtained in each construction/calculation stage.

In order to compare classical, simplified (empirical and semi-empirical) methods of estimation of surface settlements behind the excavation wall with the results of more accurate numerical FE analysis vertical displacements of the model in the stage 7, representing the final excavation, are further analyzed. The geometry of the model in the stage 7 is presented in Figure 11. The results of the analysis of stage 7 are shown at Figure 12, in form of the map of vertical displacements of the model.

The calculated value of maximum surface settlement behind the wall amounts to 7,5 mm and is located at a distance of approx. 25–30 m from the wall.

The total range of the impact zone, that is the distance to the place, in which the surface settlement disappears, may be assumed as 100–110 m. Exact zero settlement was not reached due to the model size limitation.

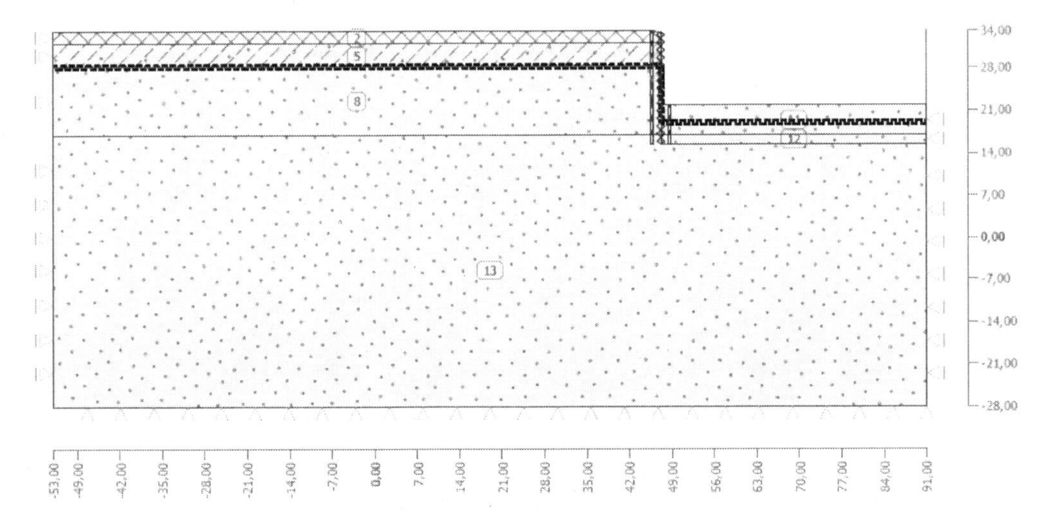

Figure 11. Geometry of the stage 7 (GEO5 FEM).

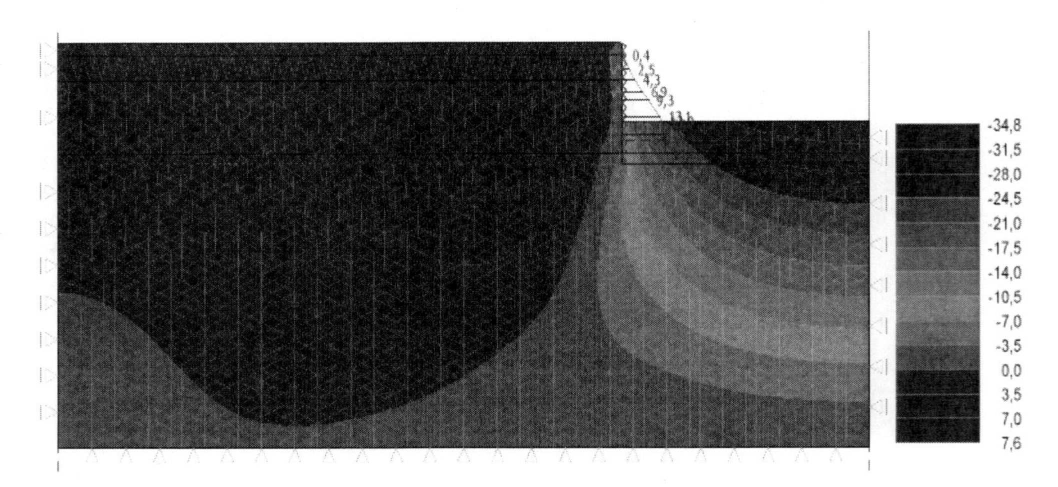

Figure 12. Construction stage 7—vertical displacements of the model [mm].

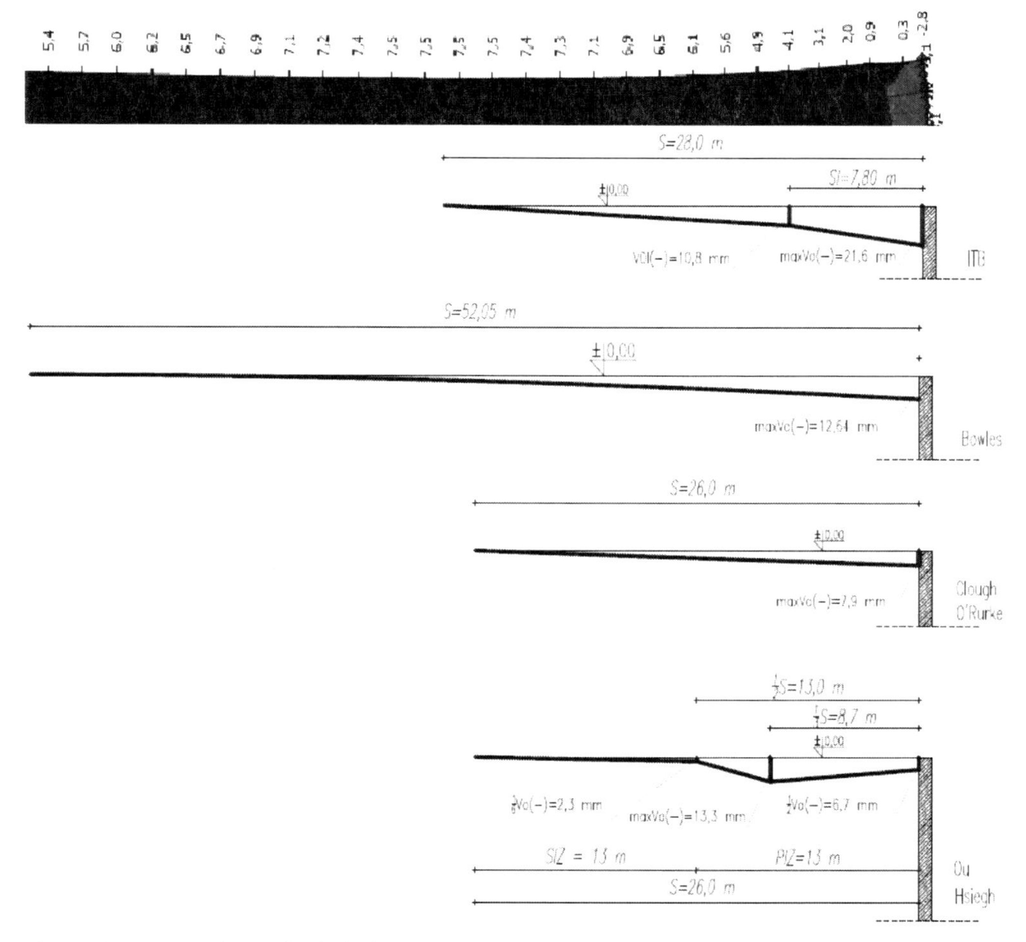

Figure 13. Comparison of calculation results.

2.4 *Comparison of results*

Second part of the case study concerned the estimation of the settlement trough behind the excavation wall applying several simplified empirical or semi-empirical methods described in chapter 1. Taking into consideration the exact geometry of the excavation (Fig. 9) maximum values of terrain settlement and the range of the settlement trough behind the wall were calculated.

The results are compiled in graphical form at Figure 13, comparing settlement curves behind the wall—in the first place the results of FE model analysis and simplified methods below.

In all cases of simplified calculations the maximum value of settlement is located in the nearest point behind the wall, with one exception of Ou and Hsieh method, in which the maximum is shifted 8,7 m from the wall.

Analyzing the results obtained by simplified methods it may be noted that the calculated range of the deep excavation impact zone is approximately 30 m, by all methods. The exception is Bowles method, which due to the substantial width of the excavation, overstates the value of the settlement trough range (52 m).

Similarly, the values of maximum vertical displacements estimated using simplified methods do not differ significantly, ranging from 7,9 mm (Clough and O'Rurke) to 13,3 mm (Ou and Hsiegh). The exception is the method of ITB, in which the maximum vertical displacement is twice higher compared to the others, that is 21,6 mm.

It should also be noted that the methods: Clough & O'Rourke's and Ou & Hsieh's, use data from the FEM calculation models (the maximum horizontal displacement of the wall and the vertical displacement of the surface at the interface point of the soil and the wall).

3 CONCLUSIONS

Taking into consideration the results of detailed finite elements analysis as well as the results of a wide range of simplified semi-empirical or empirical methods of calculation of settlements in the vicinity of deep excavation, basing on the example case described in the paper, following conclusions may be formulated:

- Estimated surface settlements behind the excavation wall calculated using simplified methods range from 7,9 to 13,3 mm (exception: 21,6 mm) and are usually located in the nearest point behind the wall or slightly shifted away from the wall (7 m). Whereas, maximum settlement resulting from FEM analysis is 7,5 mm and is located approx. 30 m away from the wall.
- The range of the deep excavation impact zone calculated using simplified methods ranges from 26 to 28 m (with one exception 52 m), while in the FEM model the impact zone is approx. 100–110 m long.
- Therefore, it may be stated, that there is a compliance in the maximum value of surface settlement but the shape and the range of the settlements trough differs significantly.
- It should be noted that significant theoretical displacements (relaxation up to 34,8 mm) of the bottom of the excavation occur in the final excavation stage (stage 7). This, probably not realistic, relaxation may result from the limitations of the constitutive soil model (modified Coulomb-Mohr model) chosen for the analysis. Overestimated relaxation of the bottom of the excavation implies unrealistic movement of the soil mass towards the excavation, which leads to the overestimation of the range of excavation impact zone. As Mitew-Czajewska (2005) proposed, further FEM analysis will be performed, with the implementation of higher values of the deformation modulus of soils below the bottom of the excavation (from 2 up to 6 times higher then in the original model), in order to better model soil behavior in this zone.

The results of analysis described in the paper and the results of additional FE modeling will be further verified and discussed basing on the on-site displacements measurements, which will be performed during construction.

REFERENCES

Bowles, J.E. 1988. *Foundation analysis and design*, 4th Ed., McGraw-Hill Book Company, New York, USA.

Buro Happold. 2013. Prosta Office Park Building permit design, Warsaw.

Clough, G.W. & O'Rourke, T.D. 1990. Construction-induced movements of in situ walls, Design and Performance of Earth Retaining Structures, *ASCE Special Publication*, No. 25: 439–470.

Fine Ltd. 2014. GEO5 User's manual. Prague.

Krajza, J. 2014. *Influence of deep excavation on surrounding structures—a case study of Prosta Office Park investment*. Warsaw University of Technology.

Mitew-Czajewska, M. 2005. *Experimental research and numerical analysis of displacements of diaphragm walls*. Warsaw University of Technology.

Ou, C.Y. 2006, *Deep excavation. Theory and Practice*, London: Taylor & Francis Group, 2006.

Ou, C.Y., Hsieh, P.G., Chiou, D.C. 1993. Characteristics of ground surface settlement during excavation. *Canadian Geotechnical Journal*. Vol. 30: 758–767.

Kuszyk, R., Tyndak, U., et al. 2013. Geology-engineering report for the Prosta Office Park investment. HydroGeoStudio. Warsaw 2013.

Siemińska-Lewandowska, A. & Mitew-Czajewska, M. 2008. *The effect of deep excavation on surrounding ground and nearby structures; Proceedings of the 6th International Symposium on Geotechnical Aspects of Underground Construction in Soft Ground (IS-Shanghai 2008), Shanghai, April 2008*. Rotterdam: Balkema.

Underground Infrastructure of Urban Areas 3 – Madryas et al. (Eds)
© 2015 Taylor & Francis Group, London, ISBN 978-1-138-02652-0

Vertical drilling depths in solid rock with the flame melting drilling technique

D.P.F. Moeller
Institute of Applied Stochastic and Operations Research, Clausthal University of Technology, Clausthal-Zellerfeld, Germany
Department of Computer Science, University of Hamburg, Hamburg, Germany

R. Bielecki
German Czech Scientific Foundation, Hamburg, Germany

ABSTRACT: The environment is already subject to enormous influence by humankind and we have to be aware of the direct and indirect consequences of activities and action scenarios of our construction planning and building. That is essential for initiating innovations and for triggering changes in people´s minds and behavior. Therefore, environmental consciousness is growing for more than a decade in construction planning and building work worldwide. One reason for this can be seen in the financial benefits of green construction work which making construction builders more willing to embrace this movement. To protect the health and the safety of the surface of the earth underground construction work becomes more and more an important issue which can contribute to protect the earth environment. Underground construction can be introduced from a very general perspective as a construction work in which at least 50% of the combined area of it is covered with a layer of earth. Looking at today´s underground construction work one can find that this is involved in the construction, rehabilitation and remediation of underground pipeline systems, covering the entire underground utilities infrastructure market including cable, gas, oil, telephone, water and wastewater, and more. In our contribution the underground construction work in much deeper depth as the foregoing mentioned underground construction work. Since we are referring to a depth of more than 1.000 meters deepness for underground construction work we introduce a new vertical drilling technique for application in solid rock.

1 INTRODUCTION

Underground construction is an important technological challenge for any contractor, municipal manager and/or professional involved in the construction, rehabilitation and remediation of underground construction work. The coverage of underground construction today extends to the entire underground utilities from underground pipeline systems for cable, gas, oil, water, and more to borehole drilling which is a narrow shaft bored in the ground, either vertically or horizontally for the usage as tunnel tubes for transportation purposes of cars, busses, trucks, metro trains, regular trains, and more.

In our contribution the underground construction work is conducted in much deeper depth as the foregoing mentioned underground construction work which mostly makes use of different techniques such as boring, blasting, drilling, hydraulic splitter, slurry-shield, wall-cover construction methods, etc. We are referring to a depth of more than 1.000 meters deepness for underground construction work which can be introduced as a deep borehole. This deep borehole is essential for the concept disposing high-level radioactive waste from nuclear power plants, and called a deep geological repository. The depth of such a deep geological repository seeks to place the nuclear waste very deep underneath the regular ground water zone, which means that drilling at the latest can be up to five kilometers beneath the surface of the earth. This depth is assumed to safely isolate the nuclear waste, stored in such a deep geological repository, from the biosphere for a long period of time, to avoid any threat to man and environment. This assumption is based on the prerequisite of the thickness of the natural geological barrier as a condition sine qua non.

The concept of a deep geological repository was originally developed in the 1970s, and a first experimental borehole has been proposed by a consortium headed by Sandia National Laboratories, USA, as reported by Fergus Gibb (Gibb 2000).

The deep borehole disposal concept consists of drilling a borehole into crystalline basement rock (typically granite) to a depth of about 5 km, emplacing waste canisters containing spent nuclear fuel or vitrified radioactive waste from reprocessing in the lower 2 km of the borehole, and sealing the upper 3 km of the borehole.

Preliminary estimates for deep borehole disposal of the entire projected waste inventory through 2030 from the current US nuclear reactors suggest a need for a total of about 950 boreholes, with a total cost that could be less than a mined repository disposal system at Yucca Mountain (Brady et al 2009).

In 2006 Christopher Ian Hog wrote a thesis at Massachusetts Institute of Technology (MIT) with the topic to design a canister for the disposal of spent nuclear fuel and other high-level waste in deep borehole repositories using currently available and proven oil, gas, and geothermal drilling technology (Hoag 2006). The proposed boreholes are also 3 to 5 km deep, in igneous rock such as granite. As reported in (Hoag 2006) the rock must be in a geologically stable area from a volcanic and tectonic standpoint, and it should have low permeability, as shown in recent data taken from a Russian deep borehole.

Looking at the techniques recommended in the publications dealing with deep borehole drilling is mostly referred to traditional drilling techniques. In (http://www.neimagazine.com/features/featuredeep-borehole-disposal-dbd-methods) Fergus Gibbs reported about a low-temperature very deep disposal, version 1, with a borehole with an inner diameter of at least 0.5 m over the lowermost 1 km of a borehole sunk 4 km into an appropriate host rock. The very deep borehole concept, reported by SKB, deployed waste packages over the bottom 2 km of a 4 km deep 0.8 m in a borehole (http://www.neimagazine.com/features/featuredeep-borehole-disposal-dbd-methods, Birgersson 1992). Another approach, the low-temperature very deep disposal, version 2, report that containers would be axially deployed over the lowermost 1 to 2 km of a 4 to 5 km deep, fully-cased borehole with a clear inner diameter of at least 0.5 m (http://www.neimagazine.com/features/featuredeep-borehole-disposal-dbd-methods).

In summary we can state that the development of underground construction space by drilling depths in solid rock represents a special construction engineering task with regard to consideration of the intrinsic complexity depending on the depth which only is solvable with a wide range of technical considerations including of following mentioned disciplines: physics, chemistry, engineering, geology, mechanics, surveying, electrical engineering, engineering, thermodynamics, hydromechanics and micro-meteorology.

2 UNDERGROUND CAVITY PRODUCTION WITH THE FLAME MELTING DRILLING TECHNIQUE

As reported by Moeller and Bielecki in (Moeller 2011) rock welding is the basic principle of a technology to sink vertical constructions or to drift horizontal driving. To overcome the energy dependent problem using the rock welding technique, a research group with scientists from the Universities Hamburg, Germany, Košice, Slovak Republic, Brno, Czech Republic, in co-operation with the German-Czech Science Foundation (WSDTI), Germany, searched and studied an option on a rock welding technique which does not need a nuclear reactor as energy source for rock melting. The resulting technical principle is deemed as flame melting technique beneath extreme high pressure, temperature, and frequency. Based on this research work a first mock up assessment was carried out in support of implementation of geological disposal based on the flame melting technique concept to melt rock indicate amendatory against present mining geological disposal concepts. The necessary exploration to test the flame melting technique to melt rock material at the laboratory scale was accomplished at the Technical University Košice, and the Slovakian Academy of Science, Slovak Republic. Therefore, for the flame melting drilling technique, the following patents are registered in recent years in Germany:

- July 3rd, 2008, DYPEN, Kosice/Slovakia; No. 10 2008 031 490 A device for inserting a drilling hole in rock with the flame melting technology (issued on August 26, 2010),
- October 2nd, 2008, Werner Foppe, D 52511 Geilenkirchen and Prof. Dr. Dr. Franz Josef Rademacher, D-89075 Ulm; No. 10 2008 049 943 A procedure and a device for melt drilling with a metal melt.

More patents for melt drilling using the flame melting technology have been registered in recent years in Slovakia and the United States.

For waterproof fulfilment of underground cavities at depths > 1,000 m the bedrock should be melted individually depending on the diameter of the cavity, based on an analysis conducted for large borehole drilling to only melt the hole edge, while the inner core will mechanically dig out, which seem to be a realistic approach for economic reasons.

The proof is to provide that the already existing cracks and fissures, caused by melting hard rock material will bring in rock melt into enough depth to close them completely that they can withstand against high pressures. Moreover, the glazed borehole wall has to be checked with regard to their long term behavior (alteration).

For this reason a principle concept as basis for a repository that permit embedding elevators in the large diameter borehole and, provided that the security barriers are arranged in a suitable way, allowing retrieval from the final repository of the containers installed in the smaller horizontal boreholes. This assumption is due to the consensus view that at first repositories will be designed for retrievable storage; but there is often a clear implication that if, after a suitable period, there are no technical difficulties and the political climate permits, the system of tunnels and access shafts will be scaled up and the repository will become a

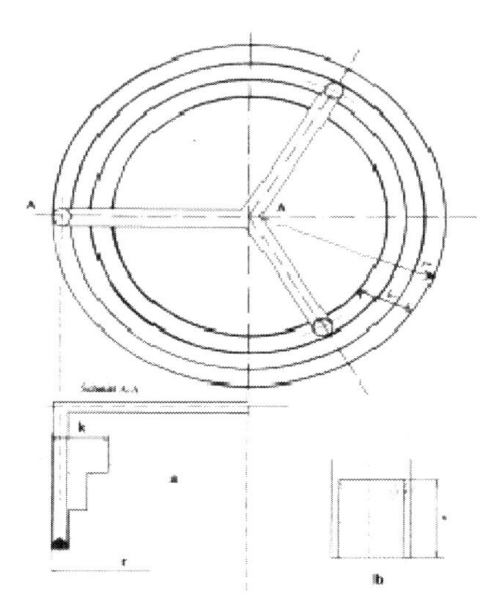

Figure 1. Physical dimension of drilling hole rim melt.

disposal. In our case big borehole diameters in rock massive only will melt the borehole border while the kernel will be mechanical removed.

Critical questions of detail with regard to melting solid rock such as granite with the flame melting technique include:

a. What responses itself occur, if large amounts of water appear during the melting process to?
   - Molten rock has a temperature of 1,400–1.800 °C.
   - Oxygen/hydrogen fusion flame has a temperature of approximately 2,530 °C.
   - Melting can be applied in spite of water inflow from cracks, which can already exist in solid rock and additionally by those which are also produced by the melting temperature.
   - Water vapor withdraws energy from the evaporation of melt and affects additionally the resulting water vapor pressure of the melt.
   - Water evaporates gradually.
   - The water vapor pushes the molt into the existing cracks and crevices during the flame (temperature approximately 2,530 °C) further burns (which can burn well below the water level).
   - Water vapor pushes the molt into the existing cracks and fissures while the flame (temperature approximately 2,530 °C) further burns (which can burn well below the water level).
   - In case of high water inflow from cracks and fissures additional sealing measures are required. In this respect sites without fissures are to select.
b. Is the glazed hydraulic sealing fringing the closed borehole rim? The answer depends from the chosen site e.g. depth.
   - The stone hem is hydraulic sealed without melt, through the selection of the rock apart from cracks, what has to be investigated by test drillings.
   - The melt seals the cracks in the rock and closes the pores of the stone hem.
   - The amount of melt can be regulated by the fusion flame.
     - To demonstrate of the depth of the crack sealing, thermal shock examinations are required. First investigations were already carried out at the Technical University of Košice and the Slovak Academy of Sciences.
c. Long-term behavior (alteration) of solidified rock?
   - The rock has a crystalline structure; the melt in contrast is amorphous, which has the same long-term behavior as the original rock, as both have the same elements (Rybar 2004).
d. What influences has the chemical milieu, i.e. the chemical composition of the rock and water gap on the solidified melt?
   - The influences on the solidified melt are the same as the influences on the original rock.
   - Aggressive substances do not influence the chemical milieu in the granite rock because they are present only in small amounts.
   - Bacteriological influences are not detectable in the desired depth.
e. How is the cooling behavior of solidified rock melting (contraction cracks)?
   - The cooling process of the melt is slow due to the low thermal conductivity (s = 0.25 to 0.73 W (m °C).)
   - The Thermo shock investigation will show detailed information for this purpose.
   - A molding of filling material in the rock area will not take place.
   - Any contractions by thermal stress should be closed by subsequent injections.
f. How deep the rocks melt enters the open surfaces in the mountain and whereof depends the depth of penetration?
   - The penetration depth of the melt depends on its viscosity and amount which can be regulated through the melting process.
   - The depth of the crack filling can be determined by measuring methods, which are not material destructive, for example, through sound measurements, supported also by drilling.
   - The production of permanently waterproof cavities in the deep bedrock so far represents a special problem. Here, a major innovation step is done with the rock melt.

3 INTERNATIONAL EXPERIMENTS WITH THE FLAME MELTING TECHNIQUE

Already in the 1970s, tests with the flame melt technique were carried out in Germany and the United States. In Los Alamos in America metal with high energy expenditure was brought initially for the rock melt by nuclear electricity glow, while in Germany the flame melting technique was favored by the engine manufacturer MBB.

For the selection of the location of a shaft system based on flame melting technique, e.g. for waste disposal, a special laboratory for the future control of complex thermodynamic processes at melting of rocks has to be establish in addition to extensive exploration.

3.1 *German attempts*

At the end of the 1970s first German drilling attempts under the auspices of the engine manufacturer's MBB have been carried out. They showed surprising results: of a rocket engine exhaust Jet that generates temperatures to 2000 °C to 2500 °C. This sudden extreme heat leads to such high strains even in toughest rocks that they shatter or melt.

In reality, drilling speed from 30 up to 90 meters per hour is achievable. A "hot gas drill" would make but always have at least 10 times faster than mechanical drills. After that time considerations, the flame melting technology could become after successful research in the underground an environment-friendly construction of traffic tunnels as well as supply and disposal pipes and laying of pipelines in hard or frozen ground through large holes to the dumping of radioactive and toxic wastes in deep (> 1 000 m) be used.

So far, such considerations remained only a vision for reasons of high development costs. Thus, the flame melting technique today is still in the laboratory or field trial stage. The research work of the engineer Werner Foppe, AT consult, Geilenkirchen, Germany, should be recommended.

3.2 *American attempts*

The use of melting technology for sinking vertical structures or horizontal driving lines is based on global test programs. Hence, underground tunneling machines based on the melting technology have already been developed in the laboratories of Los Alamos, New Mexico, in the 70th years of the last century. These machines build in the Nevada desert expanded underground caverns, the data of which have not been published. The results obtained showed that these systems have a three times better performance and the costs were 40% lower than previous classic types of technology. Despite the successes in the Los Alamos project the scientist noted that that this technological equipment cannot be used in residential areas, because the heat energy source which melt rock was based on a nuclear reactor, which would contaminate the groundwater in case of an eventual accident in the areas in which this technology is used. With regard to the result published on the Internet the scientists suggest the use of a chemical exo-thermal process instead of the use nuclear power source.

The power source of the prototype drilling systems developed in Los Alamos was an energy source with an electrical resistance, with a melting head of molybdenum. The actual melting head achieved an electric power at a temperature of up to 1,800 °C. The melting head was designed cone-shaped which result in and a clamping force which dropped the pressure in the molten rock zone on the middle of the head. This pressure generates radial cracks in the rocks and at the same time the molten rock moved into the radial cracks under his influence.

The Institute for Los Alamos worked out for cave-ins up to a diameter of 12 m for horizontal propulsion. These prototypes were working on the principle of rock melting with the special ring-shaped melt-indenter, which is up to 1,800 °C annealed and is forming a core like in the classic geological core drilling. This core was destroyed on thermodynamic tension, caused by specifically annealed needle-shaped indenters.

In this melting technology drilling head the heat energy source was transferred into the secondary circuit of the nuclear reactor of the above mentioned indenters. The medium of the transmission system of the heat energy source was liquid metal.

3.3 *Slovak attempts*

First flame injectors have been developed in the 1990 th in Slovak laboratories for the combustion of a mixture of hydrogen-oxygen gases for rock melting based on experiences originated in German and US laboratories. They have been the basis of all recent patent developments and construction plans. For this purpose a flame injector was designed based on cobalt with a 200 micrometer-thick ceramic layer of hafnium nitride which was applied to its surface by plasma technology. At the bottom of the head, the injector had a drain nozzle through which a stoichiometric balanced oxygen-hydrogen gas was applied, to melt the rock in the combustion. Under laboratory conditions drilling holes of approx. 70 mm have been bored according this diameter in different rock blocks with dimensions $0.5 \times 0.5 \times 0.5$ meters, as shown in Fig. 2. In this laboratory investigation the average penetration rate achieved was 7 mm/sec. Investigating the flame melted holes show that no dislocating pressure have occurred under the head of the jet pump system. However, radial cracks of such dimension occur that the rock boulders collapse in the final stage. Thus, the melted rock could pass through the melted chambers into the radial cracks, which has been kept in several records. Moreover, inspecting the flame melting rock boreholes show at several areas a special crust within the ambiance of the melted chambers as well as at the collapsed probes. Thermal force generates radial cracks in the direction of the free areas of the rock shoulder.

To avoid radial cracks as a result of the hydraulic pressure of the gap, it was necessary to develop a new technical principle for the mechanism forming radial cracks by means of splitting pressure under the head of the flame injector. This was to ensure that the combustion products can escape water vapor from the area under the head over the deep hole in the free atmosphere. For this reason, the shape of the flame injector was designed with an inverted

Figure 2. Lab based drilling examples.

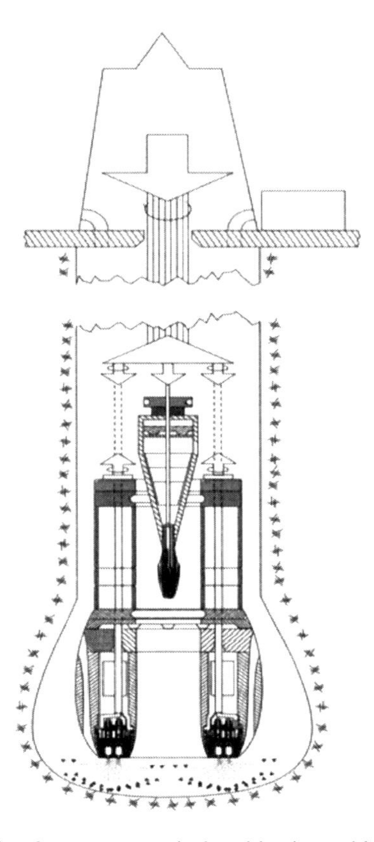

Figure 3. Principle of a machine for greater vertical and horizontal headings using the flame melting technique.

cone, developed the technology at Los Alamos. With regard to the form of the injector it became clear that it is not necessary to work with a large axial force on the axis of the injector body.

For larger cross-sections holes edge fusion are recommended for reasons of water-profess and in particular the economy—as already mentioned.

For solving the inner core problem e.g. more vertical holes with the flame melting technology or explosions or core material reduction can be performed with a boring machine developed by the professors Tobias Lazár and Felix Sekula at the Technical University of Košice. The principle of this drilling machine is shown in Fig. 3 which makes use of the flame melting technique.

4 NECESSARY RESEARCH AND DEVELOPMENT WORK

The device patented in accordance with section 2 on August 26, 2010, DYPEN enables the production of a borehole by up to 60 cm in diameter with the flame melting technique in solid rock. With this device, holes can be drilled with the flame melting technique melting the rock of the borehole a total of several 1,000 m depth. Moreover, €4 million have to be carried out more research and development work for large holes for a feasibility study to be produced over the next 3.5 years and for the practice-ready design of the flame melting technique of large holes in another 15 years approx. EUR 200 million will be required.

5 FINAL CONSIDERATIONS

The flame melting technique can be used not only for vertical drilling depths in solid rock, but also for production of horizontal line structures in crystalline rock. An advantage of the flame melting technique with regard to the degradation e.g. with roll chisel when large borehole drilling are done is to be seen, that faster and greater drilling depths can be achieved by melting and thus resolve the core in larger sections over several meters. General advantages of this technology are the construction related water resistance and efficiency of execution. For greater depths, it is necessary to use liquid hydrogen and oxygen, the specific weight of which is less compared to gaseous state. To keep with high pressure pump promoting hydrogen and oxygen in the liquid state, the mantle of their supply lines to the liquid nitrogen –195.8 °C temperature insulation is to surround. We wish all future scientists dealing with the flame melting technique success and a hearty "Glückauf".

REFERENCES

Birgersson, L., Skagius, K., Wiborgh, M. & Widen, H. 1992. SKB Technical Report 92-43.

Brady, P.V., Arnold, B.W., Freeze, G.A., Swift, P.N., Bauer, S.J., Kanney, J.L., Rechard, R.P. & Stein, J.P. 2009. Deep Borehole Disposal of High-Level Radioactive Waste, SAND2009-4401, Albuquerque, NM, Sandia National Laboratories.

Gibb G.F. 2000, Journal of the Geological Society.

Hoag, C.I. 2006. Canister Design for Deep Borehoile Disposal of Nuclear Waste, PhD Thesis, MIT http://www.neimagazine.com/features/featuredeep-borehole-disposal-dbd-methods.

Moeller, D.P.F & Bielecki, R. 2011. Storage of High Level Nuclear Waste in Geological Disposals: The Mining and the Borehole Approach, In: Nuclear Power, Ed.: P.V. Tsvetkov, Chapter 13, pp.307–330, iNTECH Publ., 2011.

Rybar, P., Lazar, T. & Hamrack, H. 2004. Studies problematiky tavenia nerastných surovin v extrémnych podmienkach, TU Košice.

Underground Infrastructure of Urban Areas 3 – Madryas et al. (Eds)
© 2015 Taylor & Francis Group, London, ISBN 978-1-138-02652-0

Analysis of selected aspects of the operation of pipelines renewed with the relining method on the basis of laboratory testing results

B. Nienartowicz
Wroclaw University of Technology, Wroclaw, Poland

ABSTRACT: The subject matter of the paper is related to the issue of strength of pipelines subjected to the renewing using the relining method. The laboratory testing conducted in 2011–2013 in the Wroclaw University of Technology laboratories are described in details herein. The specimens of concrete pipelines to a 1:1 scale, renewed with a GRP liner, placed in a ground environment were subjected to loads in the course of the testing. The basic objective of the research project was to acquire detailed information related to the specificity of the operation of pipelines subjected to renewals, in the aspect of the mating of individual elements of the renewed structure. The following phenomena were subjected to a thorough observation:

1. impact of friction appearing between the GRP liner and the grout injection on the operation conditions of the renewed structure,
2. impact of the mating of the renewed pipeline, the grout injection and the liner on the load capacity of the renewed structure,
3. impact of the quality of the grout injection applied for accomplishment on the load capacity of the renewed structure.

The fact is worth emphasizing that the model of specimens was developed so that they reflect the operation, under load, of a pipeline classified to the technical state III per ATV DVWK M127P guidelines. The paper includes the description of assumptions adopted to perform the testing and the method of their conducting, also the listing and the interpretation of the results received. A strict continuation of the laboratory testing described constituted computer analyses of the operation of structures renewed with the relining method. The paper also discusses some selected assumptions related to the computer model construction.

1 INTRODUCTION

The underground infrastructure to which all types of transfer systems are classified, such as telecommunications, power supply or fibre optic cables having the form water pipelines, gas pipelines or sewerage networks, can be laid in the underground space in two ways. Alternatively, in various types of utility tunnels, mainly providing a comfortable access to the systems located inside, or in the ground medium, which for transportation pipelines, with man-inaccessible diameters, since this paper is dedicated to that part of the underground infrastructure, considerably hinders the conducting of scientific research work, related to the identification of the structure operation characteristics. Phenomena occurring under the ground are difficult to investigate in their natural form due to their limited accessibility and the physical impossibility to observe them under actual conditions. In order to broaden the scientific cognition of the operation of the structures of pipelines intended for operation under the surface of an area it is necessary to combine the knowledge acquired in practice on operating facilities, both old and new, with the knowledge acquired during the conducting of various types of laboratory testing or computer simulations. All activities conducted in a laboratory or in front of a computer monitor have to be carried out while maintaining the due care of the faithful mapping of the underground environment conditions as well as the strength and material characteristics of the facilities under testing. In this paper the author presents selected issues from the laboratory testing, accomplished by her in 2011–2013, conducted on the specimens of pipelines renewed with the relining method to a 1:1 scale and computer simulations being their direct continuation.

2 LABORATORY TESTING

Presently, the renewals of pipelines are, without doubt, a subject matter that is worth noticing. This issue, in relation to degraded sewerage networks, is applicable to almost all countries throughout the world since the effective transportation of waste water is one of essential constituents of the functioning of our civilisation. Nowadays, the renewals of pipelines are performed using a series of differentiated technologies the exemplary classification of which has been quoted below (Kolonko et al. 2011):

- renewals using liners from continuous pipes
- renewals using liners from segment pipes
- renewals with liners from close-fit pipes
- renewals using CIPP liners
- renewals using sealing liners
- renewals using liners from spiral wound pipes
- renewals by spraying protective coatings

The complex process of the technology selection should always be based on a thorough case analysis and provide, in effect, the best fitted solutions, both in technological and business terms. The permanent development in the field of renewals, the appearance of new solutions and the uninterrupted improvement of the existing ones, takes its origin in scientific research. The results of research analyses conducted throughout the world result in the introduction of changes to computational standards used for designing.

The research work conducted by the author included the issues related to the load capacity of pipelines renewed using the relining technology and was oriented towards the analysis of degraded structure being in the technical state III in accordance with (ATV-DVWK M127P 2000), upon performance of a renewal using the GRP (Glass Reinforced Plastic) liner. The relining technology consists in the introduction, into the inside diameter of the existing pipeline, of a new pipe having the form of a liner, and the filling with an grout injection of the space formed between these elements. As a result of the renewal activities a layer structure is obtained, made up of the liner, the grout injection and the mother pipe, located in the ground medium. A schematic diagram of the renewed structure using the relining technology is presented in Figure 1.

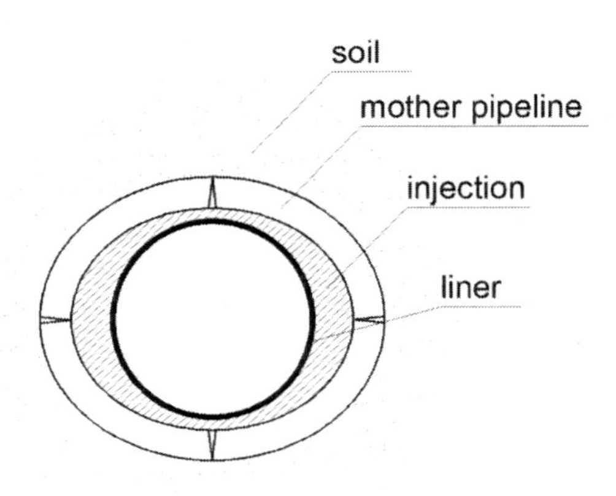

Figure 1. The schematic of the pipeline renewed with the relining technology.

In accordance with the guidelines (ATV-DVWK M127P 2000), the pipeline being in the technical state III is a stable one, however the system, pipe—ground medium, has lost its capacity for an autonomous transfer of loads, also a considerable deformation of the cross section appeared therein. The characteristics of the operation of such a facility, after the renewal, is complex and dependent on many factors, therefore during the creation of computational procedures, while aiming at their simplification and adopting the rule of risk elimination, a simplified operation diagram of the structure is adopted. One of the basic assumptions included in (ATV-DVWK M127P 2000) defines a liner as an element that, upon renewal, transfers the whole of loads acting on the duct, thus it has to be a self-supporting structure. In the opinion of the author, it would be possible, in various designing cases, to take into consideration the mating of the liner with the other two layers of the structural system, which in effect would allow to reduce the required thickness of the liner wall. When undertaking to carry out the laboratory testing, the author defined key issues being the subject matter of further analyses:

- mating of the existing pipe, the grout injection and the liner, and its impact on the load capacity of the facility after the renewal completion,
- mating between the liner and the injection, and its impact on the load capacity of the facility after the renewal completion,
- impact of the compressive strength of the grout injection applied on the load capacity of the facility after renewal.

In order to test the above-specified elements, a laboratory testing was scheduled on pipeline specimens to a 1:1 scale. A partial description of the activities conducted and of the assumptions adopted is presented by the author in referenced items (Madryas & Nienartowicz 2010; Nienartowicz 2012). Drafted below are the main assumptions of the testing schedule:

- the testing is to be conducted on specimens of pipelines damaged and renewed using the relining technology, placed in the environment of a ground medium;
- the specimens of the mother pipelines are to be prepared in such a way as to map a degraded and deformed pipeline (3%), being in the technical state III per guidelines ATV-DVWK M127P (ATV-DVWK M127P 2000);
- preparation of 15 specimen pieces in 5 different types, 3 identical specimens for each type of specimens;
- choice of parameters of the ground medium the same as for fill ground.

The testing was conducted on a test stand, providing a possibility to lay the specimens in the ground medium, with an inspection of its concentration, which allowed to achieve the operation conditions of the structure close to the actual conditions of the pipelines' functioning. The schematic diagram and the view of the test stand is presented in Figure 2 and 3.

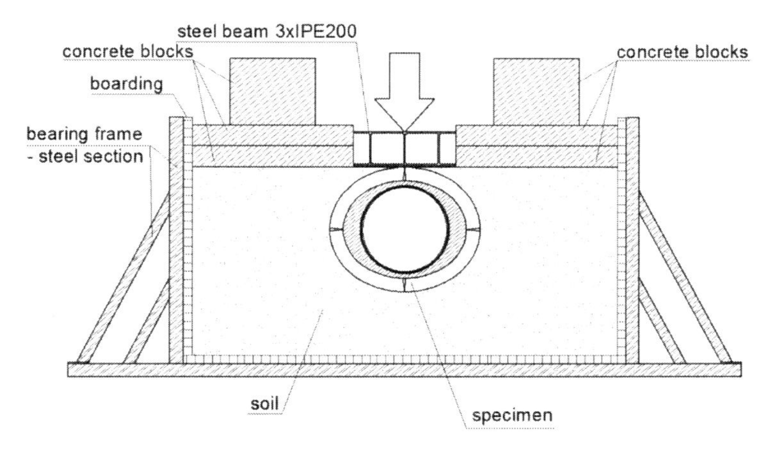

Figure 2. Schematic diagram of the laboratory testing.

Figure 3. View of the test stand prepared to conduct the testing.

Table 1. Division and characteristics of specimens.

Type	Specimen characteristics	Quantity of pieces
IA	Damaged concrete of a 600 mm diameter, renewed with the relining technology with the use of a GRP (Glass Reinforced Plastic) liner having a 500 mm diameter, with the use of an grout injection of a compressive strength equal to 1.5 MPa	3
IB	Damaged concrete of a 600 mm diameter, renewed with the relining technology with the use of a GRP liner having a 500 mm diameter, with the use of a grout injection of a compressive strength equal to 30 MPa	3
IIA	Damaged concrete of a 600 mm diameter, renewed with the relining technology with a GRP liner having a 500 mm diameter, with the use of a grout injection of a compressive strength equal to 1.5 MPa, with the application of a film between the layers, liner—injection, to exclude the mating of these elements	3
IIB	Damaged concrete of a 600 mm diameter, renewed with the relining technology with a GRP liner having a 500 mm diameter, with the use of a grout injection of a compressive strength equal to 30 MPa, with the application of a film between the layers, liner—injection, to exclude the mating of these elements	3
III	Liner of a GRP material, having a 500 mm diameter	3

15 specimen pieces were subjected to the testing. The division and a short characteristics of the specimens are specified in Table 1. In accordance with the testing schedule, the pipeline specimens were placed in the ground medium and afterwards, they were loaded longitudinally until the limit value of the vertical deflection of the liner was obtained, determined as

10% of inside diameter. The basic measured value established was that of the force required to obtain the limit value of the deflection.

Figure 4 presents the diagrams of relationship between the vertical deformation of the GRP liner's inside diameter and the force applied, for individual specimen types. The forces' limit values are listed in Table 2.

The test results allowed to assess the approximate impact of the damaged structure and the grout injection around the new liner on its load capacity, with strictly defined boundary conditions related to, among others, the degree of damages of the existing pipeline. The testing shows that the differences in forces to be applied in order to obtain an identical liner's deflection is significant in both the cases and reaches even up to 47%, if a grout injection of an increased strength is assumed for the renewal.

Also, a significant increase in the load capacity of the renewed pipeline depending on the quality of grout injection used was noticed. The choice of the grout injection having a 20 times higher compressive strength resulted in an average rise in the load capacity of the facility by 29%.

Among the three tested, previously mentioned issues, the lowest impact on the load capacity of the renewed pipeline turned out to be the provision of an appropriate mating between the layers, grout injection—liner. When using a weak grout injection, the results do not demonstrate significant changes, on the other hand, on specimens with an increased injection, they show differences of an order of 16%.

Besides, the test results demonstrated that the values of forces required to obtain the assumed deflection were several times higher than those acting under actual conditions on structures of this type and the appearance of which is allowed by the presently applied standards and guidelines.

These conclusions were drawn basing on the average results for each type of specimens. The overall analysis of all the test results obtained showed, however, essential divergences within the specimens having identical parameters and built as identical. This fact significantly affects the final interpretation of the results. Each type of the tested specimens was prepared and tested afterwards in three copies. The divergences in the test results of the specimens belonging to one type amount to 8 through 20%. This is an evidence of a high heterogeneity of specimens, both as a general concept and within individual elements. Vertical sensors of displacement were mounted on each sample at three characteristic points –10 cm away from each of the specimen ends and in the middle of the specimen's length. Figure 5 presents the results of the testing performed on a selected specimen and depicts a potential change in the strength properties of the liner over its length.

The results of liners' stocktaking, carried out after the performance of the laboratory testing, confirmed the geometrical identity of panels as guaranteed by the manufacturer. However, the values of weight measurements demonstrated considerable differences in their weight. In extreme cases, even up to 43%. The measurement results are presented in Table 2.

In the course of a further testing, in order to confirm the presumable cause for the divergences of results, a laboratory testing was conducted aiming at the determination of the composition of the GRP composite of selected liners as well as a strength testing to state their strength parameters. Using the procedures included in (ASTM D 2854 2000; ISO 1172 2002), the share of individual constituents of the composite in specimens collected at random was determined. The test results undeniably confirmed the material heterogeneity of the GRP composite. Within the scope of individual liners differences of several percents in weight ratios of individual raw materials, included in the composite, were demonstrated. However, the fact is worth emphasizing that each of the tested specimens was fully meeting the strength conditions as guaranteed by the manufacturer.

The heterogeneity of the tested panels renders difficult the interpretation and the comparison of results obtained from the laboratory testing. Nevertheless, the results obtained provide some valuable information. Since when reviewing all the data the fact is undeniable that both the use of the grout injection with higher strength parameters and the consideration of the mating between three layers of the wall structure of the pipeline subjected to the renewal may significantly affect the rise in its load capacity.

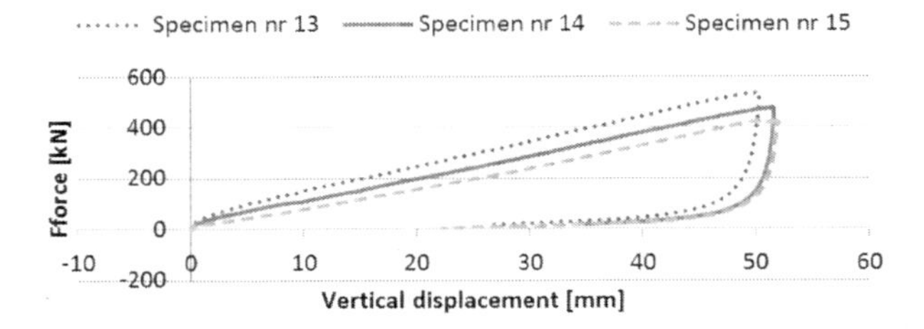

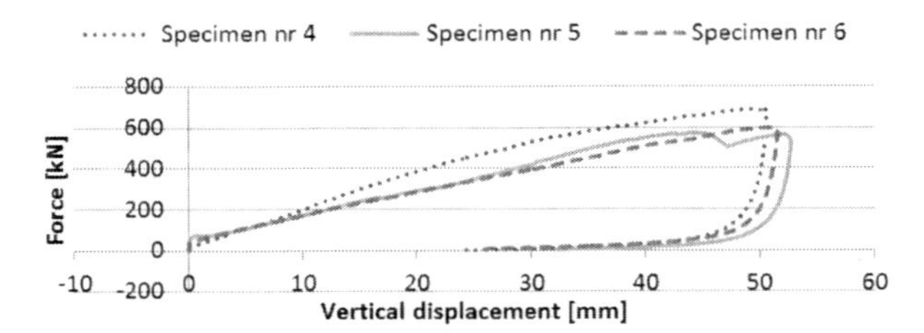

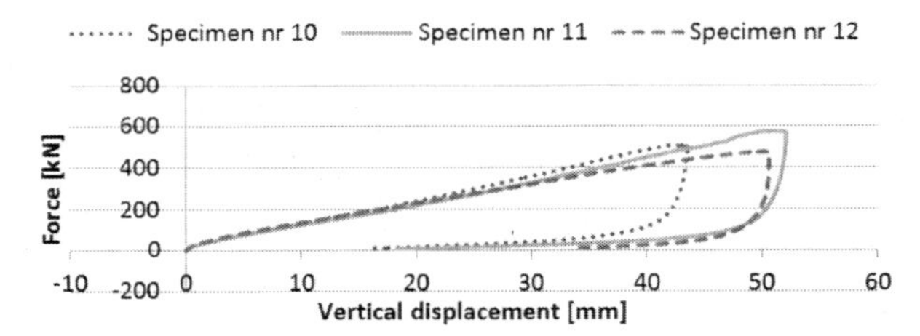

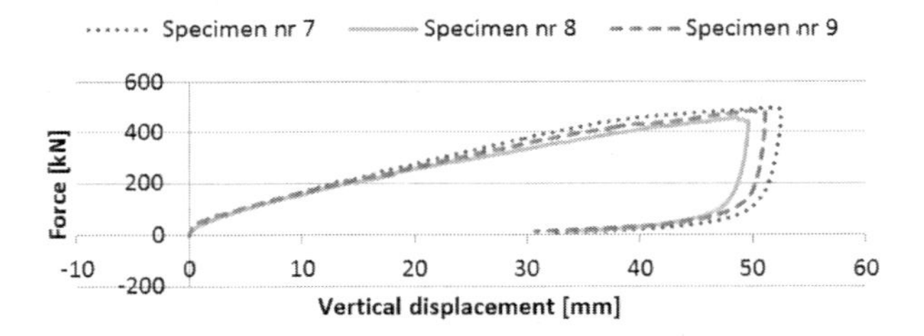

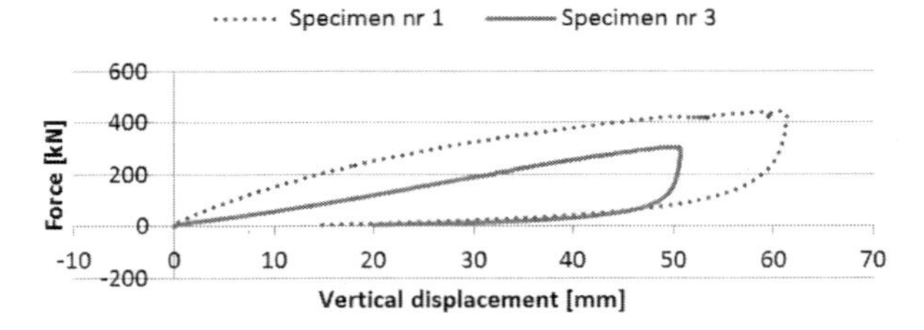

Figure 4. Relationship between the liner's vertical deformation and the force applied for all types of the tested specimens.

Table 2. Values of forces required to obtain a deflection equal to 10% of the liner's diameter as well as results of specimens' weight measurements.

Type	III	III	IB	IB	IB	IIB	IIB	IIB	IIA	IIA	IIA	IA	IA	IA
Nr	1	3	4	5	6	7	8	9	10	11	12	13	14	15
F [kN]	442	304	688	573	596	495	455	483	509	572	474	533	475	425
weight of liner [kg]	80,2	65,8	72,1	59,9	74,9	66,3	68,7	75,6	83,5	80,2	65,5	83,0	71,1	70,2

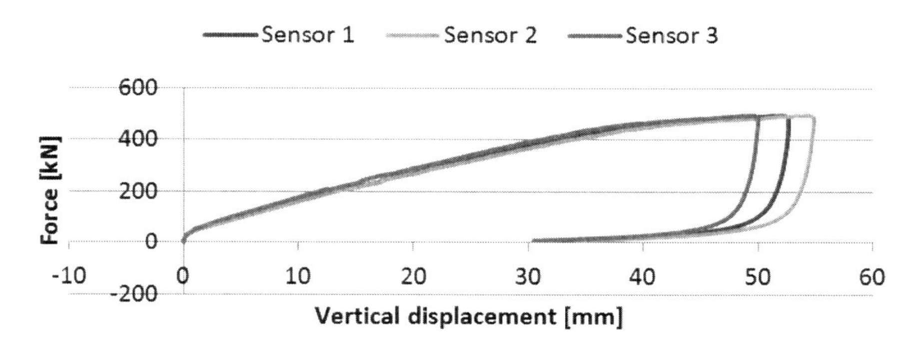

Figure 5. Measurement results of the liner's vertical deformation for specimen No. 7, type IIB.

3 COMPUTER SIMULATIONS

Under the accomplishment of the next stage of the research project, a model was created for the computer analysis of a pipeline renewed with the relining method, the correctness of the model was partially verified on the basis of the results of the laboratory testing presented.

The model of the pipeline, after the renewal with the relining technology, placed in the ground medium, was built basing on the finite elements method, in a flat state of stress. For the construction of the model assumption analogous to those for laboratory testing were adopted. The picture of the discrete model for type I specimens is presented in Figure 6. For its description rectangular, 2D elements as well as contact elements at the contact place of individual components (ground medium, concrete pipe, grout injection, liner) were used. The correct operation of the computer model, depends, in its essential part, on the correct modelling of materials. To this aim, a strength testing was conducted independently or the results of experiments conducted by manufacturers were used.

For the description of the ground medium the Coulomb-Mohr elastic-plastic model as well as the parameters corresponding to the ground used in the laboratory testing were used. The model of a concrete pipe in form of four independent symmetrical elements, defined by a precise description of the damage to the C40/50 concrete material was expressed by the mother pipeline, damaged, deformed and cracked until the loss of continuity. The grout injection material was defined based on the data obtained experimentally. The GRP liner was described on the basis of parameters delivered by the manufacturer, using the z elastic model.

The computational model constructed was used, in the first stage, to the analysis of the impact of friction at the contact place of layers, injection-liner, with the assumption of the grout injection's compressive strength equal to 1.5 MPa, adopted for laboratory specimens type I.

Figure 6. View of the discrete computer model of a type IB specimen placed in the ground model.

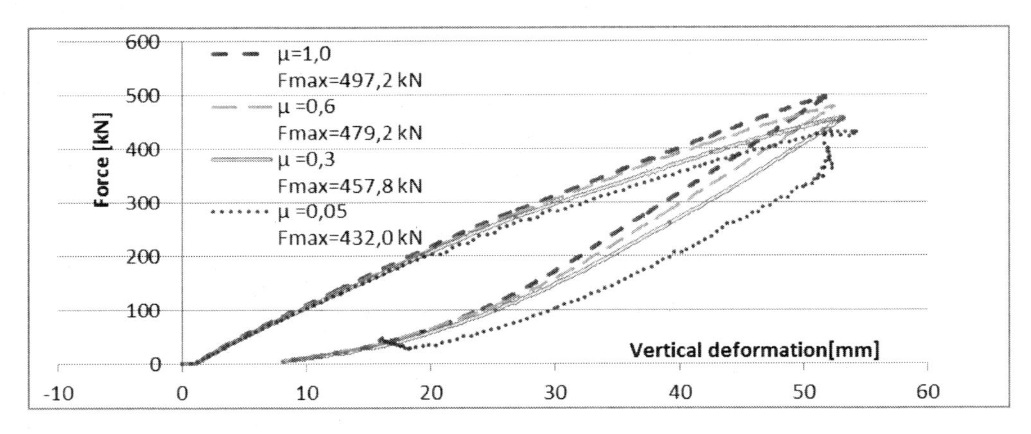

Figure 7. Results of calculations obtained in the computation model of a type A specimen.

A series of calculations was conducted for four following values of the friction coefficient at the contact place of layers, grout injection—liner: 1.0; 0.6; 0.3 and 0.05. The calculations results are presented in Figure 7.

In its assumption, the computer model has to reflect, in a realistic way, the operation mode of a structure, however, it is not possible to eliminate its partially idealized nature. A proper work-out of the model provides a certainty of the correctness of calculations made, however, in more complex situations the discussed cases without doubt are, due to, among others, the diversity of materials also the ways of their reciprocal inter-actions, it does not provide a certainty to receive results that reflect actual situations perfectly. Thus it is necessary to identify safety limits for the application of results obtained.

Additionally, the operation characteristics of linear facilities, being in the technical state III and subject to renewals may be, many a time, very difficult to model, especially in relation to the ground medium and the state of the existing facility which may change essentially or even fully over its length. As demonstrated by the laboratory testing, also the structural elements used for renewal may exhibit a high heterogeneity, which additionally renders difficult the process of an optimum designing. Thus it is necessary to consider the facility subjected to renewal in many planes.

4 SUMMARY

The results of the laboratory testing conducted and those of the still continued computer simulations provide the knowledge about the operation of pipelines renewed with the relining technology. The improvement of the computer model and the analysis of further computational cases will make possible a deeper cognition of the issues under discussion. The laboratory testing, although it does not provide unambiguous results, confirm the hypothesis about the existence of an option to reduce the required wall thickness of a liner through the consideration of the effect of its mating with the two other layers of the structural system. As none of the presently available guidelines do not take into account in their algorithms the partial mating of the existing pipeline and that of the grout injection with the liner, it is required that the designing process is based on the results of computer simulations. It is associated, each time, with a very thorough identification of the technical state of the existing facility as well as equally thorough identification of material properties of structural elements applied. The worked-out schematic of the computer model created for a separate case may be used for next ones under the condition of adapting, each time, the geometry of facilities, boundary conditions as well as strength parameters and material characteristics, paying special attention to the ground and the material of the existing pipeline.

REFERENCES

Abel, T., 2010. *Badania nośności kanałów rekonstruowanych wykładzinami ściśle pasowanymi na przykładzie technologii Trolining*. Wrocław: Praca doktorska.

ASTM D 2854, 2000. *Standard Test Method for Ignition Loss of Cured Reinforced Resins.*

ATV-DVWK M127P, 2000. *Obliczenia statyczno-wytrzymałościowe dla rehablitacji technicznej przewodów kanalizacyjnych przez wprowadzenie linerów lub metodą montażową. Uzupełnienie do wytycznej ATV-DVWK A127P.*

ISO 1172, 2002. *Tworzywa sztuczne wzmocnione włóknem szklanym. Preimpregnaty, tłoczywa i laminaty. Oznaczanie zawartości włókna szklanego i napełniacza mineralnego. Metody kalcynowania.*

Kolonko, A. et al., 2011. *Podstawy bezwykopowej rehabilitacji technicznej przewodów wodociągowych i kanalizacyjnuch na terenach zurbanizowanych*. Bydgoszcz: Izba Gospodarcza "Wodociągi Polskie".

Madryas, C. & Nienartowicz, B., 2010. *Assumptions to investigations of loading capacity of glass reinforced plastic noncircular modules used to renovation of sewerage channels*. Vancouver.

Nienartowicz, B., 2012. Rehabilitation of sewer channels. Investigations of load capacity of channels renovated with GRP liners. W: *Underground Infrastructure of Urban Areas 2*. CRC Press/Balkema, pp. 185–194.

Underground Infrastructure of Urban Areas 3 – Madryas et al. (Eds)
© 2015 Taylor & Francis Group, London, ISBN 978-1-138-02652-0

Impact of pressure control on water loses in distribution net

F.G. Piechurski
The Silesian University of Technology, Gliwice, Poland

ABSTRACT: Very often working pressure greatly exceeds required pressure. This result in worse exploitation, failures and water loses. Monitoring, control and pressure regulation allows checking and regulating water pressure that is delivered to final client. In case of leakage low pressure corresponds to low loses and high pressure corresponds with high loses. On the base of data analysis it is obvious that regulation and pressure reduction decreases leakages and reduces water loses. Pressure regulation increases no failure time intervals and reduces failure coefficients. Water company should analyze current situation, check network working conditions and introduce water monitoring and pressure control system taking into consideration economical aspects.

1 INTRODUCTION

Water monitoring system allows to continuously watching network parameters that we want to control. In analyzed company water monitoring system was limited to pressure and flow measurement in 23 locations in distribution network. Controlled reduction of water loses reduction needs to be based on software, equipment and methods of leakage detection. Leakages are found on the base of data analysis. Using water monitoring system allows checking if water monitoring system works correctly, if the whole water monitoring program works as requested and the whole water loses minimization system.

2 ACTIVE LEAKAGE CONTROL (AKW)

Analysis began from rating management method in the analyzed company. Such a change is not one day decision. Amount of work done by management and workers took long period of time. It was also money consuming. However data from 1995–2012 clearly shows that method is valid. Water loses round 25% were not acceptable in analyzed system and management decided to introduce a method allowing to reduce loses to economical minimum.

Active leakage control allows fast reaction in case of action is required (Niebel et al. 2011). Failure, increased flow, leakage needs to be fixed right away. Increased and longer flow results in greater financial loses (Piechurski et al. 2014). Previous management method allowed reacting only in case of noticing untypical behavior or after emergency call. Noticing and locating leakage in water monitoring system is not an easy issue. 23 existing points are not able to show exactly where the leakage is or what should be replaced. A part of ALC is specific equipment allowing fast and accurate leak point determination. Another important factor in minimizing water loses is pressure regulation. Pressure and its daily changes influence on failures as long as on amount of water leaked form damaged elements of network (Zheng et al. 2011). To reduce that effect company introduced pressure transmitters, controlled automatically or manually. Pressure is adduced according to citizens need. Renovation and replacement of network is an action that was started long time ago. Main steel pipes are replaced to plastic one; valves are replaced to new one to guarantee low leakage and failure rate.

3 MINIMAL NIGHT FLOW (MNF)

Network monitoring allowed much easier data analysis. One of the most efficient ways of observation and control is minimal night flow in given area. The simplest explanation of that method is continuously repeating water flow between 1:00 and 4:00 am every night (Zheng et al. 2011).

Figure 1 presents a piece of data for measurement point. The highest flow is observed round 8:00 pm and the lowest around 3:00 am. Those hours are varying during the week, working shifts, TV shows and so on. However according to norms flow characteristics are similar as long as there are no negative issues influencing network working conditions. The minimal night flow is composed of:

- water taken by citizens,
- water taken by industrial facilities,
- water loses as a result of accumulation of all small leakages,
- water loses from known and unknown leakages,
- extraordinary single water intakes.

In case of large area, there is no possibility to achieve MNF equal to zero. This is the result of the fact that water is taken all the time. Only in very small areas, where industry is not present MNF can be minimized almost to zero in case of good condition network. Analyzing Figure 1, we can say that this is ideal case. MNF is around 3 m^3/h which can be a result of having industrial company in area or leakage. It is not a good practice to look for every possible leakage for economical reasons. In other words costs of leakage detection and removing are much more expensive than generated loses. Very often these are very small leakages that are very difficult to locate in great areas.

In case of "fresh" introducing water monitoring system the situation is opposite. Night flows are much higher than required. The difference is the following. Thanks to such modern technology network company can determine what is the correct direction of network modernization to create optimized solution. On Figure 2, we can observe another area where MNF is

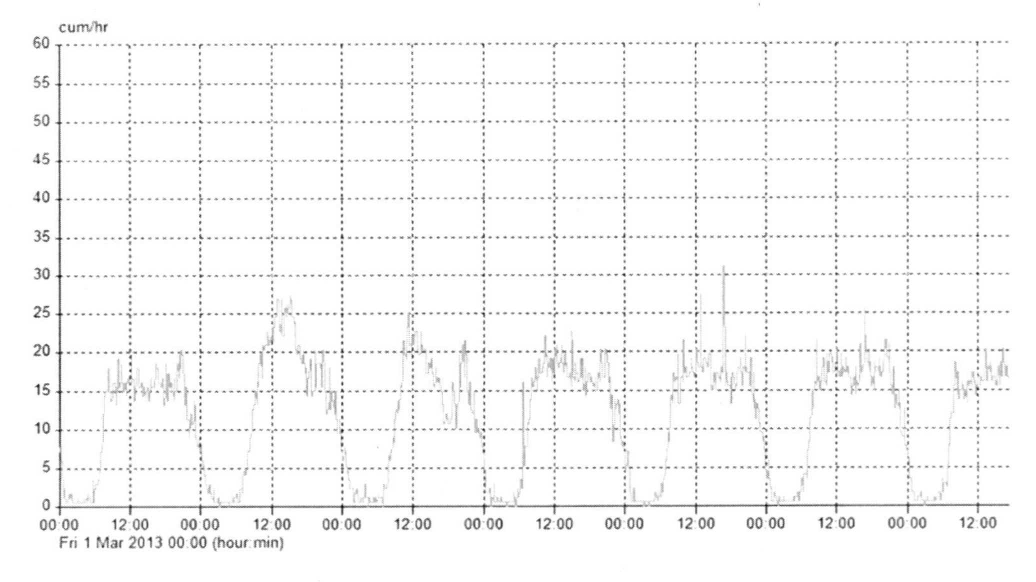

Figure 1. Flow variation according to day time.

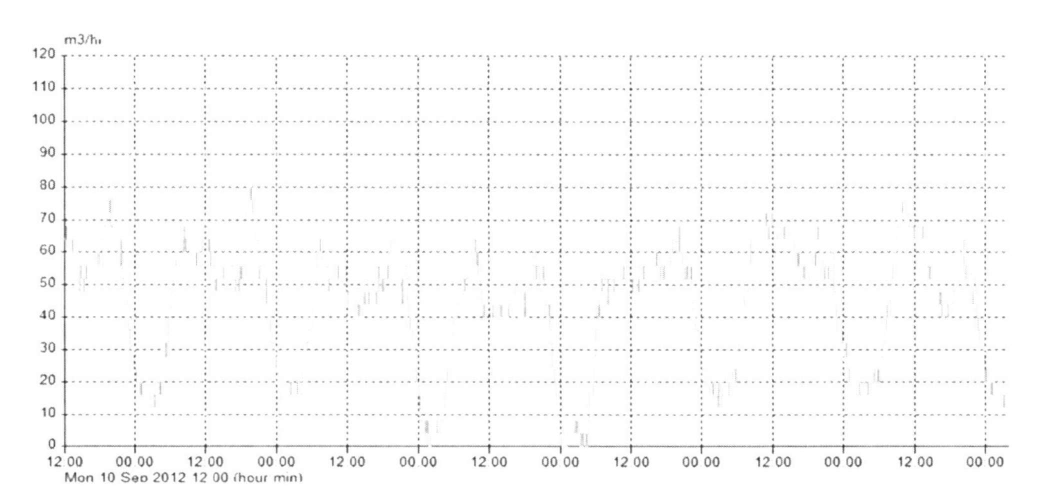

Figure 2. Night flow from another measurement point.

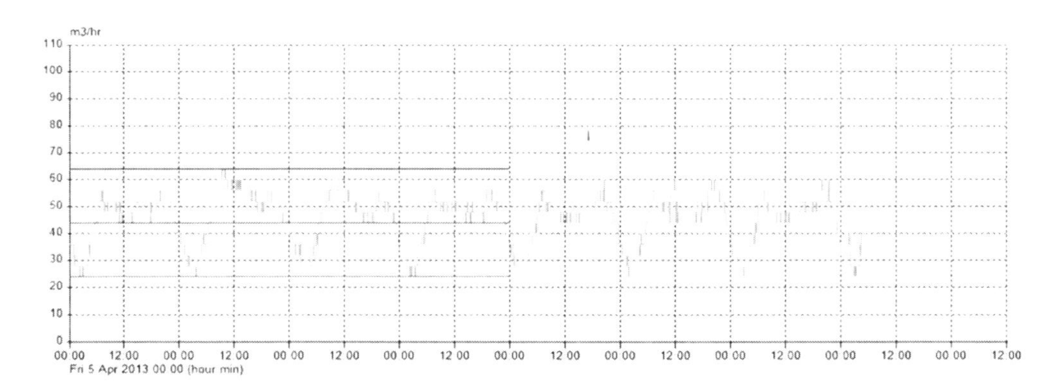

Figure 3. Minimal, maximal and mean flow—help lines used.

significantly larger. It varies 0–20 m^3/h, maximal flows are 65–80 m^3/h. This is problematic case. Not always increased night flow is a result of leakage. Sometimes other factors influence on it.

4 WATER BALANCE

Summing flows allows in very detailed way calculate water loses in analyzed area and answer the question if given area works in good condition or if they are totally noneconomic.

PMAC software allows calculating flows: hour, day, and month. Such data should not be used in billing system. Reason is very simple. A possibility of water retreat generates additional impulses in measurement equipment and this leads to errors in charging final client. Very useful is possibility of automatic reading of maximal, minimal and mean flow in exact hours. That data allows to determine network working conditions and if any corrective action is required. Analysis should be supported with proper knowledge about conditions that can generate deflections in captured data.

There are two methods to determine characteristic flow values. First, with the use of help lines (Figure 3) it is possible to read characteristic flow values in different time periods. Periods are from one hour up to whole week mean. Second and more convenient method is comparison of numerical values (Figure 4).

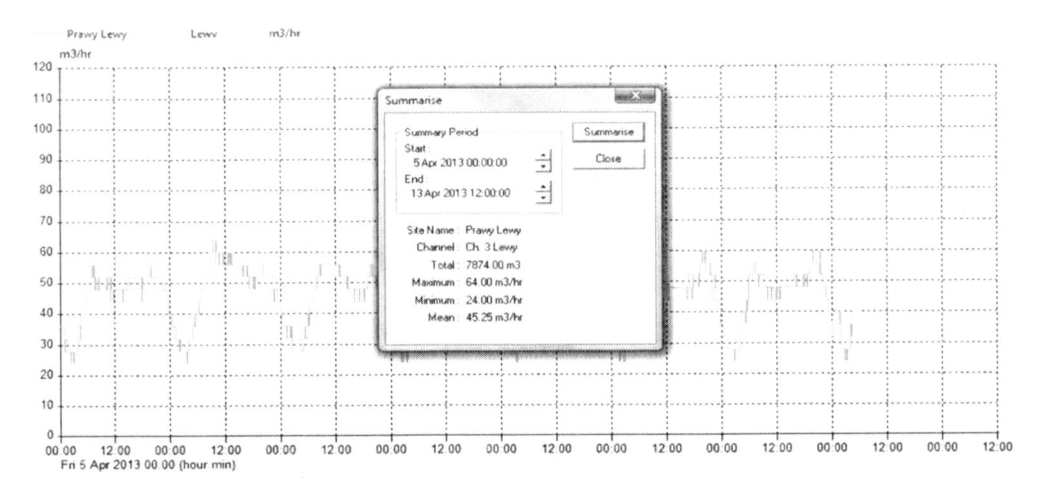

Figure 4. Minimal, maximal and mean flow—numerical values used.

During drawing analysis we can observe that network working conditions are problematic under assumption that there are no industrial companies in the area. Very high minimal flow (even 24 m³/h) and in the same time not very much bigger mean flow (45,3 m³/h). From such data it leads to result that loses can be up to 50%, but most probable is a solution that there is industrial company that continuously takes water.

5 PRESSURE MONITORING

Water network optimization means also to control and stabilize pressure in ducts. Overpressure and its variations increase amount of new failures as well as increase the amount of water leaked from existing failures. Reducing pressure leads to protection and optimization of network working conditions. Amount of lost water is also minimized (Piechurski et al. 2014).

First actions in analyzed company to regulate pressure were introduced when first problems with water delivery occurred and exploitation problems started. Pressure reducers and pumps were introduces in most critical points in city. In 2002 there were 8 pump wells and 14 pressure reducers. All that actions were reaction on variable elevation characteristic of the city, elevation of build and distance from supply points.

Introducing water monitoring system significantly increased pressure regulation and control possibilities. Pressure reducers and pump wells improved network but continuous monitoring and new technology gave wider possibilities. All equipment (Cello, Modulo, Regulo) installed in the area have possibility to read and register pressure and even its remote regulation.

6 PRESSURE MODULATION WITH RESPECT TO TIME

Pressure modulation greatly facilitates whole network operating conditions. In case of time parameters setup, on the base of experimental analysis and knowledge about elevations in network built we determine required pressure in worst point during the higher flows. In the next step we set up out device to highest pressure, the lowest one and its change time period. Such setup ensures requested pressure in given times of day and night and network will not be exposed on dynamic pressure jumps. This will result in less failures and smaller water loses.

On Figure 5 there is relationship between flow (green) at specific time and pressure change (blue). Additionally there is input pressure (red) that is supplied by GPW. Pressure drop correspond to lower water consumption by the users that from 3 am is increasing.

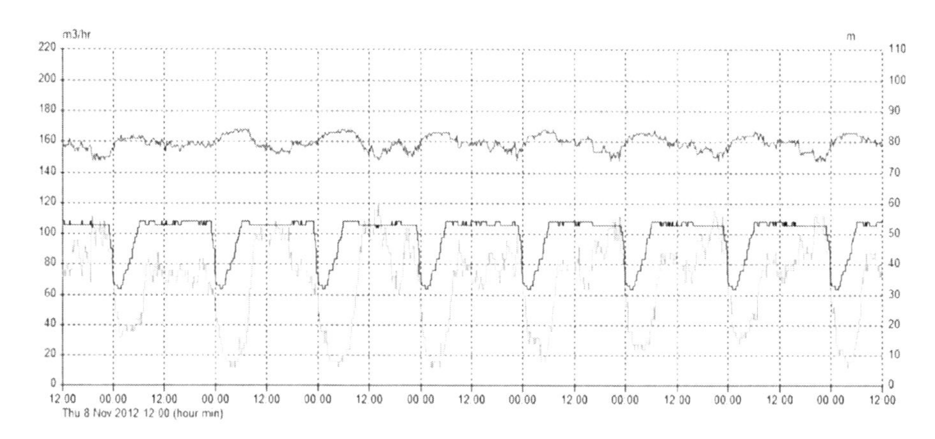

Figure 5. Pressure modulation with respect to time.

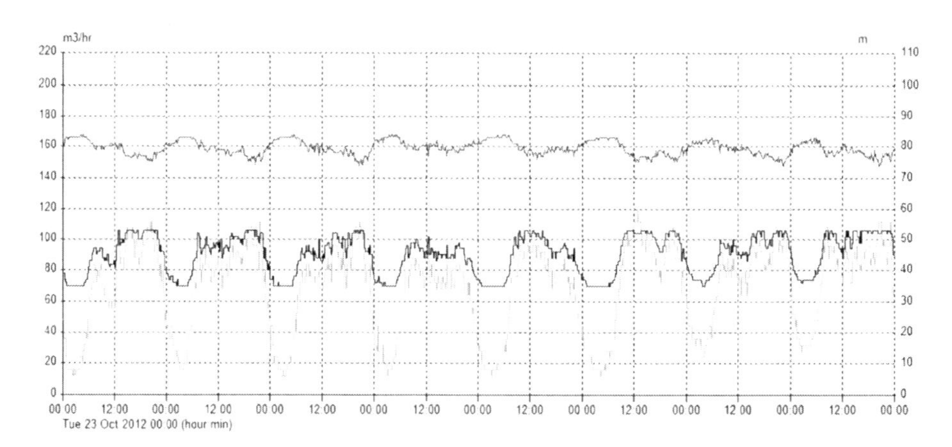

Figure 6. Pressure modulation with respect to flow.

7 PRESSURE MODULATION WITH RESPECT TO FLOW

Second way of pressure regulation in water monitoring system with use of Regulo and Modulo is modulation with respect to flow. It more efficient method because the pressure is adjusted to water flow. Unfortunately this type of modulation requires flow counter. Pressure drops are less steep and occur more often. Thanks to that we receive more effective control and network is better protected. We don't have to worry about to low pressure at any point because regulator ensures perfect work in any conditions (Figure 6).

8 FAILURE AND LEAK SEARCH METHODS

Water monitoring system is most important control element of water network but we have to remember about equipment allowing to precisely locate the failure or leakage. Water monitoring system allows determining in which area the leakage or failure is located. In some cases the area is huge so there is a need to use more precise device to fast and accurate locate the leakage.

Introducing water monitoring system determines using modern, much more effective leakage detection equipment. First and the longest time used equipment is geophones equipped with on speaker microphone produced by Palmer. Except of geophones digital correlator MicroCorr Digital produced by Palmer was bought. Correlator allows quite precise leakage point finding. This device is very useful in case of poor weather conditions, traffic, etc. Any noise has its frequency that can be cut off from data taken into analysis. Additionally to precise locate the leakage route and material that the network is built needs to be known.

9 RENOVATION AND REHABILITATION OF WATER NETWORK DUCTS

Last factor that has a great impact on water monitoring system is network renovation and rehabilitation. Water loses reduction, failure reduction and exploitation costs reducing are only a few factors advantages for water company. Old ducts under high pressure react with big amount of small cracks. The water "runs away" and such small cracks are very difficult to find. In analyzed area till 2002 most of the network was built of steel so water loses due to small leakages are obvious. Important stage was in 2002–2012 introducing PEHD pipes and renovation of existing ones (in cases where possible).

Figure 7 describes material structure of the network in 2002 in percentage. In 1998–2002 82,5 km of network was replaced with respect to age and failure coefficient. That was 20% of whole network. Of course at the beginning the most problematic ducts were replaced. In 2002–2012 company kept dynamical modernization move. Dominating material at the moment is PEHD (Figure 8). It is more resistant to mining activities than steel. Failure coefficient was greatly reduced thanks to renovation and rehabilitation activities.

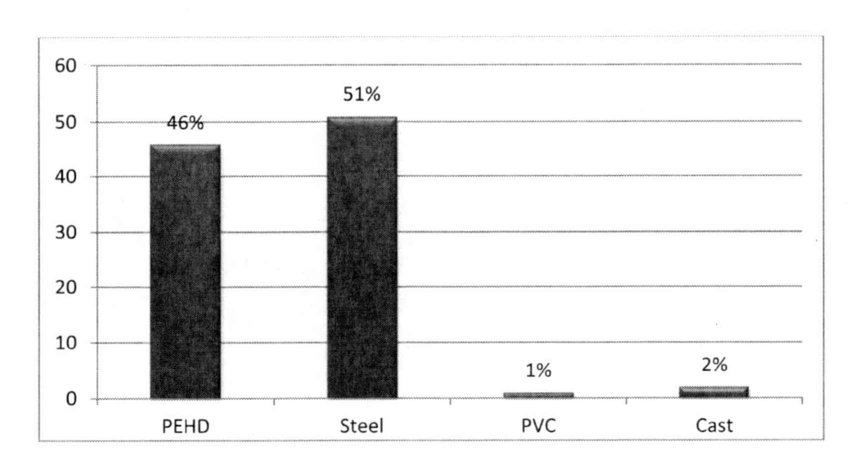

Figure 7. Material structure of the network in 2002.

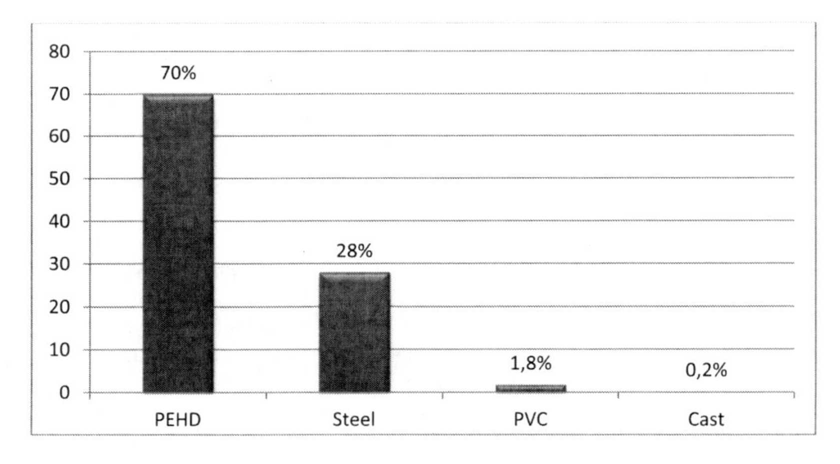

Figure 8. Material structure of the network in 2012.

In 1995–2002 water loses were about 25% Figure 9. It is estimated value because we miss exact data about water used on own needs of water company. Activities related to renovation and replacement of network parts, failure and leakage searching methods and monitoring system not only improved water loses balancing but allowed to reduce them in whole distribution system.

On the base of plots (Figures 9 and 10) we can observe great decrease of amount of lost water in distribution system in 2002 comparing to 1995. It was mainly because of reduced consumption by users and pressure drop in network. Percentage analysis shows that lose were not decreased in those years. There are no solid evidences of improvement.

In 2012 according to obtained data amount of lost water is equal to 572 296 m^3/year. Those data take into consideration water used for cleaning of the network and own company needs because they are precisely measured. After calculation the amount of lost water per day is about 1 600 m^3. Comparing to 2002 it is 4 800 m^3/day improvement. Comparing that data to total intake 19 000 m^3/day water loses were ca. 8%. That is very good result in Silesian area and great improvement in comparison to 2002 (Figures 11 and 12).

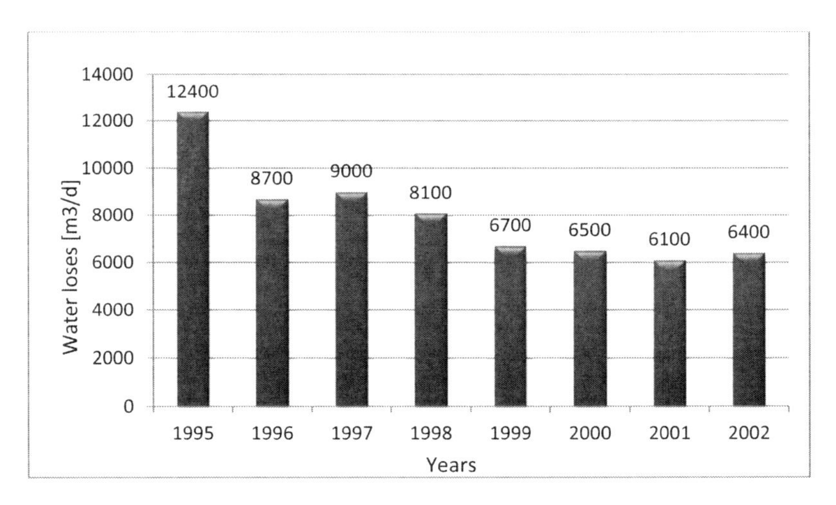

Figure 9. Water loses in 1995–2002 (m^3/d).

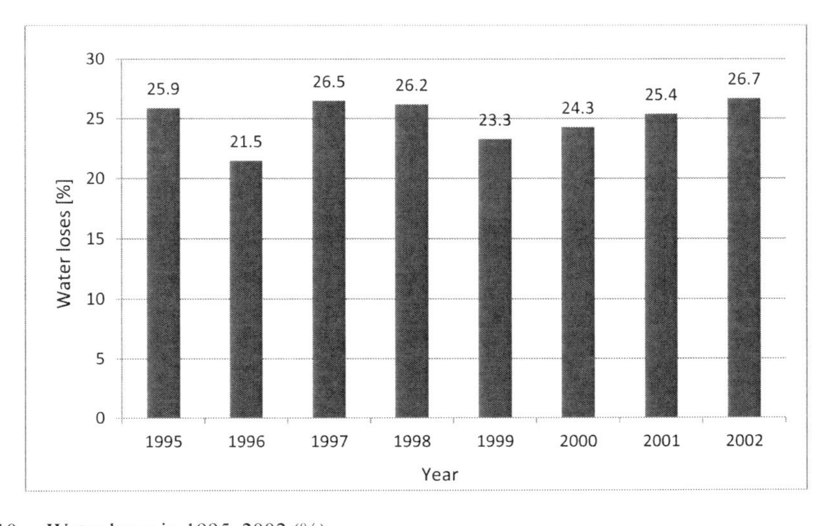

Figure 10. Water loses in 1995–2002 (%).

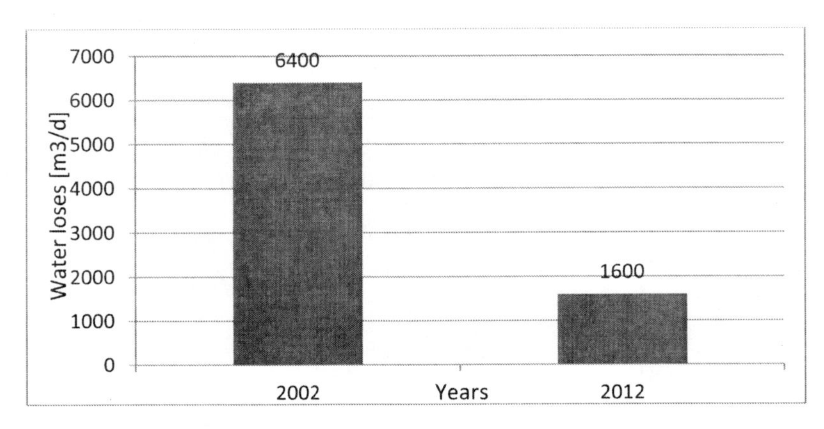

Figure 11. Water loses in 2002 in 2012 (m^3/d).

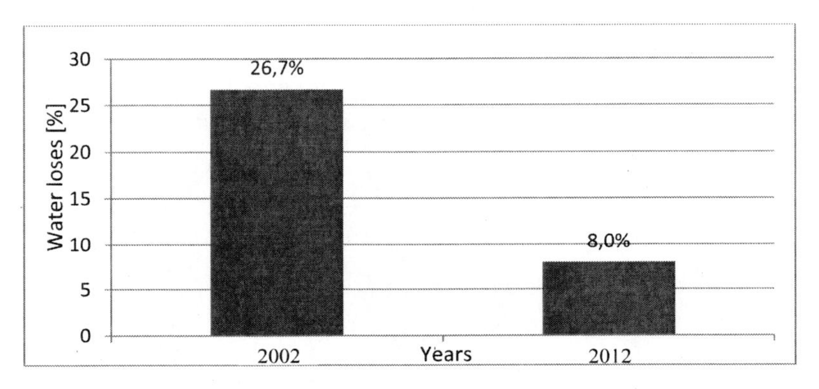

Figure 12. Water loses in 2002 in 2012 (%).

Table 1. Failure intensity coefficient in 2002 and 2012.

Year	2002	2012
Failure number	1279	447
Failure intensity coefficient [failure/km·year]	3,0	1,1

11 WATER NETWORK FAILURE REDUCTION IN 1996–2012

Very important parameter related to network condition rating is its failures rate. Great reduction of that coefficient as far as number of detected failures was improved. This is a result of introducing active leakage control. Taken parameter was 1 failure/km year as boundary condition as a result of missing data from every parts of network.

In last year you can notice significant failure decrease in analyzed distribution system comparing to 2002. All activities taken prove their effectiveness and correspond to decrease of number of failures and water loses. Main features are:

- pipes replacement and renovation—one of the most important aspects, works in this area greatly improves condition of the network, so that it is more resistant to loads.
- pressure regulation—one of high priority factors that allows to significant pressure reductions in areas that does not need high pressure.

- time of reaction—here we should mention time from leakage appearing to its detection and exact location thanks to water monitoring system. Those factors does not influence the failures rate but in case of water loses are very helpful.

12 SUMMARY

Water monitoring system in analyzed distribution network allowed easier handle of abnormal situations over the network such as: failures, rapid night consumptions, steals, night flow increase etc. Everyday monitoring data analysis allows fast response in such cases. Additionally pressure regulation with use of both time and flow methods allow reducing pressure jumps in ducts. In allows more economical water distribution in various time periods. In analyzed area final client monitoring was introduced as well. Remote reading flow meters were introduced to 80% of clients. It helps easier and more precise water balance.

According to data water loses were 25% till 2002. In 2012 it was only 8%.

Network monitoring and pressure control are intermediate methods used to minimize water loses. Very important factor is renovation and replacement of old steel pipes to new ones. In 2002 network was in 51% made of steel and 46% from PE. In 2012 70% of network was made of PE and 28% of steel. Length of the network remains the same. Additionally age of ducts was significantly decreased so the resistance was improved. As a prove we can observe failure intensity coefficient which was 3,0 in 2002 and was reduced to 1,1 failure/km year in 2012.

Other investments that were started together with water monitoring system and renewing water ducts were bought of modern equipment allowing fast and accurate leakage detection. Unfortunately water monitoring system is not capable of finding exact location of the failure. In some cases areas are so big so finding a leakage takes a few days. This is a reason why such modern equipment for fast and exact detection needs to be bought.

REFERENCES

Niebel D. et al. 2011. VAG—Guidelines for water loss reduction. A fokus on pressure management. Eschborn.

Piechurski F.G. 2014. Przykład efektów wdrożenia systemu monitoringu i sterowania ciśnieniem w systemie dystrybucji wody. Praca zbiorowa pod redakcją Kuś K. Piechurski F. Nowe technologie w sieciach i instancjach wodociągach i kanalizacyjnych. Gliwice s.279–295. ISBN: 978-83-934758-3-4.

Zheng Y.W. et al. 2011. *Water loss reduction*. Bentley Institute Press Pennsylvania.

Underground Infrastructure of Urban Areas 3 – Madryas et al. (Eds)

Key factors of success in pipe jacking, explained by means of a project realized in the Atacama desert—the first jacking project in Chile after almost 20 years: Preparation is the key!

B. Prommegger
HOBAS América Latina, São Paulo, Brazil

ABSTRACT: Chile. In one of the world's main mining countries, it took almost 20 years to apply (micro) tunneling with a TBM again. For decades, all tunnels have been made using explosives instead of modern technology since people were afraid of failure. In August 2013, French pipe jacking experts from Soletanche Bachy/Bessac teamed up with the Austrian pipe supplier HOBAS and Buildtek from Chile and convinced the owners of Sierra Gorda—one of the biggest mining projects in the world—KGHM from Poland and Sumitomo from Japan—that pipe jacking is a quick, safe, and reliable alternative, even under challenging conditions. 203 m of pipe had to be installed beneath a thermal power plant with a potential for unknown interferences in the ground, in sandy soil and time was short. HOBAS CC-GRP Pipes DE 1099 mm were jacked with an AVN 800 and one intermediate jacking station.

1 INTRODUCTION

Jacking experience in Latin America is at a different level to what is in Europa, particularly in Poland. The leading country in Latin America is Colombia with many pipe jacking projects realized, mainly in the capital city Bogota, since several years. In Brazil, mainly Sao Paulo, jacking projects up to DN 1800 have been realized for aprox. 10 years. Argentina is starting now and in Chile jacking activities are about to re-start. Panama is another "tunneling" country. However, technical expertise and especially availability of jacking machines have not yet reached the level of Europe. Jacking with pressure pipes and curvature jacking are unknown because, in lack of experience, design offices are reluctant to specify. Various players in the sector are working on promoting pipe jacking and one result is the fact the International NoDig Conference in 2016 will take place in Medellin/Colombia.

Especially in countries like Chile engineering offices and clients are reluctant to specify pipe jacking in general simply because decades ago one poorly engineered project has failed. It has taken years to re-install the confidence in this widely used technology and the aim of this paper is to enhance this process and to discuss main factors of success in pipe jacking. This will be done with the example of the first pipe jacking project in Chile after 20 years.

2 CASE STUDY

HOBAS Jacking Pipes solve the challenge of water supply in one of the largest mines in Chile connecting one of the world's largest mines with their water source. Chile is one of the world's leading regions regarding the exploitation of natural resources. Copper mining

Figure 1. Pipe jacking beneath a thermal power plant in the Atacama dessert.

is particularly important. In the north of the country, in the middle of the Atacama Desert—one of the driest regions in the world and rich in minerals, especially copper, lies the little town Sierra Gorda. It is here where one of the biggest copper mining projects is located, operated by the Polish company KGHM and Sumitomo from Japan.

Once all construction works have been completed, primarily copper and molybdenum shall be extracted by flotation. This process asks for the supply of 1.5 m^3/sec of seawater. A 142-km-long pipeline shall first bring it to the thermic power plant owned by Suez Energy where it is used as cooling water. The water is subsequently pumped to the mine, 1626 meters away from the plant and 1626 m above sea level. A part of the line traverses the plant in 8 meters depth in sandy soil, with possible unknown interferences, and there was no time to lose.

The client began to analyze different options, and soon realized that only a trenchless installation method would be feasible. Open trench installation was not possible beneath the plant. Pressed for time, the client needed a quick and safe solution for this section. Trenchless jacking installation presented itself as the rescue at hand. The decision makers based their choice mainly on the following requirements:

- The pipe material should be resistant to sea water, have a long service life, and be suitable for jacking.
- The space at the construction site was limited, so that only one starting and reception were feasible.
- Microtunneling through sandy desert soil with an AVN 800 jacking machine.
- Longtime, extensive experience in pipe jacking was a must.
- Delays should be kept to a minimum.
- In order to avoid unexpected interference, it was decided to jack 8 meters depth.
- Jacking had to be precise since the beginning and final point of the line were already installed.

At the end of August 2013, HOBAS was awarded the contract together with their local partner Buildtek and the French contractor Soletanche Bachy/Bessac. HOBAS delivered a total of 203 m of jacking pipes De 1099, SN 100000, PN 1, with a wall thickness of 51 mm. The jacking pipes were designed for a maximum jacking load of 3,348 kN which never had to be fully used—perfectly suitable for the installation beneath the power plant.

Figure 2. Interferences, one potential danger in pipe jacking.

Apart from pipes, HOBAS also delivered 86°-bends, couplings, and reducers to connect the pipeline to the system. The construction works were completed within 3 weeks only and to the complete satisfaction of all parties involved. The client very much appreciated the professional project handling and the timely deliveries and was more than happy with the HOBAS Products and their fast and simple installation. Thanks to the perfect cooperation of Soletanche Bachy/Bessac, HOBAS, and Buildtek nothing stands in the way of starting up the mine very soon. So the first tunneling project in Chile was realized after nearly 20 years at the turn of 2014 in less than 3 weeks.

Figure 3. HOBAS Jacking Pipes in the starting chamber.

Figure 4. HOBAS bend in the receiving pit.

Year	2013/2014
Construction time	3 weeks
Length	203 m
Diameter	1099 mm external
Pressure	PN 1
Stiffness	SN 100000
Installation method	Sierra Gorda
Jacking Customer	
Construction company	Bessac/Soletanche Bachy
Advantages	Fast installation, low risk, less likely to have errors, less excavation, small work area.

3 FACTORS OF SUCCESS IN PIPE JACKING

Experience, good preparation and good teamwork of all parties involved will lead to a successful jacking project.

Coming back to the decisive factors of success in a—or every—jacking project:

1. Knowing the geological conditions well
2. Knowing interfering structures in the ground
3. Experienced contractor
4. Working with reliable material(s)
5. Working with optimal materials—not every material is suitable for all conditions
6. Adequately designed pipe for the individual project
7. Flexible pipe material
8. Adapting the pipes to the tunnel boring machine
9. Well-thought-out lubrication concept
10. Experienced operator for the tunneling machine
11. Optimized concept of intermediate jacking stations and pipes before and after the stations
12. Monitoring system if needed
13. Cooperation and communication of all parties involved

3.1 *Knowing the geological conditions well*

This is a must in any underground project, and especially in pipe jacking projects where you have sometimes pushes of up to 1 km between two shafts. According to geological conditions, experts design not only the Tunnel Boring Machine (for soft soil, mixed oil or rock), but also the pipes (stiffness needed according to abrasion resistance), the lubrication system and the

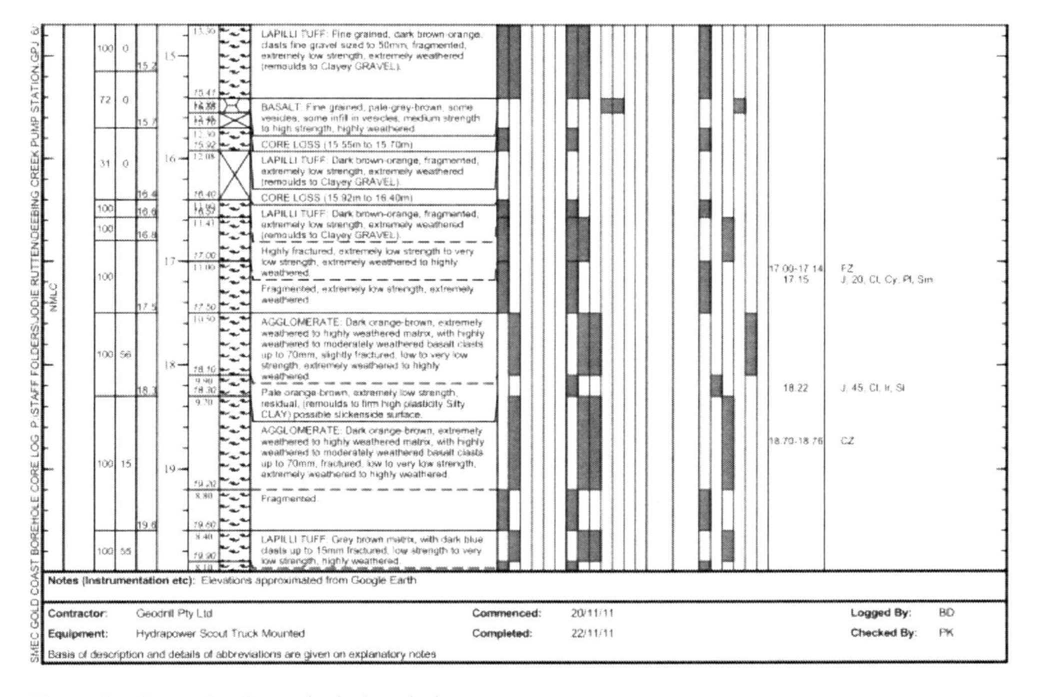

Figure 5. Example of a geological analysis.

need for Intermediate Jacking Stations. Very often—especially speaking about LatinAmerica—investors are reluctant in investigating in geological surveys before defining the engineering and the final budget of a project. However, this is indispensable for the success of a project. There are engineering offices specialized in this topic and the results, their interpretation and impact on the project design is crucial for its success. The results of these studies show the consistence/composition of the soil, the hardness of the rock if any, the friction to expect, water level to expect, and other characteristics.

3.2 *Knowing interfering structures in the ground*

Cables, pipes, housing structures, concrete walls, sheet piles, or other structures can be expected or unexpected interferences in the ground. Also in very experienced pipe jacking countries in Europe, unexpected interferences have caused severe problems in jacking projects; this can go until losing a TunnelBoringMachine because it gets stuck. One possible solution is—like it was the case in the project in Chile I presented earlier—to go deeper with the jacking pipe. In Sierra Gorda, none of the involved parties knew where the interferences were (cables and pipelines of the existing thermal power plant) and therefore the jacking depth was increased to 7 m—needless to say, no problem for the technology. In the city of Sao Paulo/Brazil, most project specifications place the risk of unknown interferences in hands of the contractor—with the result of a considerable cost increase in construction costs due to the risk premium, which makes many jacking projects unfeasible… .

3.3 *Experienced contractor*

An experienced contractor cannot only assess best all these risks mentioned here, but also has the most experienced personnel to handle the machine, to supervise the lubrication, and manage all the other factors.

Figure 6. Underground interferences.

3.4 *Working with the right equipment and material(s)*

These refers to mainly all equipment and material on the construction site, but to mention the most important:

a. Pipes
b. Tunnel Boring Machine
c. Inter Jacking Station(s)
d. Hydraulic Drives
e. Seperator

Materials and equipment should be reliable and apt for the pipe jacking method, but also ideal for the respective project. Not every material or equipment is suitable or ideal for all conditions and a thorough selection process should be done on a project by project base.

3.5 *Adequately designed pipe for the individual project—Flexible Pipe Material*

A flexible pipe material like GRP can much better accommodate the constant—and inevitable—movements in steering the MiniTBM during the pipe jacking project than a rigid pipe material that will produce a gap in the joint or—in extreme cases—a pipe failure. Due to this flexibility, GRP pipes do not need any wooden rings between the joints. This flexibility becomes especially important in curvatures jacking.

3.6 *Adapting the pipes to the tunnel boring machine*

3.7 *Well-thought-out lubrication concept*

Bentonite is used as a lubrication material and also to later one, when it has cured, to fill the small annular space between the bigger diameter of the bore head and the outer diameter of the pipe. The lubrication concept should be a teamwork between contractor, producer of the

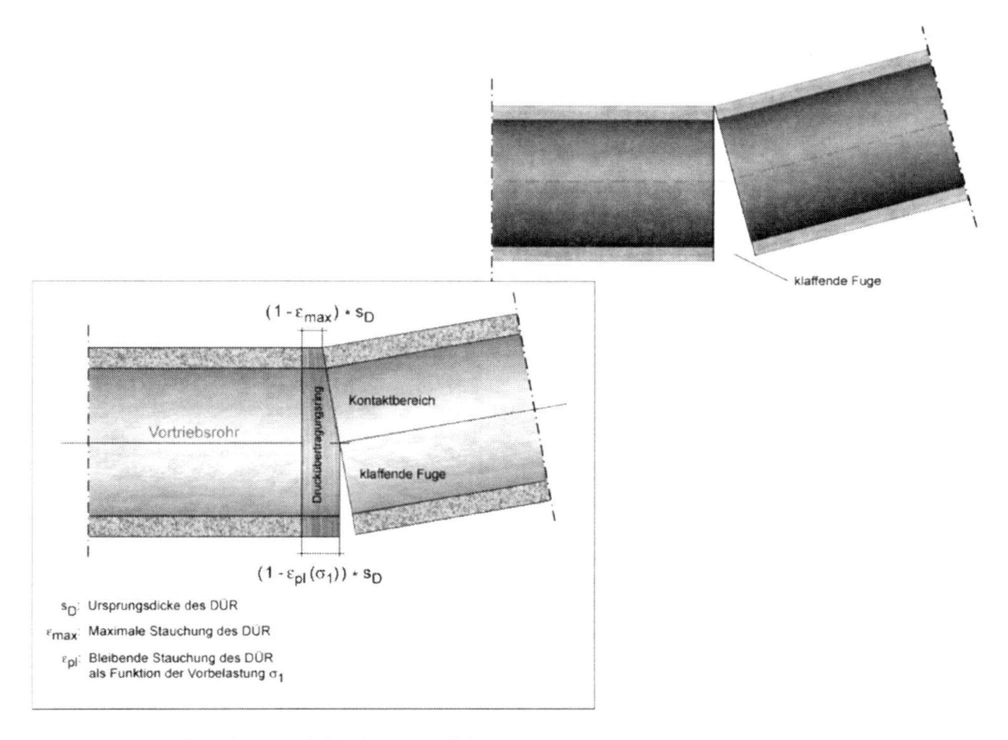

Figure 7. Flexible pipe vs. rigid pipe material.

tunnel boring machine and pipe supplier considering the geologic conditions and the friction of the outer pipe diameter. HOBAS Pipes, due to their smooth outer pipe wall, have a lower friction coefficient than other pipe materials. It is usually recommended to use injection nozzles every 2–3 pipes, depending on the soil type, the diameter and the pipe length. HOBAS supplies tampons to fully seal the injection nozzles after installation has been finished.

3.8 *Experienced operator for the tunneling machine*

An experienced operator of the MiniTBM will not only reduce the movements in the steering process to a minimum, but he will also anticipate changes in geology and breaking out of the machinery. Experience, feeling, hearing the noise of the excavated material and feeling how the machine behaves play an important role and thus a well experienced operator will be able to make up for deficiencies in at least some of the before and later mentioned factors.

Figure 8. Injection nozzle and sealing tampon.

Figure 9. Schematic view of a jacking site with one intermediate jacking station.

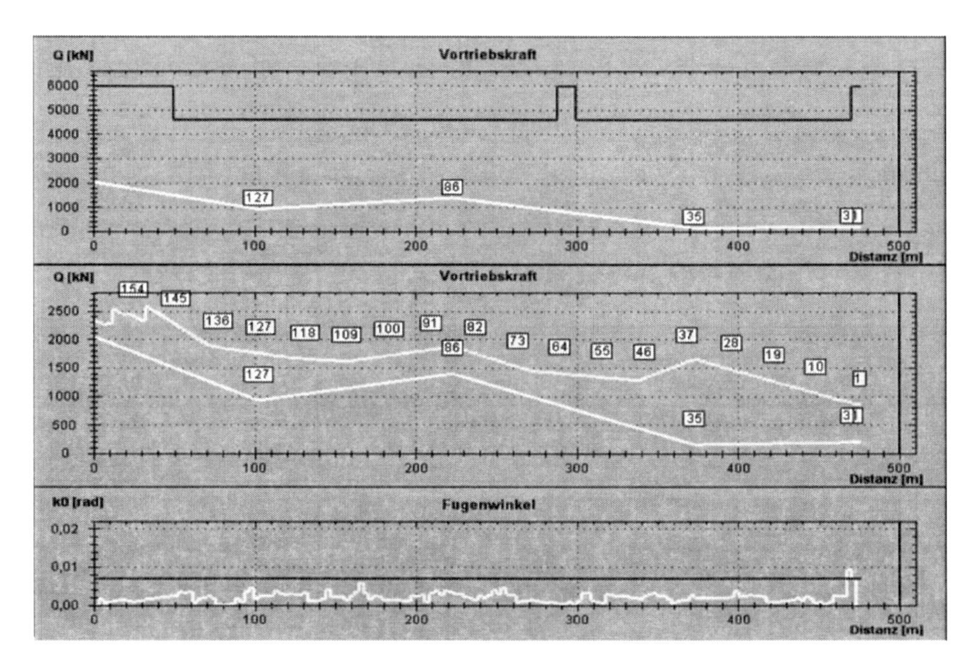

Figure 10. JackControl—Controlling and monitoring system.

3.9 *Optimized concept of intermediate jacking stations and pipes before and after the stations*

Intermediate Jacking Stations are used to distribute the force of the jacking machine. Depending on the geological conditions and the coefficient of friction of the outer pipe wall as well as the pushing length, contractors usually apply an InterJackStation every 100–150 m. They also give an additional security for the jacking process. Experience with HOBAS have shown that very often these InterJackStations do not have to be put into operation, thanks to the smooth outer surface of HOBAS Pipes and the reduced jacking force needed. This helps in reducing fuel consumption, but also to speed up the process since every InterJack in operation reduces the installation speed. After finishing the installation process, those InterJacks are very often replaced by pre-fabricated shafts using installation couplings, or they are partly left in the ground if accessibility is limited.

3.10 *Monitoring system if needed*

Monitoring systems monitor, control and record the movements of the machine and the jacking forces applied, compare them to the design and refer to lubrication and other auxiliary processes.

4 CONCLUSIONS

After all, when considering above mentioned factors, and with a good cooperation with all parties involved

- Design engineer
- Pipe supplier
- Jacking company
- Equipment supplier

every jacking project will be a success and I am convinced that this technology has not yet reached its limits.

Underground Infrastructure of Urban Areas 3 – Madryas et al. (Eds)
© 2015 Taylor & Francis Group, London, ISBN 978-1-138-02652-0

Trenchless technologies for the construction of small diameter water supply and sewer networks—selected issues

B. Przybyła
Technical University of Wroclaw, Wroclaw, Poland

ABSTRACT: The paper presents technologies for trenchless construction of water supply and sewer pipes, which enable the realizations for small diameters. Realizations of this type concern mainly to connections pipes and the end sections of the network, moreover pressure sewer systems serving rural areas or new real estate developments, often scattered over a large areas. These are situations frequently occuring in Poland, furthermore trenchless methods for small diameters pipes are characterized by specified constraints which are presented in this paper.

1 INTRODUCTION

A great number of trenchless technologies has been on the market for a long time, though, often being significantly different from one another. The fundamental assumption in order to classify them was to distinguish trenchless methods for underground pipes installations from trenchless methods for pipelines technical rehabilitation.

The division of trenchless methods applied in the construction of sewer lines is presented in BS EN 12889. This division—Figure 1—is universal and may be also referred to the construction of water supply pipelines. This standard describes a summary of assumptions of particular methods according to the adopted division and the rules for controlling the pipelines after their implementation.

The logical and frequently quoted is the division of pipeline construction technology proposed by *International Society for Trenchless Technology* ISTT which distinguishes following groups of methods:

- Impact Moling,
- Impact Ramming,
- Boring and Guided Boring,
- Directional Drilling (Horizontal Directional Drilling),
- Pipe Jacking,
- Microtunnelling.

This classification does not take into consideration the possibility of constructing pipelines using methods which are customarily reserved for commucation tunnels like SM and TBM discs, the methods which are also applied for the construction of the objects of network infrastructure. It particularly refers to deeper situated collectors with bigger diameters and special purpose tunnels, for example retention and cumulative ones. Moreover, in the course of time new technologies may appear which will have to be distinguished in the presented classifications. The *Direct Pipe* technology introduced in recent years by the company Herrenknecht AG which combines the elements of microtunnelling and directional drilling might be one of the examples.

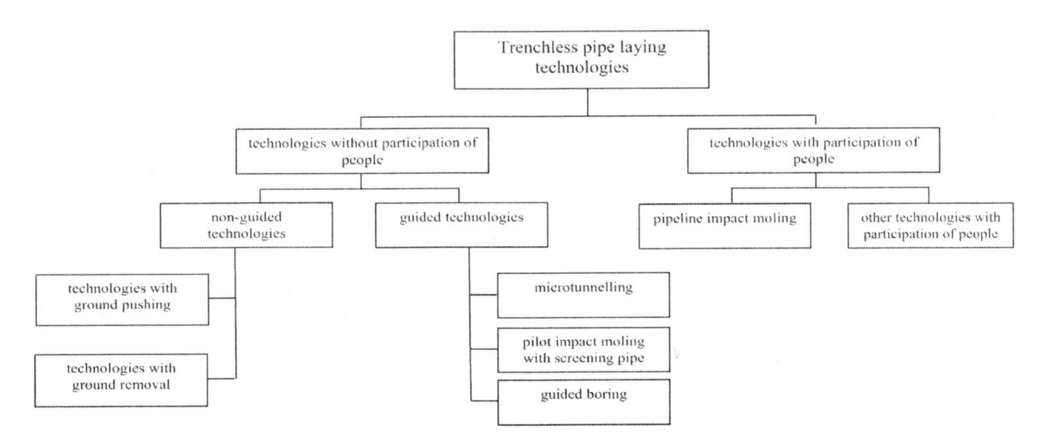

Figure 1. The division of trenchless technologies according to BS EN 12889:2003.

While analyzing the above mentioned classifications we may observe that according to BS EN 12889 the pipe cracking method and the pipe replacement method (a variant of microtunnelling) belong to the pipeline construction technology, while the same methods according to a different standard (EN ISO 11295:2010) are also classified as trenchless methods for pipelines technical rehabilitation.

Trenchless technologies are applied in the construction of pipelines both in the areas of intensive urban building development and in rural areas. In Poland the construction of new networks in rural areas is normally carried out in a traditional way—using a trench system, at the same time applying the trenchless technologies is profitable or even necessary because of frequent crossing the natural and artificial obstacles (rivers, roads, protected areas). A surface development of cities and towns brings results in the natural and urgent need for constructing new pipelines. Laying new pipelines is mostly planned together with building new streets, in trenches, but crossing existing communication arteries brings about the practical necessity of applying trenchless technologies.

The expansion or reconstruction of sewage and water supply systems in the city centers e.g. connected with the intensification of the construction constitutes the next group of trenchless enterprises.

Trenchless technologies are prevalent in this case and their application is indispensable while implementing the subsequent, deeper levels of urban media supply systems.

The construction of small diameter water supply and sewage systems in Polish conditions is a popular issue and first of all refers to constructions in rural areas, it is also connected with surface development of the cities and towns. In case of rural built-up areas or already existing small settlement units the fundamental issue is a construction of a new or reconstruction of the existing sewage system network (often as the pressure system which demands small diameters). Providing a new sewage and water supply the infrastructure is updated because of intensive occupation of new areas (mostly those which have been hitherto used in agriculture) for scattered housing residential units. This mostly negative phenomenon is referred to as "urban splashing".

In the above mentioned implementations the pipes of dimensions not bigger than 500 for sewage systems and 200 (300) mm for water supply systems are prevalent.

While analyzing the issues of constructing new pipelines of small diameters, it is necessary to mention so called narrow-trench methods of implementation which are an alternative for traditional implementations carried out in non-urbanized areas or areas which were earlier used in agriculture. The cable plough and milling technologies belong to narrow trench methods. They constitute a separate group, independent from trenchless or traditional methods.

Considering trenchless methods which are suitable for water supply and sewage systems construction it is necessary to take into account the requirements for these pipelines, first of all the precision of their implementation and materials used. Such requirements are established by investors to meet technical requirements of COBRTI INSTAL (Płuciennik & Wilbik 2001, Płuciennik & Wilbik 2003) and BS EN 805. Some of the above mentioned technologies give the ability to meet these requirements, some do not.

According to (Płuciennik & Wilbik 2001) for the water supply pipelines, following accuracies are assumed: aberration in the plan—max. 0.10 m for plastics and 0.02 m for other materials, decline aberration—+/–0.05 m for plastics and +/–0.02 m for other materials. Requirements for sewage systems are determined by the necessity of maintaining the declines superimposed in the design in accordance with the ordinate of inspection chambers. According to (Płuciennik & Wilbik 2003), the permissible aberration of the ordinates of laid pipeline should not exceed +/–1 cm whereas aberration in the plan of the pipeline axis should not exceed +/–2 cm. While using trenchless technologies, these requirements are difficult to meet.

2 TRENCHLESS TECHNOLOGIES FOR CONSTRUCTING PIPELINES OF SMALL DIAMETERS

It is possible to claim that besides pipe jackings, other groups of technologies mentioned by ISST enable the pipelines to be installed, with the diameters from DN 100 to DN 500. Implementation of the smallest diameters exclusively (basically to DN 200) pneumatic impact molling technology is provided. Other technologies also enable bigger diameters to be constructed, they are in a way designed for it—e.g. microtunnelling.

Main assumptions of selected technologies which enable the implementation of water supply and sewage systems, focused on small diameters are mentioned below. In there an elaboration (Madryas, Kolonko & Przybyla 2010) has been partially used.

The principle of microtunnelling is to drill a horizontal hole (or a hole with a small decline) between two previously made chambers (a staring one and a finishing one). The ground between both chambers is mined with a special drilling head and transported to the starting chamber. The excavated material is transported by using a washing system, a helical conveyor or pneumatically. From the starting chamber some pipes are driven into the ground space, they are pushed there together with the mining head using the station of hydraulic servomotors (Fig. 2). In order to diminish the friction between the external surface of moved pipes and the ground, a lubricant is interposed by suitable nozzles located on the perimeter of the pipes. After getting to the finishing chamber, the mining head is raised onto the surface of the ground. The finishing chamber may become a starting chamber for the next stage, drilling from the starting chamber in two opposite directions is also possible. In this technology special pipes with thickened walls are used, they may carry big normal stresses which arise during pushing the pipes.

Microtunnelling technology may be applied in different ground conditions, from hard rocks to hydrated and unstable grounds. Also the straight sections or curved sections of the diameters in the range of several hundred meters may be implemented. Maximum lengths of implemented sections are more than 500 m.

According to the information given by the equipment manufactures, (e.g. Herrenknecht AG), it is possible to make pipelines with diameters from DN 250 to about DN 4000. Also smaller values (from 100 mm) are given, however the author has not found any description of a microtunnelling device for such implementation. Since implementation of big diameters with the usage of this technology does not bring any doubts, implementing the smallest diameters is rarely met but not excluded. In the most common situations when the application of this technology is justified—straight implementation of short sections in easily-mined grounds, overcoming obstacles and communication embankments-microtunnelling finds its natural competitors in horizontal drilling, especially in a guided version.

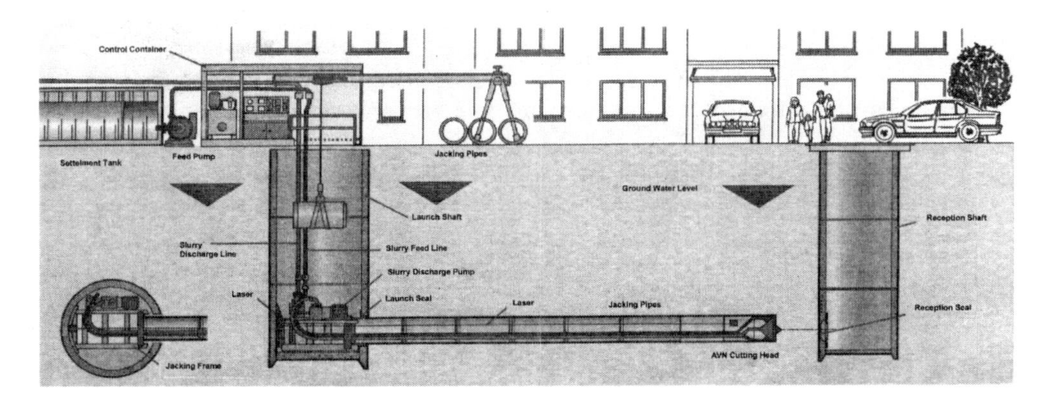

Figure 2. Implementation of the pipeline with the microtunnelling method (based on Herrenknecht AG—handouts).

For implementation of small diameters with the usage of this technology the possibility of transportation of excavated material using merely the helical conveyor is the basic factor. It gives the possiblity of implementing the pipeline by sections of the length of several dozen metres (up to about 80 or 120 m, depending on the source of information).

Horizontal guided boring is used first of all for laying pipelines under such obstructions like roads, railroads and embankments, mostly in a situation when the remaining part of the works is implemented in open trenches. It is a simple method allowing the installation of pipes with diameters from DN 100 to DN 2000 in the sections of length of mostly up to several dozen metres. In this method mechanical boring devices are used and the boring is made from the starting to the finishing trench together with simultaneous hydraulic introduction of steel screening pipes. The excavated material is then moved to the starting trench using the helicon conveyer. Next, from the side of the starting trench target pipes are introduced and screening pipes are removed gradually in the finishing trench.

Horizontal boring is distinguished in variants with or without the possibility of guiding the boring trajectory. Applying horizontal guided boring devices (in Polish: WPS) is particularly useful while implementing sewage system pipelines which must be implemented with ensuring the designed decline. The application of a special guiding head enables the boring devices to monitor permanently the boring trajectory and its continual corrections.

Guided borings are divided into 2-stage ones and 3-stage ones. In Fig. 3 a scheme of a 3-stage guided boring is presented, as well as an example of a boring device manufactured in Poland. Figure 4 presents a 2-stage version in which behind the guiding boring device a continual pipeline is pulled from the finishing trench; this pipeline is produced of plastic, it has a diameter which suits the rod diameter. It is also possible to build in short pipeline modules in this variant of the method and this is why it is suitable for the implementation of the smallest diameters which are considered in here.

Additionally, it is necessary to mention a variant called "blind boring" in which finishing trenches do not exist and which is used for constructing the connections with a direct joining the new pipeline with the existing one, with a bigger diameter.

Directional drilling and horizontal directional drilling—technology which is applied for laying the cables, pressure pipelines and (less often) gravitational ones under the obstacles of a natural or artificial character (e.g. rivers, hills, roads, railroads). In general, the method is based on making a guide hole under the obstacle, then enlarging it to a required diameter and pulling a designed pipe or cable into the hole prepared as above. The scheme of procedure is presented in Fig. 5. To fulfill the task, it is necessary to have an access and suitable working space on both sides of the crossed obstacle. In the works, proper drilling devices are used, they are characterized by basic parameters like: drawing load and torque—the values which decide what the maximal length of the drilling and planned diameter of the pipeline

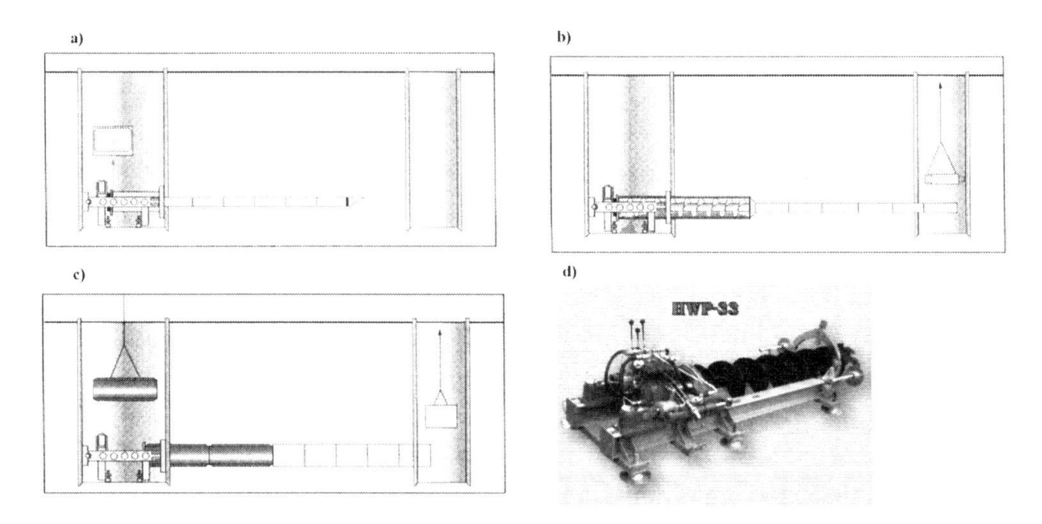

Figure 3. Implementation of the pipeline with the method of guided, 3-stage boring (based on Bohrtec handout): a) stage 1—guided boring; b) stage 2—boring together with simultaneous introduction of steel screening pipes and removal of the guide rod; c) stage 3—introducing the pipes together with simultaneous removal of screening pipes; d) an example of a boring device of the HWP-series (Wamet).

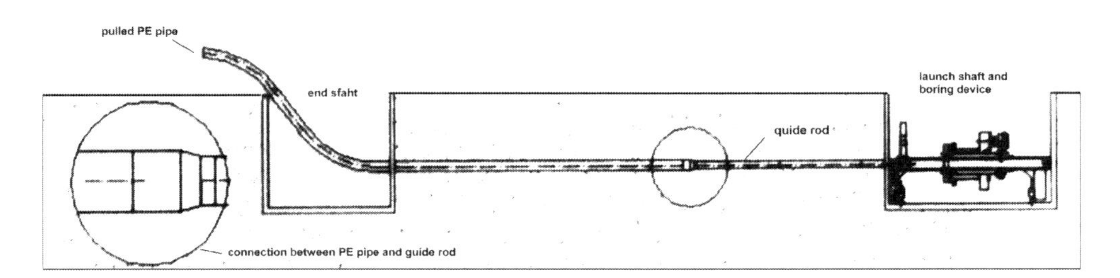

Figure 4. The 2-stage variant of the horizontal directional drilling: the scheme of pulling-in the PE pipe after finishing the first stage—guided boring (based on Perforator handouts).

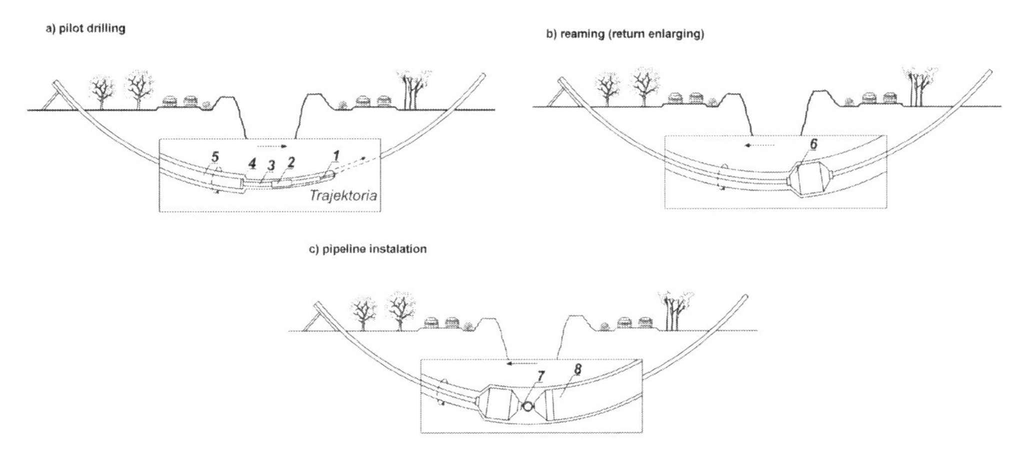

Figure 5. The scheme of activities in the technology of directional drilling (Madryas et al. 2006): 1—drilling head, 2—measuring probe, 3—pilot pipeline, 4—enlargement bit, 5—drilling line, 6—enlargement device, 7—swivel, 8—technological (e.g. sewage) pipe.

should be. The drilling device is situated on one side of the obstacle, on the other side the target pipeline is prepared to be pulled in. There are also other variants in which two drilling devices are used, they work on opposite sides of the obstacle.

The first stage of drilling—the guide drilling sets out the course (axis) of the pipeline; during this stage there is a possibility to guide the boring head and a proper correction of its course. Enlarging the hole during the second stage may be single or multiple. In both stages drilling fluid plays an important role (usually a bentonite wash). The liquid helps the drilling process and stabilizes the enlarged hole securing it from falling in while the target pipe is introduced. The target pipe, which may be a screening or technological pipe, is introduced in the third stage.

This method is used for laying the cables and pipelines with the diameters from DN 100 to DN 1500, with the length exceeding even 2000 m. With reference to the analyzed problem, in Poland this method is frequently used—particularly in the situation when water supply pipelines are laid under communication lines and there are no embankments or they are not high enough to apply alternative methods (e.g. for horizontal drilling) without using starting and finishing chambers.

Pneumatic impact moling method enables the installation of pipelines with diameters from 25 to 200 mm, with the length up to about 60 m. According to BS EN 12889:2003 this method is referred to as impact drilling and is classified as a method without ground removal. Polish specialists of Gdansk University of Technology should be univocally considered founders of this method (or maybe of first devices of this type). In this method the ground is pushed and thickened by a pneumatic punch (called "the impact mole") moving in it which is driven by compressed air. While moving in the ground, the punch pulls in cables, plastic pipes or steel pipes with diameters suitable to its diameter. It is possible to install pipes, of the same length as the length of the implemented section (in the case of plastic pipes or cables—also unwound from a cable winch), moreover it is possible to install short plastic pipe modules behind the head.

Since the dynamic head thickens the ground while moving in it, in the grounds of naturally high level of density—e.g. in equigranular gravel there may appear difficulties in covering the designed route. The loose ground may start to subside as a result of vibrations and it can cause an uncontrolled movement of the head in the ground and falling in the surface of the ground.

In general the moles are non-guided devices which move in the straigth line therefore correct setting and directing the device at the target in the starting trench is of a great importance.

In spite of this, the heterogeneity in the ground may cause deviations from the designed route. There is a possibility to monitor the exact position of the punch from the surface of the ground and in case of unpredicted behaviour the device may be withdrawn to the initial position (modern devices have also an ability to move backwards). The scheme of the method of impact moling is presented in Fig. 6.

There is also a device whose route is possible to be steered (it can move along the curve) (Tracto Technik GmbH). However, this solution does not give any possibility to install straight sections.

This method is suitable to construct the crossings under communication obstacles or connections. In case of construction for the need of gravitational sewage systems the basic problem, which may disqualify the method, is a small precision of the crossings. However, for the implementation of water supply pipelines—especially for final sections of the network and for connections, this method is exceptionally useful because of its simplicity and economical effectiveness of the trenchless implementation.

In case of pneumatic impact ramming, steel pipes are driven into the ground by means of pneumatic hammers (rams), also punches (moles) may be used in pneumatic moling. The pneumatic hammer is placed in the starting trench on a special bed or (with grater diameters) only attached to the impacted pipe. The pipe is gradually driven into the ground as a result of hammering(ramming). The whole pipe is placed in the guide rails (the length of the pipe is the same as the construction length) or its sections are welded together. The scheme of the method is presented in Fig. 7.

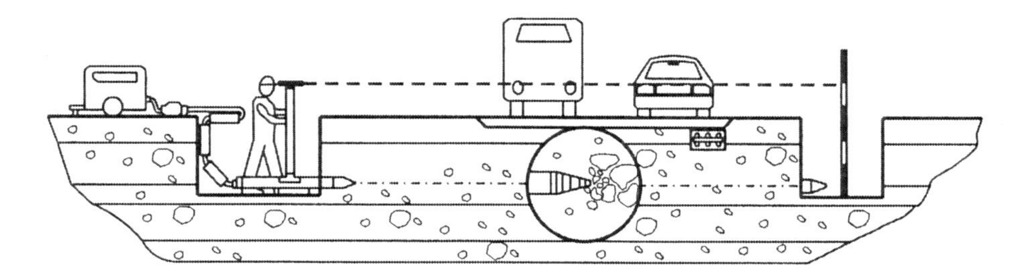

Figure 6. The scheme of the impact moling (based on Tracto Technik GmbH handouts).

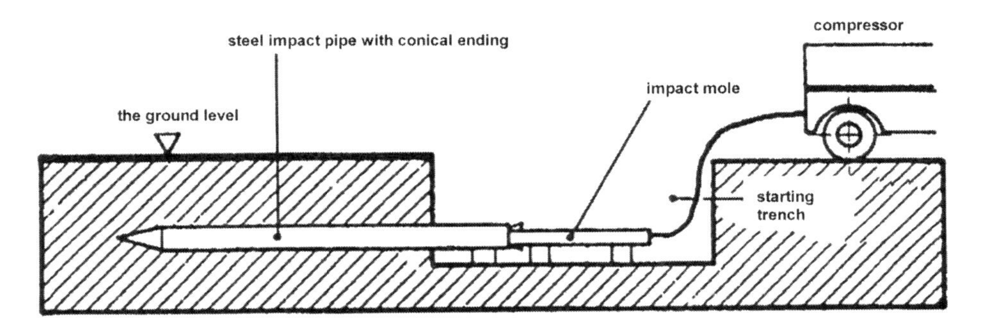

Figure 7. Impact ramming of closed-face pipes (Zwierzchowska 2006).

It is possible to install the sections of the length of 20–50 m at one time depending on the diameter of the pipe and ground conditions. In suitable conditions even the distances of 100 m are reached. The range of the diameters of installed pipes is from 110 to 2000 mm and possibly more.

Two variants of this method are used: open-face pipe ramming and closed-face pipe ramming. The second variant refers to the pipes with the smallest diameters—from 110 to about 300 mm but—because of the danger of neighbouring installations—in general the diameters of 200 mm are not exceeded. A cone shape head is used to close the pipe or ragged and bent inside endings of a pipe are welded.

In the variant of open-face pipe ramming, after reaching the final trench the ground is removed from inside of the pipe. It is done with the usage of compressed air, water under pressure, helical conveyor or even by hands (with greater diameters).

To avoid the danger of destroying the road surface, the minimum depth of pipe installation is assumed. More problematic is the variant with the closed-face pipe and for diameters from 130 to 200 mm this depth should not be smaller than 1.3 m and for the diameters of 200 to 250 mm not smaller than 1.8 m (Zwierzchowska 2006).

The method of pneumatic impact ramming of steel pipes is characterized by simplicity and short time of installation. Assessed installation of the pipe of length of 15–20 m lasts in suitable conditions half an hour only (Madryas et al. 2006). Comparative times of installation are reached in the method of impact drilling. In Polish conditions the method is not used too often because the most of installations of small diameters are done by means of impact moling. However, in the range of small diameters justifiable and economically effective may be the application of this method for installing steel screening pipes—where it is necessary.

The cable plough method belongs to narrow-trench methods, and it is a way of installing cables and pipes with diameters to DN 350 for plastics (PE) and DN 300 for steel pipes. This method was also used in case of laying water supply pipelines of spheroidal cast iron (Beer, Sans & Ertelt 2006). In this method a cable plough is used, its share cuts away the ground making a narrow

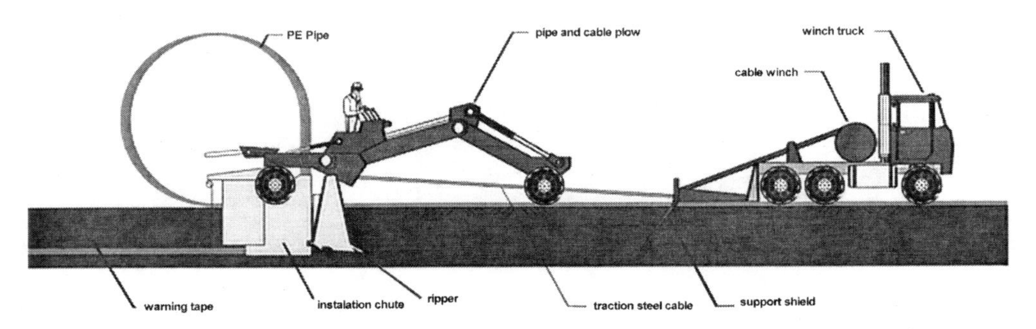

Figure 8. Laying the cables and plastic pipelines up to DN 200 with the cable plough method (based on Foeckersperger GmbH handouts).

space in which a previously prepared pipeline is laid. The plough is pulled on a rope by a winch mounted on a wheel or caterpillar tractor. The plough places the pipeline simultaneously in the post-share space and the ground is poured intristically covering it. In the ground over the pipeline a warning tape is developed automatically, it is also possible to fill the space in the ground with sand from a feeder or bentonite as a lubricating medium in the situation when the variant of the method is used while a laid pipeline is pushed through the ground.

The surface of the ground after laying the pipeline may be levelled by a bulldozer or left without the intervention which in case of laying the pipelines across agricultural grounds is necessary. This technique enables even 5 km of the pipelines to be laid a day, on the depth referring to the length of the share. In the situation of carrying out the works in stone grounds or when the pipeline is laid in a road strip, a similar method of milling is applied. In this case the trench is made behind the tractor with the usage of a mill which cuts the surface of the road and the ground below. Thereafter the surface may be reproduced immediately (provided that the leaktightness of the pipeline is previously controlled).

The scheme of the cable plough method is presented in Fig. 8. In Polish trade press more information about it may be found in (Roszkowski 2005, Janiak & Roszkowski 2006); this method is also described by a German guideline ATV-DVWK-M 160.

3 CAUTIONS REFERRING TO PIPELINE CONSTRUCTION

While analyzing the materials which may be used for construction water supply or sewage pipelines with trenchless methods, it is necessary to pay attention to limitations made by the technologies themselves, particularly in connection with requirements placed for the installation of pipelines. Because of the considered range of diameters, the choice of products available on the market may be crucial. In this paper water supply pipelines with internal diameters from 100 to 200 mm and sewage systems with diameters from 200 to 500 mm are analysed.

In the guided boring technology and in microtunnelling, steel pipes and impact pipes—stoneware, reinforced concrete, basalt, polymer and GRP pipes are used (also of other materials, rarely applied). The smallest available diameters of impact moled pipes are DN 100 (basalt), DN 150 (stoneware, polymer-concrete), GRP, DN 300 (reinforced concrete).

In the method of two-stage guided boring, the method of pneumatic impact molding, in the method of guided boring and the cable plough method drawn-in pipes are applied. They are first of all suitable types of PE pipes, steel pipes and pipes of spheroidal cast iron. PE and steel pipes are available also in diameters smaller than DN 100, there are also available pipes of spheroidal cast iron with the system of connections transferring the axis forces of DN 100.

While constructing the water supply systems using the impact pipes there is a problem with leaktightness connected with a big number of inadequate connections between particular pipes. Therefore constructions made of those pipes are in general screening pipes for driving a technological pipe inside. It is mostly justified by the investors' demands regarding laying the water supply pipelines under obstacles. On the other hand, they can constitute an independent sewage pipeline in case when the regulations do not demand a screening pipe.

The basic problem connected with applying the technology which assumes drawing-in pipes is the danger of destroying the external insulating layer for the steel and cast iron pipes whereas for plastic pipes there is a danger of scratching during drawing into the ground medium and occurence of point stresses which may lead to the phenomenon of slow increase of the cracks or to the local plasticizing (flowing) of the material. Therefore it is recommended to apply multilayer pipes, at least of PE 100 together with a protecting layer, resistant to point stresses as well as pipes of PE 100 RX in different configurations.

In sewage system pipelines also short modules of PVC or (less often) PP may be applied as drawn-in which is used in the method of pneumatic moling or of two-stage guided boring.

4 FINAL REMARKS CONCERNING APPLICATIONS OF TRENCHLESS TECHNOLOGIES AND THEIR ACCURACY

For particular, analyzed here trenchless technologies the following characteristics should be mentioned:

- the possibility of installation of sections of particular length and range of diameters;
- material solutions applied;
- limitations connected with ground and water conditions;
- minimal heights of covering the pipelines with the ground over the top of the pipeline;
- accuracies of building-in the pipelines.

Above mentioned characteristics do not constitute a full list of limitations. Some of them mentioned above have been given with a summary description of the methods in Chapter 2.

Having compared average maximum lengths it may be assumed that possible to be installed in one stage of works are:

- for pneumatic impact moling—up to 35 m (for non-guided impact moling);
- pneumatic ramming of steel pipes—up to 35 m for a closed-face pipe and up to 120 m for open-face pipes;
- non-guided boring with the transportation of excavated material with a helical conveyor—up to 60 m;
- guided boring (with the pilot boring) and the transportation of excavated material as above—up to 80 m;
- directional drilling—up to 2000 m;
- microtunnelling—up to 120 m with the transportation of excavated material with a helical conveyor (200 m with the pneumatic transportation of excavated material, 500 m with the washing transportation of excavated material—however, it refers to bigger diameters of the pipelines than analyzed here).

However, it is necessary to mention that given distances may be impossible to obtain in case of bad ground conditions.

Considering possibilities of applying analyzed technologies in case of high level of ground water, it is possible to state that basically the only usage of impact moling and open-front pipes is here rather excluded. In other cases—if it is only possible to lower the level of ground water locally in the area of the starting and finishing trench, the works may be carried out after taking a suitable configuration of applied technology.

In the range of accuracy of constructed underground network aroused differences in the ordinates in the profile are analyzed as well as decline from the axis in a plan (projection) differing from the design documentation.

The greatest inaccuracies occur in technologies in which there is no possiblity to monitor continually the situation of the tool (head) which moves in the ground. If such a possibility exists, the next thing to take into consideration will be a chance to correct the route of the tool currently. The greatest accuracies are obtained in technologies in which laser or gyroscopic methods of steering and controlling are used, simultaneously with a high level of automation of the process of pipeline construction. In Table 1 there is a summary of maximum values of decline deviation and deviation in the plan of axis of a constructed pipeline from the predicted design assumptions.

The laser system of control is applied in microtunnelling, the teleoptical system in horizontal drilling with steering, radiolocation systems in guided boring as well as with pneumatic impact moling (not always).

In general, the stricter demands concerning accuracy of implementation are placed before pipelines carrying the medium gravitationally—i.e. sewage system. Because of the lack of the government guidelines concerning accuracy of implementation of pipelines with trenchless methods, it is possible to apply the German instruction ATV A 125. Permissible deviations according to this instruction are presented in Table 2. It is necessary to pay attention to the fact that these requirements are smaller than given to sewage system pipelines and (in most of cases) water supply systems (included in chapter 1) in instructions [3, 4].

The issues of required accuracy of the works should be stated precisely by the investors in tender specifications.

It is necessary to notice that in the work (Zwierzchowska 2006) an interesting suggestion of the method of choice of trenchless technology of constructing the pipeline was presented and above mentioned characteristics were taken into consideration.

The author does not possess any detailed information concerning the accuracies obtained while laying the pipeline by means of narrow-trench technologies. Detailed information on this subject has not been presented in professional literature in Poland in the course of last years and practical experiences emerge from implementations described in (Roszkowski 2005, Janiak & Roszkowski 2006). Despite potential possibilities that the technologies may be useful in implementation of sewage systems under pressure and of water supply systems in non-urban areas, in Polish conditions they were applied only occasionally. With such a destination this method was used much broader both in Europe and in North America.

Table 1. Maximum values of the decline deviation and the deviation in the plan of axis of a constructed pipeline from the predicted design assumptions (according to Zwierzchowska 2006).

Kind of steering and control system	Decline deviation	Deviation in the plan of the axis
Laser	+/−20 mm	+/−25 mm
Teleoptical	+/−20 mm	+/−25 mm
Radiolocation	Not given	Not given
Without steering	1–2% of the length of constructed section	1–2% of the length of constructed section

Table 2. Admissible accuracies required for gravitational pipelines (according to Zwierzchowska 2006, based on ATV A 125).

Nominal diameter, DN, mm	Admissible decline deviation of the pipeline, mm	Admissible deviation of the pipeline in the axis, mm
DN < 600	+/−20	+/−25
600 ≤ DN ≤ 1000	+/−25	+/−40
1000 < DN ≤ 1400	+/−30	+/−100
DN ≥ 1400	+/−50	+/−200

5 SUMMING UP

The usage of trenchless technologies for implementation of water supply systems or small diameters sewage systems must be anticipated by a detailed analysis of their profitability in relation to the implementation by means of traditional methods. Unfortunately, in local Polish conditions we have often to deal with the negligence of other costs than immediate costs of implementation and the situations, not exceptional ones, in which newly constructed local roads are destroyed.

The application of trenchless technologies mostly is determined by the necessity of crossing road and railroad obstacles, less often by the water flows. Those situations generally demand a necessity of using the screening pipes which may be introduced as steel pipes with the method of pneumatic ramming of the steel pipes or—less often—using the mole ploughs on short sections. The latter may be also an alternative for using trenches in case of short passages for water supply pipelines—especially while implementing the house connections.

Horizontal non-guided boring may be efficient while building-in the screening pipes, first of all the steel ones, also for the passages of water supply networks under communication obstacles.

For constructing the sewage system pipelines the methods with guiding are the most suitable, no matter whether the screening pipe is necessary or not. So, it is necessary to mention the guided horizontal boring and microtunnelling, however for small diameters and short sections using them is doubtful because of the costs.

Constructing the sewage system using the methods without guiding may apply to pressure pipelines, in addition of indication to short sections.

The method of guided boring may be treated as a base for implementation of sewage air-trap passages under the obstacles. It is also applied for constructing the water supply pipelines. It gives a possibility to overcome particularly long sections in practically all ground conditions and with particularly difficult obstacles, e.g. in mountain conditions. Numerous examples of spectacular implementations are present in professional magazines.

REFERENCES

ATV A 125: 2008. Rohrvortrieb und verwandte Verfahren. Deutsche Vereinigung für Wasserwirtschaft, Abwasser und Abfall e.V., Hannef.

Beer, F., Sans, C. & Ertelt, S. 2006. Układanie rur z żeliwa sferoidalnego przy użyciu pługa rakietowego w Adelsheim i Öhringen. Wyjątki stają się regułą. *Nowoczesne Budownictwo Inżynieryjne* (5).

Janiak Ł. & Roszkowski A. 2006. Płużenie—realizacja projektu w Osiecznej koło Leszna. *Inżynieria Bezwykopowa* (6).

Madryas C., Kolonko A. & Przybyła B. 2010. Efektywność techniczna, ekonomiczna i ekologiczna technologii bezwykopowych. *Gaz, Woda i Technika Sanitarna* (9).

Madryas C., Kolonko A., Szot A. & Wysocki L. 2006. *Mikrotunelowanie*. Wrocław: Dolnośląskie Wydawnictwo Edukacyjne.

Płuciennik S. & Wilbik J. 2002. Warunki techniczne wykonania i odbioru sieci wodociągowych. Wymagania Techniczne COBRTI INSTAL. Warszawa: COBRTI INSTAL.

Płuciennik S. & Wilbik J. 2003. Warunki techniczne wykonania i odbioru sieci kanalizacyjnych, Wymagania Techniczne COBRTI INSTAL, Warszawa: COBRTI INSTAL.

PN EN 12889:2003. Bezwykopowa budowa i badanie przewodów kanalizacyjnych. Warszawa: PKN.

PN EN 805:2002. Zaopatrzenie w wodę—Wymagania dotyczące systemów zewnętrznych i ich części składowych. Warszawa: PKN.

PN EN ISO 11295:2010. Zalecenia dotyczące klasyfikacji i projektowania systemów przewodów rurowych z tworzyw sztucznych stosowanych do renowacji. Warszawa PKN.

Roszkowski A. 2005. Układanie nowych rurociągów i kabli w gruncie metodą płużenia. *Inżynieria Bezwykopowa*, (5).

Zwierzchowska A. 2006. Technologie bezwykopowej budowy sieci gazowych, wodociągowych i kanalizacyjnych. Kielce: Wydawnictwo Politechniki Świętokrzyskiej.

Underground Infrastructure of Urban Areas 3 – Madryas et al. (Eds)
 ISBN 978-1-138-02652-0

Realistic classification of stoneware sewerage network technical condition

A. Raganowicz
Zweckverband zur Abwasserbeseitigung im Hachinger Tal, Taufkirchen, Germany

J. Dziopak & D. Słyś
Department of Infrastructure and Sustainable Development, Rzeszów University of Technology, Rzeszów, Poland

ABSTRACT: The paper presents a proposal concerning realistic classification of results obtained from full-range optical inspection of the stoneware domestic-household sewerage network transporting domestic-household sewage from the catchment area of the highly urbanised Bavarian municipality Unterhaching situated east of the city of Munich. The realistic classification meeting requirements related to technological and organisational specificity of renovation work is a necessary complement to the priority classification that allows to perform statistical studies on empirical material. In turn, such studies constitute a preliminary phase of work aimed at network recognition for the purpose of developing a technical and operating condition forecast for the sewerage infrastructure in question. Analysis of results of statistical studies performed in the framework of examination of the currently operated network has proved that the prognosis concerning ageing of a sewerage network can be successfully based on the two-parameter Weibull probability distribution. The model is also frequently used in studies concerning reliability of operation of building structures, mechanical devices, electronic circuits and other systems.

1 INTRODUCTION

The priority classification of results obtained from a full-range optical inspection has been carried out with respect the currently operated stoneware sewerage network channelling domestic-household sewage from the catchment area of the Bavarian municipality of Unterhaching that was carried out in accordance with German guidelines (Merkblatt ATV-M 149 1999). In the year 2000, when a television survey was performed, the network at issue was composed of 1,162 segments with nominal diameter in the range DN 200–400 mm and total length of 39,970 m. Analysing results of the developed classification it has been found that the German guidelines did not provide any definition that would allow to determine unambiguously a limit after which the examined segments should be subjected to renovation work. In fact, it is based on the assumption that the mots serious damage is decisive for the priority class of the analysed sewerage network segment.

The situation in which the examined sewer has only one damage of the worst class (KU1) occurs very frequently. Such a case results in assigning to such sever automatically the worse, i.e. the first priority class (KP1) which, as a consequence, means the necessity to carry out its renovation. In many instances such isolated damage can be effectively removed by performing a low-cost maintenance operation. If, however, the priority classification will be adopted as the sole base for planning renovation investment, its results should be always subjected to a meticulous analysis aimed at possible adjustment of the final assessment. In such situation there is a need to carry out another classification that to a larger extent would refer to technological and operational specificity of renovation works. Such new strategy should allow to analyse the damage occurring in sewers in their longitudinal cross-section, and not in the transverse section as in the case of the priority classification. Only reliable and correct (in the sense of adequate professionalism) classification providing the actual picture of structural condition of the sewers, can constitute a base for further studies aimed at development of a statistical forecast concerning technical and operating condition of the examined network. The concept of the realistic classification put forward by the present authors is based on long-term experience in the area of planning, executing, and accounting for the actually performed renovation operations.

2 REALISTIC CLASSIFICATION OF RESULTS OF A FULL-RANGE OPTICAL INSPECTION OF THE EXAMINED SEWERAGE NETWORK

Renovation is the largest component of costs related to operation of any sewerage network. The cost of embodying one metre of a self-supporting repair sleeve into a sewer with diameter DN 250 was about 350 euro in 2009/2010 prices. Therefore the outlays for renovation of a dozen or so network segments may reach the value of several hundred thousand euros. Such amount will constitute a significant item in the budget of any network operator. The sewerage network operating costs can be divided generally into two groups—the fixed costs known also as the running expenses, and the investment costs. The first group includes mainly the expenditure incurred on hydrodynamic cleaning of the sewers, optical inspection, and maintenance operations.

On the other hand, the second group encompasses costs related to network extension and renovation. The above breakdown of operating costs in calculation for tariffs for domestic-household sewage disposal. From the studies and analyses carried out in Germany (Pecher 1994, Pecher *et al.*1995) it follows that the fixed cost component in fees is 40% whereas the remaining 60% is earmarked for investment undertakings. Determination of such fixed network operation costs that could be applied generally is a difficult task as sewerage systems may differ significantly in length, age, type, and longitudinal slope. One of the related studies (Pecher 1994) presents an analysis based on calculation of domestic-household sewage collection fees carried out by 352 German towns operating sewerage networks with the total length of 92,600 km. As a result of the analysis it has been determined that the annual per capita fixed sewerage network operation costs varied very significantly and ranged from 15 euro to 85 euro. 40% of these figures represented the cost of maintaining the personnel. As related to the unit length of the analysed networks, the costs turned out to be 8.6 euros per metre. The Ministry of the Environment of the German federal state of Brandenburg, Germany, has determined the approximate fixed sewerage network operating costs that confirmed results of previous studies and analyses (Ministerium für Umwelt, Naturschutz und Raumordnung des Landes Brandenburg 1996).

In view of great significance of renovation investments in the grand total of operating costs, renovation work plans must be based on reliable forecasts concerning construction and operating condition of the network that in turn should be established on the grounds of realistic classification of results of full-range television surveys (Madryas and Wysocki 2008). Among many proposals of such classification there is a model based on the value of the network's construction substance. In the framework of this model, the so-called renovation profile is determined for each network segment by assigning to it the renovation class ranging from 1 to 5 whereas, similarly to the German guidelines (Merkblatt ATV-M 149 1999), the first class is the worst one. The fundamental difference between the priority classification and the realną classification consists that in the first case, the condition class assigned to a given network segment is based on the most serious of the identified damages while the cumulated renovation length of all damages in the segment is decisive for the second type of classification. All technical details concerning classification of actual sewerage network operated by the municipality of Unterhaching are a result of long-term experience gathered by the present authors in the course of planning, execution, and financial accounting for the performed renovation.

Based on data concerning completed renovation project it was possible to compare costs with the obtained operating effects. A decisive moment of these comparative studies and analyses was determination of a specific value determining cost-effectiveness of a segment renovation project. Typical damages occurring in stoneware sewers had usually the point-like and thus local nature. The actual length of these cavities fall in the range 0–0.50 m which allowed to analyse them only in the network's cross section. The cumulated net length of damages occurring within an analysed network segment should not be adopted as a ruling determinant decisive for necessity of renovation. On the grounds of the cost analysis performed with respect to the renovation it has been found that embodying the repair sleeve even as short as 0.50 into the sever in order to repair a minor point-like damage generates fixed costs that are roughly the same as for installing sleeves up to 5.0 m long. Such fixed cost components included various activities, such as:

- construction site and workplace organisation;
- hydrodynamic cleaning of sewer conduits;
- performing optical inspection before and after renovation;
- carrying out leakproof tests;
- compiling the post-renovation documentation.

After summing up the costs and comparing them with the costs of renovation of the whole sewer segment it was possible to determine the minimum renovation length that turned out to be from 3.0 to 8.0 m. As an indicative minimum renovation length, the value $l_{od(min)}$ = 5.0 m has been adopted. Each individual damage was assigned a minimum renovation length and thus the cumulated renovation lengths were determined. In case when damage length with different classes overlapped, then the class of the whole renovation length was decided by the most serious damage. Another important quantity used in this classification is the gross cumulated renovation length and its ratio to the total segment length. As a result the question has arisen what value of the ratio of the cumulated renovation length to the total segment length would be considered a ruling threshold indicating the need for renovation. When establishing the value of this important determinant, once more the reference was made to experience gathered in the course of realisation of former renovation undertakings, and first of all, accounting for their costs. Analysis of the renovation costs allowed to arrive at the conclusion that installation of several short sleeves on a given network segment is not necessarily more cost-effective with respect to the option of providing the repair sleeve along the whole of it length. Finally, the limiting value of 30% has been adopted which means in practice that the damage class for each the gross cumulated damage length will reach 30% of total length of the analysed network segment is decisive for its ultimate qualification to the renovation class.

Details concerning the realistic classification are presented using as an example the network segment between inspection chambers 304690029 and 304690030, and situated in the town of Unterhaching at Professor-Huber-Str. Tabular interpretation of the classification is presented in Table 1, while its graphical representation is shown in Figure 1. From the above example it follows that the cumulated gross renovation length reaches 30% of length of the

Table 1. The tabular method of establishing the renovation profile and the renovation class for a network segment with length of 54.1 m.

	Gross damage length				
Damage class	KU 1	KU 2	KU 3	KU 4	KU 5
Single (m)	5.00	0.00	18.00	5.00	26.10
Cumulated (m)	5.00	5.00	23.00	28.00	54.10
Cumulated (%)	9.24	9.24	42.51	51.76	100.00
Renovation class	KO 1	KO 2	KO 3	KO 4	KO 5

KU—Damage class; KO—Renovation class.

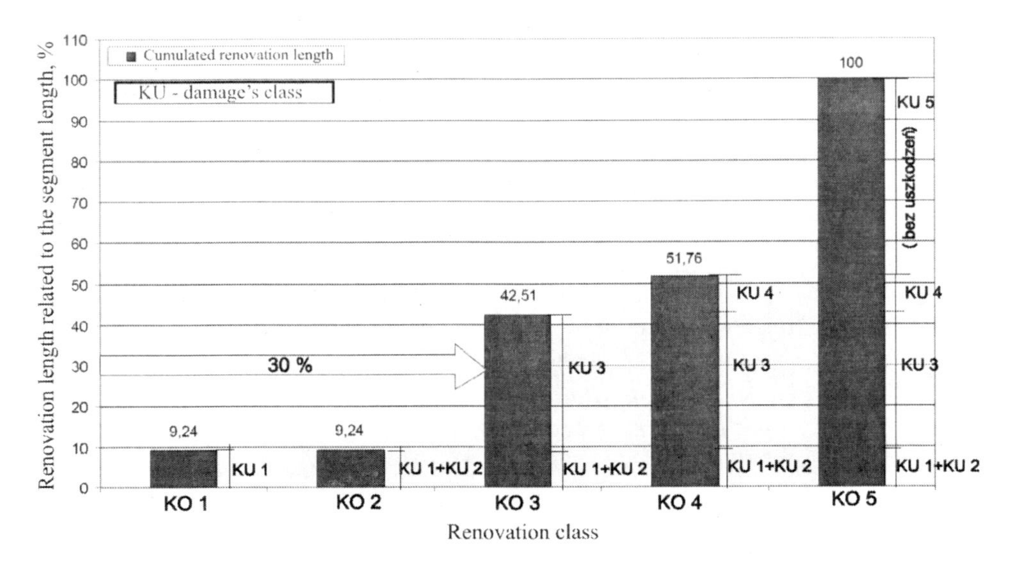

Figure 1. Graphical method of establishing the renovation profile and the renovation class for a network segment with length of 54.1 m.

whole segment only in the area of the third damage class, which means that the segment is assigned the third class of renovation (KO 3).

The realistic classification of the examined network was carried out in such a way that each network segment was assigned a gross renovation length and based on this value, the corresponding renovation class, from the best fifth class (KO 5) to the worst first class (KO 1). Comparing results of the realistic classification with those of the priority classification it is easy to notice significant shifts in renovation classes. Many of network segments that in the framework of the priority classification were rated among the first class items, in the case of the realistic classification were assigned the second or the third renovation class. A fragment of the realistic classification carried out for the domestic-household sewerage network channelling sewage from the municipality Unterhaching are presented in Table 2. Within each of the five renovation classes, the segments have been sorted by their age, from the youngest (newest) to the oldest. In order to simplify calculation operations, the following age groups have been created: 5, 10, 15, 20, 25, 30, 35, 40, and 45 years. The age group width B can be determined from the formula

$$B = \frac{t_{\max} - t_{\min}}{n_k}, \tag{1}$$

where $t_{\max}$ is the maximum technical service life of given sample (in years), $t_{\min}$ is the minimum technical service life of given sample (in years), $n_k = \sqrt{n}$, and n is the numerical strength of the sample.

The width of a given age group depends directly on the numerical strength of the whole random sample. With increasing number of elements constituting given random sample, the number of age groups increases and their width decreases accordingly. Percentage distribution of renovation classes referred to the number of network segments as well as to the network length is presented in Figure 2. The differences resulting from which of these two quantities the class distribution was related to are small and do not exceed several percentage points. Analysing the results obtained by means of these two categorisation methods it can be concluded that the realistic classification creates, to a certain degree, more favourable picture of technical condition of the network.

Table 2. Realistic classification of the domestic-household sewerage network operated by the Unterhaching municipality.

Network segment	Street name	Segment length (m)	DN (mm)	KP	Year of construction	Age (years)	Foundation depth (m)	KO
313090019–313090020	Biberger Str.	39.54	250	1	1967	33	4.51	1
313190038–313190039	Bürgermeister-Prenn-Str.	58.97	250	1	1973	27	2.54	2
308890061–308890062	Deisenhofener Weg	58.96	250	1	1982	18	3.28	4
313290036–313290037	Finsinger Weg	54.88	250	1	1966	34	3.39	3
397690011–397690012	Frühlingstraße	28.22	250	1	1966	34	2.97	1
308990015–308990016	Grünauer Allee	50.02	250	1	1984	16	2.00	4
317290011–317290012	Grünwalder Weg	15.84	250	1	1985	15	3.46	1
394790033–394790037	Habichstr.	42.83	250	1	1967	33	3.00	2
309090060–309090061	Hallstattweg	9.94	250	1	1972	28	2.30	1
300890023–300890024	Jägerstr.	43.14	250	1	1965	35	3.31	3
300890025–300890026	Jägerstr.	44.32	250	1	1965	35	3.16	3
308790033–304790012	Karwendelstr.	47.79	250	1	1967	33	2.66	2
397690019–397690020	Parkstr.	56.90	250	1	1957	43	3.10	1
304790040–304790037	Pittingerplatz	18.13	250	1	1966	34	2.66	1
304790021–304790047	Pittingerstr.	48.67	250	1	1967	33	2.84	4
304690029–304690030	Professor-Huber-Str.	54.10	250	1	1967	33	2.65	3
300990040–300990034	Sommerstr.	58.38	250	1	1966	34	3.23	1
309090041–309090042	Truderinger Str.	27.30	250	1	1967	33	1.75	1
309090042–309090043	Truderinger Str.	59.48	250	1	1967	33	2.10	1
313090007–313090008	Utzweg	53.69	250	1	1967	33	3.02	5
308990038–308990039	Von-Staufenberg-Str.	41.76	250	1	1966	34	3.16	1
308990039–308990040	Von-Staufenberg-Str.	41.66	250	1	1967	33	3.19	1
397690007–397690008	Waldstr.	38.83	250	1	1966	34	2.71	4
304790078–308790036	Zugspitzstr.	27.04	250	1	1971	29	3.60	3
308890044–308890045	Adejeweg	41.84	250	2	1991	9	4.32	4

(continued)

Table 2. (Continued).

Network segment	Street name	Segment length (m)	DN (mm)	KP	Year of construction	Age (years)	Foundation depth (m)	KO
394790017–394790005	Alfred-Lingg-Str.	49.00	250	2	1956	44	4.05	2
308790010–308790011	Am Klosterfeld	37.02	250	2	1972	28	4.69	4
301091014–301091013	Am Sportpark	57.71	250	2	1988	12	3.28	3
312990011–312990013	Annastr.	43.08	250	2	1968	32	3.61	3
309090075–309090076	Bajuwarenweg	25.89	250	2	1972	28	2.34	4
313090016–313090017	Biberger Str.	43.37	250	2	1967	33	4.36	2
313090016–313090019	Biberger Str.	38.55	250	2	1967	33	4.56	3
313190030–313190037	Bürgermeister-Prenn-Str.	50.05	250	2	1973	27	2.40	4
313190039–313190040	Bürgermeister-Prenn-Str.	38.59	250	2	1973	27	2.66	4
394790020–394790013	Carl-Duisberg-Str.	42.23	250	2	1956	44	3.89	3
394790019–394790020	Carl-Duisberg-Str.	42.36	250	2	1956	44	3.71	3

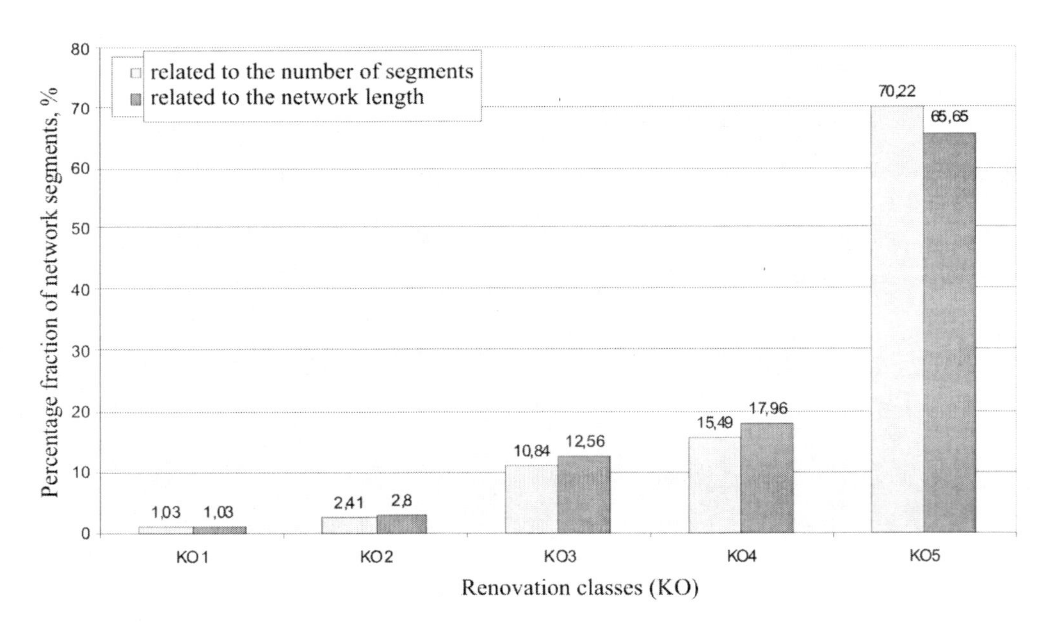

Figure 2. Percentage distribution of renovation classes for a given system related to the total number of segments and the network length.

Only 3.44% of network segments were classified to the first and the second renovation classes, compared to 70.22% of segments categorised as belonging to the fifth class, which is a confirmation of a generally good technical condition of the stoneware domestic-household sewerage network operated by the municipality of Unterhaching. Network segment that were assigned the first or the second renovation class need be urgently repaired. Comparison of

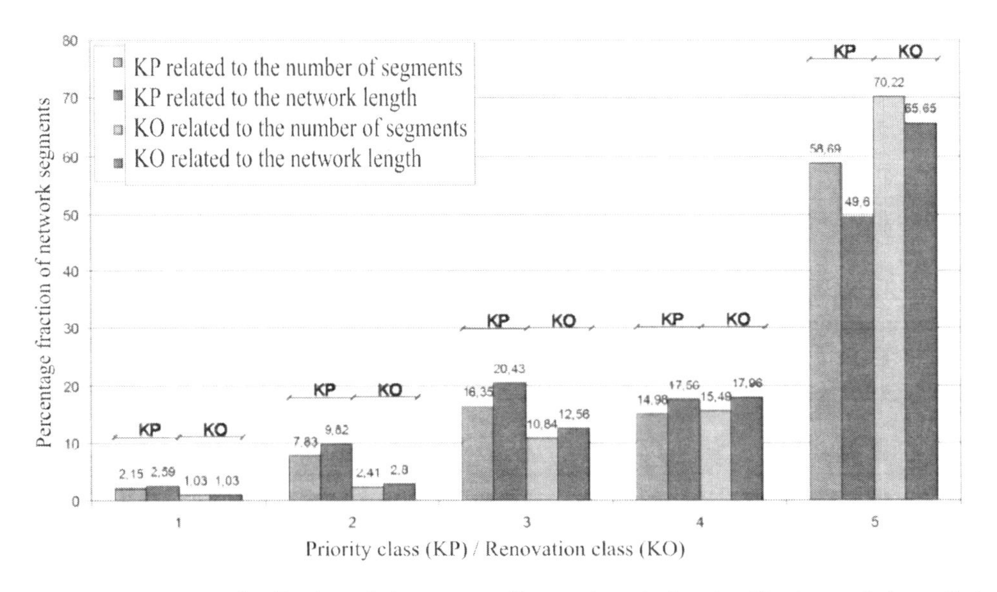

Figure 3. Percentage distribution of classes according to the priority classification and the realistic classification related to the total number of segments and to the network length.

the two classification schemes (Fig. 3) shows a distinct shift of the segment percentage fraction towards better classes when the realistic classification is employed.

Adoption of the 30% threshold defined as the ratio of the cumulated renovation length to the total length of the segment is the effect of, first of all, experience gathered in the course of renovation work, and also the result of selection of a rational renovation strategy based on specific operation targets.

The strategy allowing to avoid excessive deterioration and depreciation of a sewerage network in a long-term perspective by means of appropriate preventive actions is based on maintenance of the structure's construction substance. Performing conservative renovation of a network increases definitely its construction substance current value and thus also the relative construction substance determinant value (WSB_{rel}) that is a quantity representative for the structure's technical condition (Merkblatt DWA-M 143–14 2005). The determinant, defined as the ratio of the network construction substance value to the structure reconstruction value, taking into account the renovation costs, after completion of such renovation should fall in the range of at least 0.3–0.5.

The results of classification concerning technical condition of the stoneware domestic-household sewerage network operated by the municipality of Unterhaching allow to single out five different renovation classes that constituted a base for statistical studies leading finally to working out the forecast concerning technical condition of the examined sewerage network. This required one more classification operation consisting in setting up five transition classes determining boundaries between different states of technical condition attributable to network segments. This in turn allowed to describe the network ageing process in a continuous manner. Segments of the fifth renovation class (KO 5) constitute a base that allows to establish a limit value between the fifth and the sixth renovation class. In the case of transition from the fourth to the third technical condition state, the data base comprised the segments included in the fifth and the fourth renovation class (KO 5–4).

Analogously, other transition classes, namely KO 5–3, KO 5–2, and KO 5–1 have been constructed. The last and thus the worse transition class (KO 5–1), that determines the critical condition limit, includes all segments of the examined network. Sorted and ordered data concerning network segments included in the fourth transition class (KO 5–4) are presented in Table 3.

The class representing transition from the fourth to the fifth renovation class (KO 5–4) included 996 network segments that revealed rather minor damages. In the case of each

Table 3. Segments included in the fourth transition class (KO 5–4) of the stoneware domestic-household sewerage network of the municipality of Unterhaching.

Segment age (years)	Number of segments n_i	Frequency n_i/n	Vulnerability function $F^*(t)$ (%)	Transition function $R^*(t)$ (%)	Failure intensity function $f^*(t)/R^*(t)$
5	55	5.52	5.52	100.00	5.52
10	97	9.74	15.26	94.48	10.31
15	89	8.94	24.20	84.74	10.55
20	61	6.12	30.32	75.80	8.08
25	107	10.74	41.06	69.68	15.42
30	440	44.18	85.24	58.94	74.96
35	135	13.55	98.80	14.76	91.84
40	0	0.00	98.80	1.20	0.00
45	12	1.20	100.00	1.20	100.00

$n = \sum = 996$; (*)—empirically determined function.

segment included in this class, the cumulated length of fifth and fourth class damages accounted for at least 30% of the total length of the segments.

From the five transition classes identified this way, five random samples were created that included from 816 to 1,162 segments of the analysed stoneware network segments sorted by their age. Next, for each of the random samples, empirical functions describing probability density, vulnerability, reliability and failure intensity were determined.

3 EMPIRICAL FUNCTIONS OF PROBABILITY DENSITY, VULNERABILITY, RELIABILITY, AND FAILURE INTENSITY

Construction of empirical functions of probability density, vulnerability, reliability, and failure intensity was presented above and subjected to a detailed analysis with the use of the example of the transition class corresponding to transition from the fourth to the fifth renovation class (KO 5–4). Determination of each of the transition functions represents a separate statistical problem in view of different size and characteristics of each random sample. The basic parameter and the point of reference is the time that denotes the age of individual segments correlated with specific technical and operating condition.

At the outset of statistical studies, the histogram was created of absolute numbers n_i of segments included in successive age groups as well as the histogram of the failure occurrence frequency defined as n_i/n. In the case of sewers the term "failure" is interpreted as reaching a precisely determined renovation class by a network segment or a set of segments. The frequency histogram in the block form that, in this case, does not show the features of the normal distribution, carries the information on how many segments are included in each of the age groups within given renovation class. By joining upper vertices of the value bars with a continuous line it is possible to pass directly to the empirical probability density function $f^*(t)$ representing likelihood of occurrence of a failure (Fig. 4). The function determined this way represents the relative probability density representing the likelihood that the network segments will pass from the fourth to the fifth renovation class. In case when the number of segments included in the fourth transition class tends to infinity, the empirical probability density function evolves into the theoretical function $f(t)$. If the relative transition frequencies will be considered as values of the abscissas, then the surface area under the line representing the probability density function density will take the value equalling one. The function represents the number of segments of the stoneware domestic-household sewerage network, included in individual age groups that move from the fourth to the fifth renovation class. The form of the probability density distribution for the fourth transition class (KO 5–4) is not

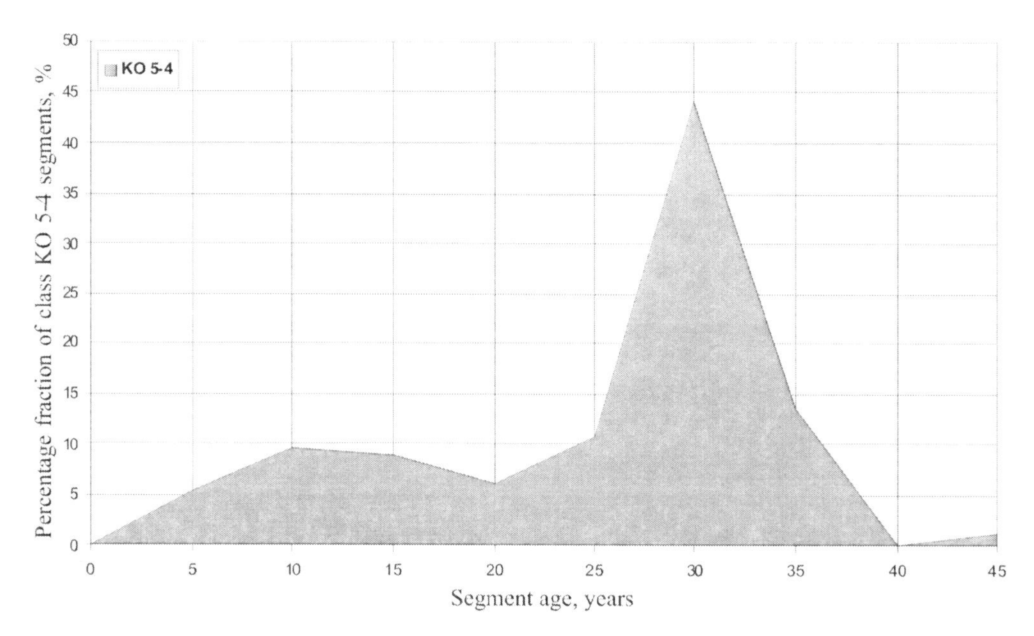

Figure 4. The empirical probability density function representing the likelihood of the network segments transition from the fourth to the fifth renovation class.

symmetrical. A distinct peak of the function occurs for segments 30 years old that represent in total 44.18% of the whole population. On the other hand, zero value of the probability density function in segments operated for 40 years corresponds to the fact that segments of this very age group are not represented in the fourth transition class.

In the next phase of the statistical studies, the failure occurrence frequency $H(m)$ was summed up according to the formula

$$H(m) = \sum_{i=1}^{n} h_{rel}(i) \tag{2}$$

where i denotes the age group number.

The histogram representing the sum of failure frequencies obtained this way provides the answer to the fundamental question can be reduced to determining what is the number of segments that, at given point of time of operation of the network, have reached the technical condition corresponding e.g. to the third renovation class (Fig. 5).

A characteristic feature of the histogram is the significant increase of the frequency of reaching the third renovation class by segments that are operated for 30 years. This is the most populated age group in the fourth transition class. Transformation of the histogram into an empirical function is similar to this shown above in the case of the failure occurrence probability density function.

In case when the numerical strength of the examined sample $n \to \infty$, it is possible to proceed from the empirical distribution of the sum of failures $F^*(t)$ to the continuous theoretical function $F(t)$. The plot of the vulnerability function starts at the point $F(t) = 0$ and then increases monotonically along with adding successive failure frequencies tilt the moment of time when all segments of the network in question reach a predefined technical condition in the case of which $F(t)$ takes the vale equalling the unity.

Examining statistically the event consisting in passing of a given set of segments from one class to another and taking into account the fact that the set is a large random sample ($n \to \infty$), the failure distribution function can be represented in the form of the integer of the probability density function (Wilker 2004, Wilker 2010):

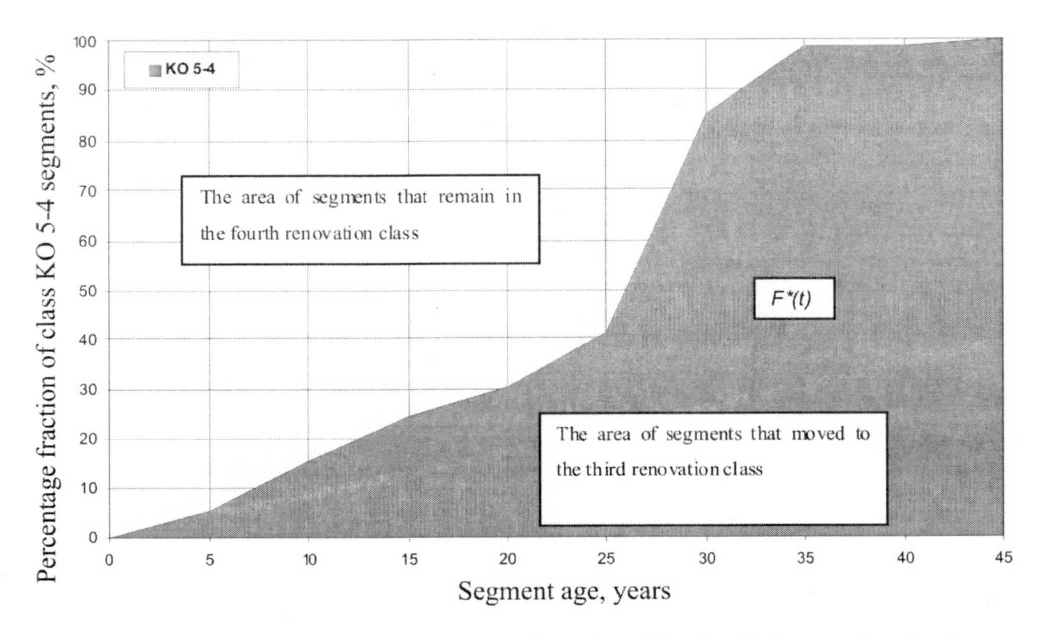

Figure 5. Empirical function representing probability of reaching the third renovation class by segments of the examined network.

$$F(t) = \int f(t) \cdot dt. \tag{3}$$

On the other hand, by differentiating the failure distribution function described by equation (3), one obtains the probability density function (Wilker 2004, Wilker 2010) in the form

$$f(t) = \frac{dF(t)}{dt}. \tag{4}$$

The distribution function $F(t)$ describes the probability of occurrence of an event consisting in that the sewerage network segments included in the fourth class will reach, at the specific moment of operation, the technical and operating condition defined as the third renovation class. The above is the base for a frequently used definition of the term 'failure' as the probability that a specific technical condition is reached. In case when the sought quantity are the network segments that did not fail until a given instant of the operating time, solution of the related problem is illustrated by means of the survival frequency histogram.

On the grounds of the above, the empirical survival function is constructed which is frequently called the reliability function (Fig. 6). The empirical survival curve $R(t)$ is a function complementary with respect to the distribution function $F(t)$ that can be expressed by means of the following relationship (Wilker 2004, Wilker 2010):

$$R(t) = 1 - F(t). \tag{5}$$

The characteristic inflection point of the empirical transition function occurs for these segments of the examined sewerage network that were operated for 30 years, similarly as in the case of the probability density function and the mortality function. From that point in time of operation of the network subjected to the analysis, the number of network segments that will remain in the fourth renewal class rapidly decreases.

The sum of segments that, at any moment of operation and within any selected age group, either failed or did not fail, is 100%. The plot representing the reliability function starts at the moment $t = 0$ before any of the segments has not been damaged yet which means also that

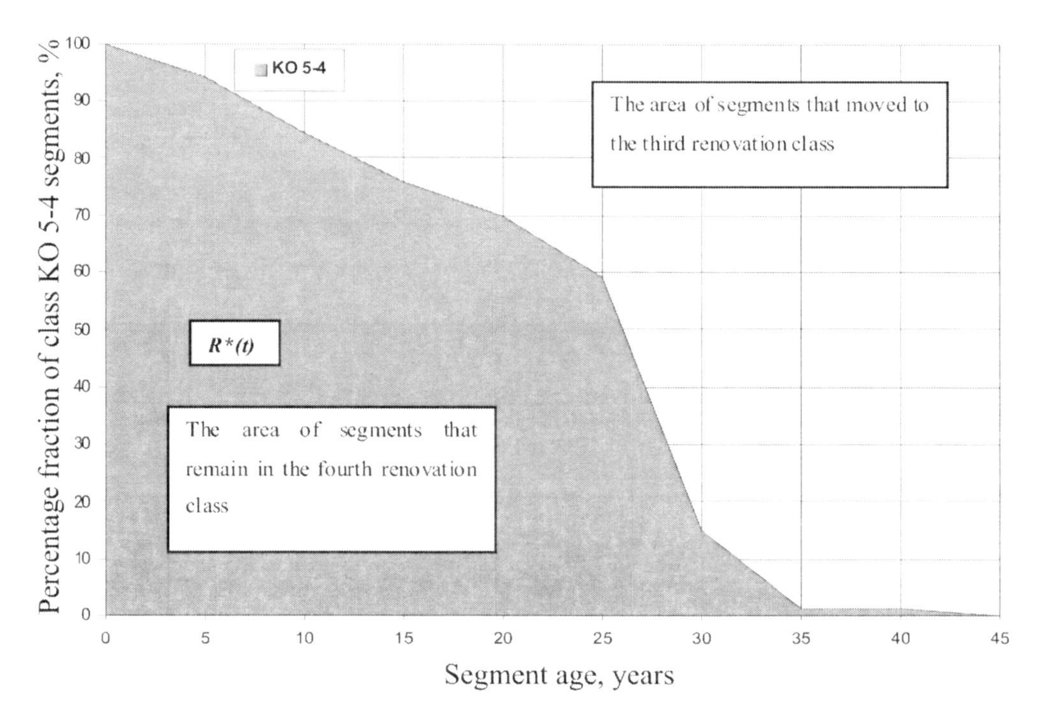

Figure 6. The empirical probability density function representing the likelihood that segments of the examined network will remain in the fourth renovation class.

none of them moved to the next renovation class and therefore $R(t) = 100\%$. The function tends to the point $t = t_n$ in which most of the segments reach the worst renovation class.

Another important element of the statistical analysis is the failure intensity $\lambda(t)$, that represents vulnerability of a network segment operating still without any failure at the moment of examination. The failure intensity can be defined as ratio of the number of units of a sewerage system damaged at the time t of its operation to the number $n_s(t)$ remaining still in good working order (Wilker 2004, Wilker 2010) and written in the form

$$\lambda(t) = \frac{1}{n_s(t)} \cdot \frac{dn_s(t)}{dt}. \tag{6}$$

Empirical distribution of the failure intensity for the fourth renovation class is presented in Figure 7. If the sum of failures at the moment t will be described by means of the density function $f(t)$, and the sum of undamaged units by the survival function $R(t)$, then the failure intensity can be expressed by means of the following relationship (Wilker 2004, Wilker 2010):

$$\lambda(t) = \frac{f(t)}{R(t)}. \tag{7}$$

The failure intensity referred to the instant of time t carries information about the number of network segments undamaged before will move to the next, i.e. the worse renovation class. The highest probability of transition from the fourth to the third renovation class show these network segments that are operated for 30 and 35 years. There are two most populated age groups in the fourth transition group (KO 5–4). Zero value of empirical function $\lambda(t)$ for segments aged 40 years is a result of the fact that this age group is absent in the fourth transition class and therefore function $f^*(t)$ also assumes zero value in this point.

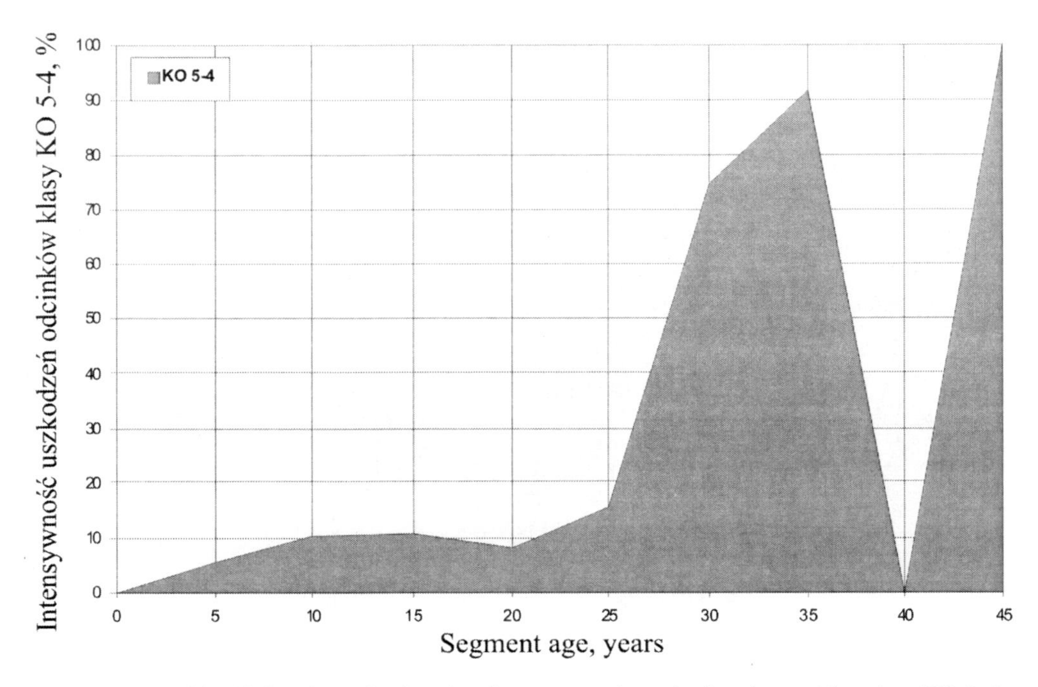

Figure 7. Empirical failure intensity function for segments from the fourth transition class (KO 5–4).

The failure intensity can be also interpreted as the probability that a failure will occur at the moment t of operation. The parameter can be used to assess the changes occurring in vulnerability of building structures or mechanical devices in time.

Using results of numerous studies (Wilker 2004, Wilker 2010) and statistical analyses, empirical failure intensity curve has been constructed which has the characteristic form of a bathtub and can be divided into three areas with respect to the failure intensity expressed as a function of the operating time (Fig. 8), namely:

- Area A—failures occurring in the initial period of operation of the structure,
- Area B—failures occurring in the course of normal operation,
- Area C—failures occurring as a result of long operation of the system.

The empirical function $\lambda(t)$ determined for the network segments included in the class corresponding to transition from the fourth to the third renovation class shows in general an increasing trend that allows to locate it in the third area of the bathtub plot. Main causes of the failure intensity increase in this phase include:

- changes occurring in the materials used to construct the sewers as a result of long operation, e.g. because of worn sealing elements,
- construction material fatigue,
- unprofessionally executed or inadequate maintenance of the system,
- essential changes in the network operating conditions such as the road traffic intensity, renovation of other municipal infrastructure components, variable ground water level, etc.

Statistical studies of the same scope as this presented in the example concerning transition class KO 5–4 were also carried out for the remaining four classes, i.e. KO 5, KO 5–3, KO 5–2, and KO 5–1.

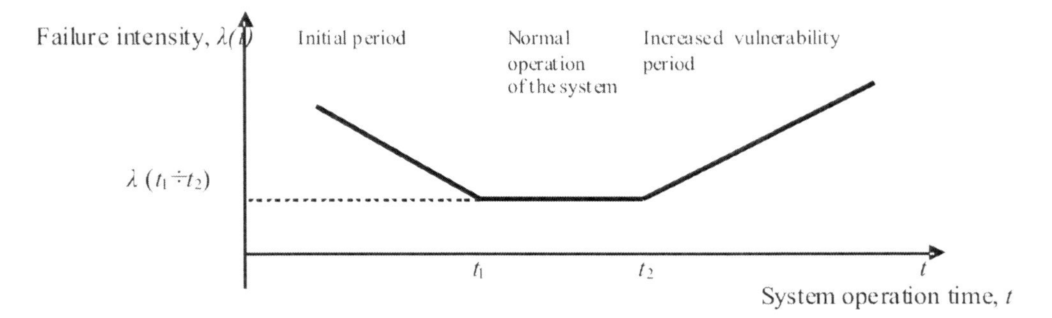

Figure 8. Schematic representation of the failure intensity curve for building structures and mechanical devices.

From analyses of the failure occurrence probability density distributions it followed that they were not of the normal Gaussian distribution type. A characteristic shift observed in these distributions suggested presence of features typical for the Weibull distribution which is an empirical two-parameter exponential distribution particularly useful in reproduction of results obtained in studies on technical service life of building structures, mechanical devices, and electronic systems. Estimation of parameters is the essential problem that must be solved in the case of using a distribution of that type for further statistical studies. In practice, also a three-parameter version of the Weibull model is used when the examined unit of an assembly of units remains undamaged in the initial phase of operation. Assumption of that kind in not true in the case of sewers, as a damage in such structures can be originated as early as in the investment project realisation phase. Other publications of the same cycle present graphical and analytical methods used to estimate parameters of the Weibull distribution in order do determine theoretical functions determining fail-safe operation of sewers. These functions, called also the transition functions, describe transition of a network segment or a group of segments from one to another, worse renovation class.

4 CONCLUSION AND FINAL REMARKS

Realistic classification of results obtained from full-range optical inspection of a sewerage network in the version proposed by the present authors takes into consideration the two essential operating aspects, namely the prevention-oriented renovation strategy guaranteeing full availability of the network throughout the whole of its technical service life and technological and operational conditions of conducting renovation works. The classification of technical condition of an existing network carried out this way will form a base for further statistic studies aimed at working out a rational forecast for ageing of the examined domestic-household sewage collection system.

On the grounds of preliminary statistical studies it could be found that the observed probability density distributions were not symmetrical. Further analyses suggested appropriateness of developing a planned forecast concerning technical condition of the stoneware sewerage network operated by the municipality of Unterhaching on the grounds of a two-parameter Weibull model.

The most important objective of the developed forecast is to determine the quantitative scope of necessary renovation operations. The obtained technical condition assessment results become an indispensable instrument of rational management exercised over any sewerage network. For each sewerage network operator, the knowledge of actual needs in the area of renovation investments is the starting point for rational planning repair work and budgeting financial resources in the long-term perspective.

REFERENCES

Madryas C, Wysocki L. 2008. Renovation of brick interceptor sewers. Tunnelling and Underground Space Technology 23 (6), 718–726.

Merkblatt Atv-M 149. 1999. Zustandserfassung, -klassifizierung und –bewertung von Entwässerungssystemen außerhalb von Gebäuden.

Merkblatt DWA-M 143-14. 2005. Sanierung von Entwässerungssystemen außerhalb von Gebäuden, Teil 14: Sanierungsstrategien.

Ministerium für Umwelt, Naturschutz und Raumordnung des Landes Brandenburg. 1996. Abwasserentsorgung in Brandenburg—Orientierungswerte für den Kostenaufwand bei der Abwasserableitung und –behandlung.

Pecher, R. 1994. Bau- und Betriebskosten bestehender Anlagen zur Abwasserentsorgung in der Bundesrepublik Deutschland, Korrespondenz Abwasser (41), Nr. 12.

Pecher, R. Dudey, J. Lohaus, J. Esch, B. 1995. Ergebnisse der ATV-Umfrage „Abwassergebühren", Korrespondenz Abwasser (42), Nr. 1.

Raganowicz A. Metodyka prognozowania stanu technicznego sieci kanalizacyjnych. Wrocław University of Technology. Raport CPRE 3/2010, Wrocław.

Wilker, H. 2004. Weibull-Statistik in der Praxis, Leitfaden zur Zuverlässigkeitsermittlung technischer Produkte, Verlag: Books on Demand GmbH, Nordestedt.

Wilker H. 2010. Weibull-Statistik in der Praxis, Leitfaden zur Zuverlässigkeitsermittlung technischer Komponenten, Verlag: Books on Demand GmbH, Nordestedt.

Underground Infrastructure of Urban Areas 3 – Madryas et al. (Eds)
© 2015 Taylor & Francis Group, London, ISBN 978-1-138-02652-0

Models of level diversification assessment of Water Supply Subsystem

J.R. Rak & A. Włoch
Rzeszow University of Technology, Rzeszow, Poland

ABSTRACT: In the paper a new method for assessing the diversification degree of water supply for human consumption from independent sources was presented. The Shannon-Wiener index provides such possibility and allows for objective assessment and comparison between CWSS with different final water production. Also an author's indicator based on the polynomial function was suggested. The method can constitute also supplementing of deepened financial analyses of the decision about water purchase. Reality of the market economy more and more often conveys to the possibility of the water purchase from other producers. This is the second element of the importance of the level of water supplies diversification assessment. The third factor of the validity of water supply diversification regards crisis situations connected with undesirable events of technical failures or terrorism acts. For two models water supply diversification indices were determined in 22 cities.

1 INTRODUCTION

Most generally the collective water supply system (CWSS) can be divided into the water supply subsystem (WSS) and the water distribution subsystem (WDS) (Wieczysty 2001).

The annular nature of the water supply system provides multi-directional water supply of each ring and each object supplied from the distribution network is capable of bidirectional water supply. Thus, the diversification of water supply in the WDS results from the methodology of water supply system design (Wieczysty 2001).

In small water supply systems generally you can find one source of water supply, however, group municipal water pipelines can be found too. Large metropolises are supplied with water from two or more independent sources. The CWSS safety can be analysed in terms of the protection against threats and the vulnerability to threats (Rak 2012). The issue of WSS diversification belongs to the study of CWSS vulnerability to threats.

The analysis of literature shows a lack of deeper analyses related to the diversification of water supply for large urban and industrial agglomerations.

The only premises as to the diversification of water supply result from the reliability theory. Using the Bernoulli distribution, functioning of the so-called threshold structure "m-k" out of "m" is analysed. In this way the probability of operating states is determined, with failure $k = 0, 1, ..., m - 1$ Water Supply Systems (WSS) (Rak 2005). A limitation in this case is the need to assume equal probabilities describing operating states of "m" WSS.

The method taking into account the various probabilities of operating states of individual WSS was developed by Krakow's research team, led by the late professor Arthur Wieczysty and is called "the analysis of the expected value of water scarcity" (Wieczysty 2001).

There are three variants of this method for different CWSS: excessive $Qd > Qdmax$, sustainable $Qd = Qdmax$ and deficit $Qd < Qdmax$. This method has been adapted by the research team in Rzeszów to determine the risk of lack of water for diversified WDS (Rak 2005).

Against this background, the aim of this paper is to present a new methodology for the analysis and assessment of diversification of water supply from WSS to urban and industrial agglomeration.

2 EXAMPLES OF CRITICAL SITUATIONS

During the weekend 18–19 February 2012 in Jaworzno occurred 4 major failures in the water supply system, which forced the temporary shutdown of the water intake for the pipeline repair. Alternatively, the residents were supplied with water from water carts and got 10 litre aseptic water bags. To get water the residents could go directly to the intake with their own water containers. Water supply was resumed on Monday, February 20.

At night from 24 December to 25 December 2013 at about 1a.m. a failure in the water supply in Jastrzębie occurred. The residents of several blocks of flats and their relatives did not have water after returning from Midnight Mass till Christmas morning hours.

At the end of 2013 and the beginning of 2014 Kosovo was affected by a severe drought. The tank Batllawa, which was a reservoir for water production for capital city Pristina and city of Podujew, was almost completely dry. More than 400,000 residents of these cities could not have tap water for 3 months. The authorities have introduced drastic restrictions on water consumption. The crisis lasted until mid-February 2014.

On January 9, 2014 took place the technical failure and the associated with it leakage from tanks in Chemical Company "Freedom Industries" located in the State of West Virginia in the USA. Oily liquid with an odour of mint, under the trade name "Crude MCHM", whose production is based on methanol, got into the river Elk in the amount of 20,000 litres. This river is a source of water supply for 300,000 people in the region. Water in water supply network was contaminated. Symptoms associated with the consumption of tap water or contact with it include: nausea, vomiting, diarrhea, dizziness, rash and skin irritations. The state has not reached the data about the degree of toxicity of the liquid and had to wait until the tide of polluted water flows. The population was provided with bottled water by the National Guard. During three days one million litres of bottled water was delivered.

3 EXAMPLES OF DIVERSIFICATION OF WATER SUPPLY SOURCES

Jastrzębie Zdrój community covers nearly 80% of the demand for water by importing relatively cheap water from the Czech Republic. The town of Oława purchases about 50% of the amount of water from the Mokry Dwór waterworks company in Wroclaw. The town of Myślenice covers about 10% of its needs for water purchasing water from WTP in Dobczyce, which is subject to the Water Supply and Sewerage Company in Krakow (Iwanejko 2006).

In March 2013 an agreement was reached between the Water Supply and Sewerage Company in Racibórz and the Water Supply and Sewerage Company in Lubomia on wholesale water sale from Racibórz to Lubomia community. It became possible after commissioning the 11.2 km long pipeline transmitting drinking water from Racibórz to Syrynia.

In mid-2013 the residents of Dąbrowa Górnicza organized an action to influence the town authorities in order to diversify their sources of water supply. The Water Supply and Sewage Company is a Polish-German partnership. The residents believe that the water price level is too high for their financial capabilities. At the same time they indicate the possibility of creating a second independent source of water supply to the city from the direction of the tank in Kuźnica Warężyńska administered by the Regional Water Management Board in Gliwice.

In practice, sometimes the local water supply pipelines are connected and one intake is shut out of service—anti diversification. The pipeline Niewiesz and Chropy, located in Poddębice, Łódź Voivodoship, is such an example. The local authorities consider a variant to shut off the intake in Chropy which is situated in the industrial plant. The basic argument for such a decision is, observed for several years, a significant decrease in water consumption. A similar situation took place in Nowa Sól situated in the Lubuskie Voivodoship, where in 2010 one of the three water intakes were excluded from the operation.

4 METHOD FOR ASSESSING THE DIVERSIFICATION DEGREE OF WATER SUPPLY

In the literature there are many non-parametric measures of diversity, some of them are listed here.

1. Simpson index of diversity (Simpson, 1949)

$$\sum_i p_i^2 \tag{1}$$

where p_i is a proportional contribution of option i and $\sum_i p_i = 1$.

2. Shannon index of diversity (Shannon & Weaver 1962)

$$\sum_i p_i \ln p_i \tag{2}$$

where option p_i has analogous meaning as in the Simpson index.

3. Shannon Evenness index of diversity (Pielou, 1969)

$$\frac{-\Sigma_i p_i \ln p_i}{\ln S} \tag{3}$$

where S is a number of options.

In the papers (Hill 1973, Hurlbert 1971, Weitzman 1992) also other indices of diversification was considered.

Practice and economy give some preferences in choice of suppliers with a fixed value of supply. These preferences should be taken into consideration in the construction of the diversity index.

The starting point in the construction of the diversification index is the choice of the preference function. The preference function is defined in the closed interval <0,1>, it has a positive values and, additionally, it takes the value 0 at 0 and 1. It should reflect experts knowledge, practice and system analyses. They determine the most desirable size of supply for CWSS. This value should be a local maximum of the preference function.

So there is an infinite number of choices for the preference function. In this work two diversification indices are considered. First one is the Shannon-Wienner index (Shannon & Weaver 1962) based on the function:

$$y = -x\ln(x) \quad \text{for} \quad x \in (0,1>, \text{ and } y(0) = 0 \tag{4}$$

This function has a local extremum at the point 1/e, so it promotes the suppliers with shares close to 1/e.

Additionally, if the preference function is continuous on the closed interval <0,1>, then it is natural to choose a polynomial as the preference function.

The best size of supply in the system is approximately equal to 1/3. There is infinitely many polynomials with maximum at the point 1/3 and value 0 at the points 0 and 1.

It is not difficult to construct a polynomial satisfying the above conditions. The second preference function considered in this work is:

$$x^3 - 2x^2 + x \quad \text{for} \quad x \in <0,1> \tag{5}$$

As it was mentioned earlier, it is not difficult to define a polynomial which possess an extremum in a fixed point. For example polynomials $(x - x^2)$ and $(x^2 - x^3)$ have extremes at 1/2 and 2/3, respectively.

The graphs of preference functions are presented in Figure 1. The function (4) (dashed line) promotes the suppliers with shares close to 1/e, while the function defined by the formula (5) (solid line) suppliers with shares 1/3.

The advantage of the polynomial preference functions is the ease and accuracy of calculations.

The diversification indices constructed on the function (4) and (5) are as follows:

$$ds = \sum_{j=1}^{m} u_j \ln(u_j) \tag{6}$$

$$dw = \sum_{j=1}^{m} \left(u_j^3 - 2u_j^2 + u_j\right) \tag{7}$$

where m is a number of WSS (water supply systems); u_j is the share of the jth WSS in a daily structure of production capacity of the whole CWSS.

The share of the jth WSS is defined by the formula:

$$u_j = Q_{dmaxj}/Q_{dmax} \tag{8}$$

where in:

$$u_j \in (0,1> \text{and} \sum_{j=1}^{m} u_j = 1$$

For one WSS, which means $u_j = 1$, the values $ds = dw = 0$, and for a hundred WSS with $u_j = 0.01$ the values are: $ds = 4.605$ and $dw = 0.98$.

The index ds is called the index of Claude Shannon and Norbert Wiener (Shannon & Weaver 1962; http://en.wikipedia.org) and has been used, among others, by James A. Stirling to determine the level of safety of supply electric power systems with energy carriers (Stirling 1999; Leszczyński 2012).

For both indices for assessing the diversification degree the rule saying that the higher the value the more favourable the diversification degree of water supply, is applied.

Table 1 summarizes the numerical values of the index for m = 2 independent WSS (a very common situation in domestic CWSS).

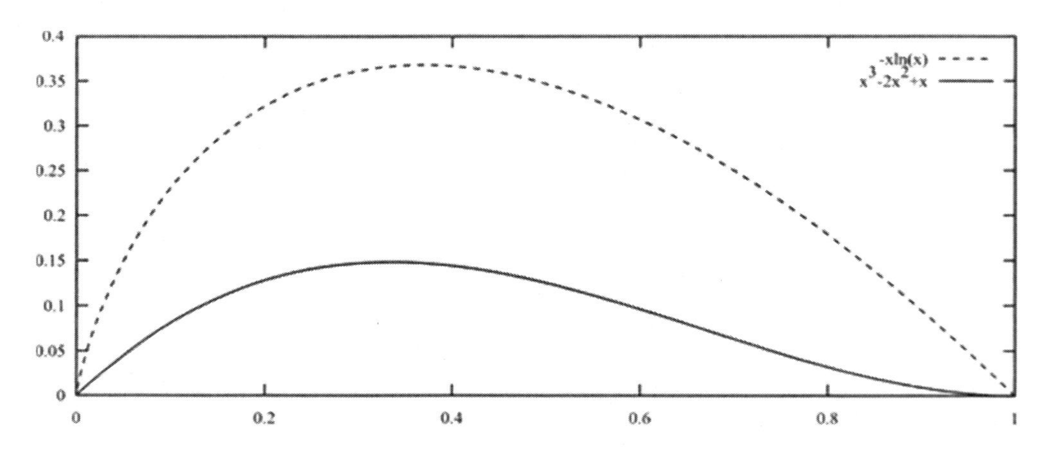

Figure 1. The course of the preference function for assessing the diversification degree of water supply.

The data contained in the table 1 show that the indices are the most favourable when the structure of water supply is sustainable (shares of the two sources are 50% each).

Table 2 summarizes the numerical values of indices for sustainable WDS with different number of WSS.

Fully sustainable indices can be used to objectively compare the degree of diversification in the groundwater intakes, where each of the wells can be treated as a source of water.

In the tables 3, 4 and 5 the exemplary values of indices for different number of independent WSS and various water supply shares, are presented.

The analysis of the indices shows that the degree of diversification for fairly sustainable WDS (similar shares) depends on the number of WSS.

Table 1. The values of the indices for two independent sources and different shares.

m = 2	$u_1 = 0.5$ $u_2 = 0.5$	$u_1 = 0.6$ $u_2 = 0.4$	$u_1 = 0.7$ $u_2 = 0.3$	$u_1 = 0.8$ $u_2 = 0.2$	$u_1 = 0.9$ $u_2 = 0.1$	$u_1 = 0.95$ $u_2 = 0.05$
ds	0.693	0.673	0.611	0.500	0.325	0.198
dw	0.250	0.240	0.210	0.160	0.090	0.048

Table 2. The list of indices for sustainable WDS.

m	2	3	4	5	10	20
u_j	0.50	0.33	0.25	0.20	0.1	0.05
ds	0.693	1.097	1.386	1.609	2.303	2.996
dw	0.250	0.444	0.563	0.640	0.810	0.903

Table 3. The values of indices for m = 3 independent sources of water supply.

m = 3	$u_1 = 0.33$ $u_2 = 0.33$ $u_3 = 0.33$	$u_1 = 0.4$ $u_2 = 0.3$ $u_3 = 0.3$	$u_1 = 0.5$ $u_2 = 0.3$ $u_3 = 0.2$	$u_1 = 0.6$ $u_2 = 0.3$ $u_3 = 0.1$	$u_1 = 0.6$ $u_2 = 0.2$ $u_3 = 0.2$	$u_1 = 0.7$ $u_2 = 0.2$ $u_3 = 0.1$	$u_1 = 0.8$ $u_2 = 0.1$ $u_3 = 0.1$
ds	1.097	1.089	1.030	0.898	0.950	0.802	0.639
dw	0.444	0.438	0.400	0.324	0.352	0.272	0.194

Table 4. The values of indices for m = 4 independent sources of water supply.

m = 4	$u_1 = 0.25$ $u_2 = 0.25$ $u_3 = 0.25$ $u_4 = 0.25$	$u_1 = 0.3$ $u_2 = 0.3$ $u_3 = 0.2$ $u_4 = 0.2$	$u_1 = 0.4$ $u_2 = 0.3$ $u_3 = 0.15$ $u_4 = 0.15$	$u_1 = 0.5$ $u_2 = 0.3$ $u_3 = 0.1$ $u_4 = 0.1$	$u_1 = 0.6$ $u_2 = 0.2$ $u_3 = 0.1$ $u_4 = 0.1$	$u_1 = 0.7$ $u_2 = 0.1$ $u_3 = 0.1$ $u_4 = 0.1$
ds	1.386	1.366	0.297	1.168	1.089	0.940
dw	0.562	0.550	0.508	0.434	0.386	0.306

Table 5. The values of indices for m = 5 independent sources of water supply.

m = 5	$u_1 = 0.2$ $u_2 = 0.2$ $u_3 = 0.2$ $u_4 = 0.2$ $u_5 = 0.2$	$u_1 = 0.3$ $u_2 = 0.3$ $u_3 = 0.2$ $u_4 = 0.1$ $u_5 = 0.1$	$u_1 = 0.4$ $u_2 = 0.3$ $u_3 = 0.1$ $u_4 = 0.1$ $u_5 = 0.1$	$u_1 = 0.5$ $u_2 = 0.2$ $u_3 = 0.1$ $u_4 = 0.1$ $u_5 = 0.1$	$u_1 = 0.6$ $u_2 = 0.1$ $u_3 = 0.1$ $u_4 = 0.1$ $u_5 = 0.1$
ds	1.609	11505	1.418	1.359	1.227
dw	0.640	0.584	0.534	0.496	0.420

In the case of large differences in the shares of the particular WDS this rule does not apply.

The degree of diversification of a smaller number of independent WSS that are sustainable may be higher than in case of a higher number of WSS with very differentiated shares.

For example:

$m = 4$: $u_1 = 0.7$ $u_2 = 0.1$ $u_3 = 0.1$ $u_4 = 0.1$; $ds = 0.940$, $dw = 0.306$.

$m = 3$: $u_1 = 0.5$ $u_2 = 0.3$ $u_3 = 0.2$; $ds = 1.030$, $dw = 0.400$.

which gives $1.030 > 0.940$ and $0.400 > 0.306$. Similarly:

$m = 5$: $u_1 = 0.6$ $u_2 = 0.1$ $u_3 = 0.1$ $u_4 = 0.1$ $u_5 = 0.1$; $ds = 1.227$, $dw = 0.420$

$m = 4$: $u_1 = 0.4$ $u_2 = 0.3$ $u_3 = 0.15$ $u_4 = 0.15$; $ds = 1.297$, $dw = 0.508$

and also:

$m = 3$ $u_1 = 0.8$ $u_2 = 0.1$ $u_3 = 0.1$; $ds = 0.639$, $dw = 0.194$

$m = 2$ $u_1 = 0.6$ $u_2 = 0.4$; $ds = 0.673$, $dw = 0.240$

The analysis of the CWSS in Krosno, Krakow, Warsaw and Wroclaw are as follows:

- Krosno—m = 3
  - ZW Iskrzynia 17 000 m^3d^{-1} – $u_1 = 0.28$
  - ZW Sieniawa 36 000 m^3d^{-1} – $u_2 = 0.60$
  - ZW Szczepańcowa 7 000 m^3d^{-1} – $u_3 = 0.12$
  - Total: 60 000 m^3d^{-1}
  - ds = 0.917, dw = 0.334
- Kraków—m = 4
  - ZW Raba 180 000 m^3d^{-1} – $u_1 = 0.62$
  - ZW Dłubnia 32 000 m^3d^{-1} – $u_2 = 0.11$
  - ZW Rudawa 55 000 m^3d^{-1} – $u_3 = 0.19$
  - ZW Bielany 24 000 m^3d^{-1} – $u_3 = 0.08$
  - Total: 291 000 m^3/d
  - ds = 1,057, dw = 0,369
- Warszawa—m = 3
  - ZW Centralny 400 000 m^3d^{-1} – $u_1 = 0.54$
  - ZW Północny 240 000 m^3d^{-1} – $u_2 = 0.32$
  - ZW Praski 100 000 m^3d^{-1} – $u_3 = 0.14$
  - Total: 740 000 m^3d^{-1}
  - ds = 0.973, dw = 0.366
- Wrocław—m = 2
  - ZW Mokry Dwór 120 000 m^3d^{-1} – $u_1 = 0.48$
  - ZW Na Grobli 130 000 m^3d^{-1} – $u_2 = 0.52$
  - Total: 250 000 m^3d^{-1}
  - ds = 0.692, dw = 0.250

These calculations determine the ranking of CWSS in those cities consistent with respect to the both indices.

- Kraków: ds = 1.057; Kraków: dw = 0.369
- Warszawa: ds = 0.973; Warszawa: dw = 0.366
- Krosno: ds = 0.917; Krosno: dw = 0.334
- Wrocław: ds = 0.692; Wrocław: dw = 0.250

A proposal for categorization and determination of standards for assessing the diversification degree of water supply in CWSS for the Shannon-Wiener index and the polynomial index introduced in this work.

Table 6. List of the diversification indices ds and dw for selected cities in the country.

	Shares—u_j										Index	
City Number UZW	u_1	u_2	u_3	u_4	u_5	u_6	u_7	u_8	u_9	u_{10}	ds	dw
Lublin	0.31	0.04	0.11	0.11	0.17	0.09	0.07	0.06	0.03	0.01	2.00	0.70
m = 10	Diversification category: satisfactory											
Radom	0.01	0.02	0.12	0.16	0.12	0.04	0.01	0.44	0.06	0.02	1.71	0.58
m = 10	Diversification category: satisfactory											
Krynica	0.42	0.23	0.05	0.10	0.06	0.08	0.04	0.02	–	–	1.66	0.58
m = 8	Diversification category: satisfactory											
Kraków	0.62	0.11	0.19	0.08	–	–	–	–	–	–	1.06	0.37
m = 4	Diversification category: sufficient											
Opole	0.51	0.26	0.07	0.16							1.17	0.44
m = 4	Diversification category: sufficient											
Andrychów	0.54	0.35	0.11	–	–	–	–	–	–	–	0.94	0.35
m = 3	Diversification category: sufficient											
Częstochowa	0.32	0.62	0.06	–	–	–	–	–	–	–	0.83	0.29
m = 3	Diversification category: sufficient											
Kielce	0.58	0.30	0.12	–	–	–	–	–	–	–	0.93	0.34
m = 3	Diversification category: sufficient											
Krosno	0.28	0.60	0.12	–	–	–	–	–	–	–	0.92	0.33
m = 3	Diversification category: sufficient											
Łańcut	0.37	0.28	0.35	–	–	–	–	–	–	–	1.09	0.44
m = 3	Diversification category: sufficient											
Nowy Sącz	0.26	0.59	0.15	–	–	–	–	–	–	–	0.95	0.35
m = 3	Diversification category: sufficient											
Poznań	0.53	0.29	0.18	–	–	–	–	–	–	–	1.0	0.38
m = 3	Diversification category: sufficient											
Tarnów	0.61	0.21	0.18	–	–	–	–	–	–	–	0.94	0.34
m = 3	Diversification category: sufficient											
Warszawa	0.54	0.32	0.14	–	–	–	–	–	–	–	0.97	0.36
m = 3	Diversification category: sufficient											
Bydgoszcz	0.48	0.52	–	–	–	–	–	–	–	–	0.69	0.25
m = 2	Diversification category: average											
Brzesko	0.95	0.05	–	–	–	–	–	–	–	–	0.20	0.05
m = 2	Diversification category:: little											
Myślenice	0.9	0.1	–	–	–	–	–	–	–	–	0.32	0.09
m = 2	Diversification category:i: little											
Rabka Zdrój	0.28	0.72	–	–	–	–	–	–	–	–	0.59	0.20
m = 2	Diversification category: average											
Rzeszów	0.43	0.57	–	–	–	–	–	–	–	–	0.68	0.24
m = 2	Diversification category: average											
Szczawnica	0.55	0.45	–	–	–	–	–	–	–	–	0.69	0.25
m = 2	Diversification category: average											
Wrocław	0.48	0.52	–	–	–	–	–	–	–	–	0.69	0.25
m = 2	Diversification category: average											
Zielona Góra	0.60	0.40	–	–	–	–	–	–	–	–	0.67	0.24
m = 2	Diversification category: average											

– lack of diversification ds = 0; dw = 0
– little diversification 0 < ds ≤ 0.325; 0 < dw ≤ 0.090
– average diversification 0.325 < ds ≤ 0.693; 0.090 < dw ≤ 0.250
– sufficient diversification 0.693 < ds ≤ 1.386; 0.250 < dw ≤ 0.563
– satisfactory diversification ds > 1.386; dw > 0.563

Table 6 summarizes the values of diversification indices for 22 cities.

5 CONCLUSIONS

Dimensionless values of the indices *ds* and *dw* predispose them to assess absolutely the level of WDS diversification and give the possibility to compare the units with different demand for water in that regard.

Market economy conditions make it possible to diversify water supply through its purchase from various manufacturers operating in neighbouring administrative units or the economic operators.

The numerical values of the indices *ds* and *dw* can be used in the decision making process on the purchase of water for the various economic operators.

Diversification of supply sources for CWSS has a particularly positive role in crisis situations.

Every transition from one to two sources of water supply gives an increase in the diversification degree measured by the indices *ds* or *dw*.

Any combination of the transition from two WSS to their larger number gives an increase in the diversification degree if they have fairly sustainable shares in the global demand for water. When the shares of WSS are considerable differentiated an increase of their number does not always lead to an increase of the diversification degree of water supply to the settlement unit.

REFERENCES

Hill, M. 1973. Diversity and Evenness: a Unifying Notation and its Consequences. *Ecology* 2(54): 427–432. http://en.wikipedia.org

Hurlbert, S. 1971. The Non-concept of Species Diversity: a critique and alternative parameters. *Ecology* 4(52): 577–86

Iwanejko, R. 2006. Koszty bezpieczeństwa a wybór optymalnej decyzji zakupu wody przez przedsiębiorstwo wodociągowe. *Czasopismo Techniczne* 2-Ś: 143–156.

Leszczyński, T.Z. 2012. Wskaźniki bezpieczeństwa energetycznego. Rurociągi 2: 3–10.

Rak, J.R. 2005. *Podstawy bezpieczeństwa systemów zaopatrzenia w wodę*. Wydawnictwo Komitetu Inżynierii Środowiska PAK: Lublin.

Rak, J.R. 2012. Ochrona systemów wodociągowych w ujęciu securitologii. *Instal* 7/8: 38–41.

Rak, J.R. 2014. Problematyka dywersyfikacji dostaw wody. *Technologia wody* 1(33): 22–24.

Shannon, C. & Weaver, W. 1962. The Mathematical Theory of Communication. University of Illinois Press: Urbana.

Stirling, A. 1999. *On the economics and analysis of diversity*. SPUR Electronic Working Paper Series. Paper No. 28.

Weitzman, M. 1992. On Diversity. *Quarterly Journal of Economics* 107(2): 363–405.

Wieczysty, A. red. 2001. *Metody oceny i podnoszenia niezawodności działania komunalnych systemów zaopatrzenia w wodę*. Monografie Komitetu Inżynierii Środowiska PAN: Kraków.

Underground Infrastructure of Urban Areas 3 – Madryas et al. (Eds)
 ISBN 978-1-138-02652-0

Construction of metro station underneath the existing road tunnel

A. Siemińska-Lewandowska
Warsaw University of Technology, Warsaw, Poland

ABSTRACT: In the paper the construction of a metro station underneath the existing road tunnel is presented. The station consists of two parts at the both sides of the road tunnel. In order to connect the platforms and the tracks the connection under the road tunnel, consisting of three tunnels, with a length of about 40 m is built. Due to the vicinity of river geotechnical conditions of the site are very complex. That caused substantial problems in the construction process of the connection tunnels. During the works with use of NATM, the temporary lining of the tunnel collapse and the west part of the station was flooded with water and soil. To complete the works the grouting, jet grouting and ground freezing was used.

1 INTRODUCTION

The central part of the 2nd metro line in Warsaw is 6.3 km long, consist of 7 stations and 6 running tunnels. It passes the center of the city from east to west crossing the Vistula river close to "Powiśle" Station. The "Powiśle" station, which is the subject of analysis, is located entirely on the left bank of the Vistula River under existing road tunnel (Fig. 1).

The station consists of two parts (AGP 2010): west (Fig. 1) and the east (Fig. 2). Their length are respectively 65.00 m and 43.00 m, and the width 21.60 m. The station is built using the top & down method, where elements of structure of the station are 1,4 m thick diaphragm walls, four intermediary slabs, foundation slab and piles. The maximum depth of excavation is approximately 24.00 m from the upper level of retaining walls for the western part and 22.50 m for eastern part. Between the western and eastern part of the station the

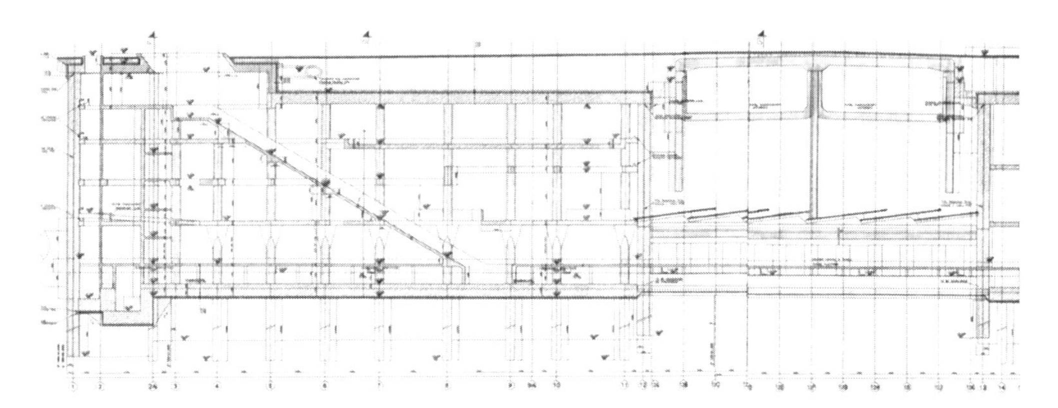

Figure 1. Cross section of the western part of the "Powiśle" station and the road tunnel.

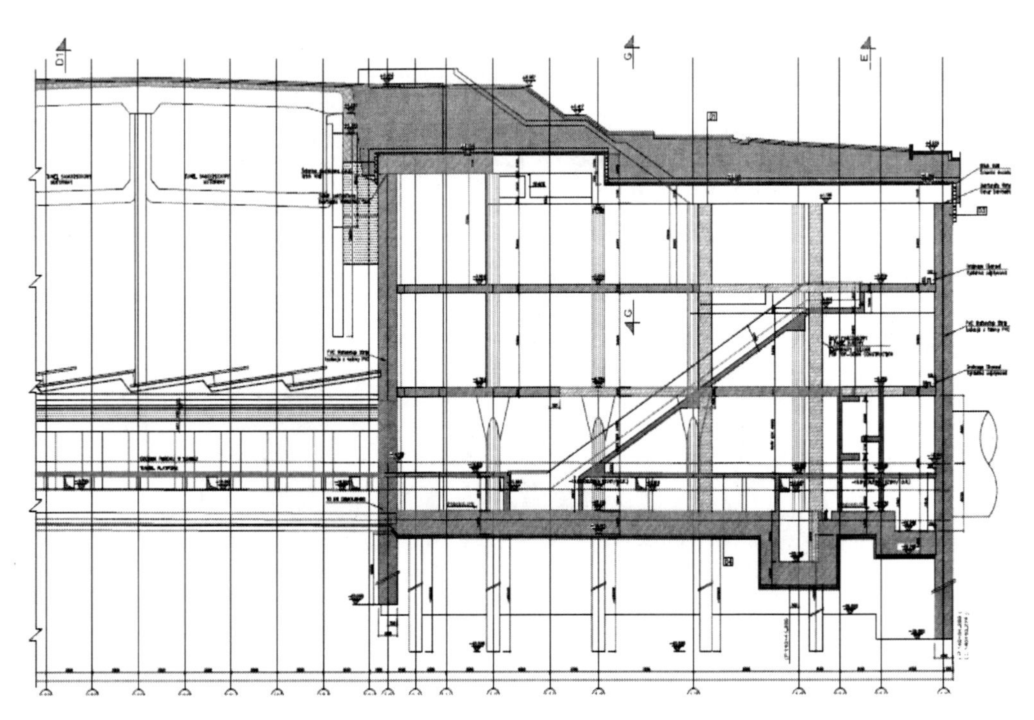

Figure 2. Cross section of the east part of "Powiśle" station.

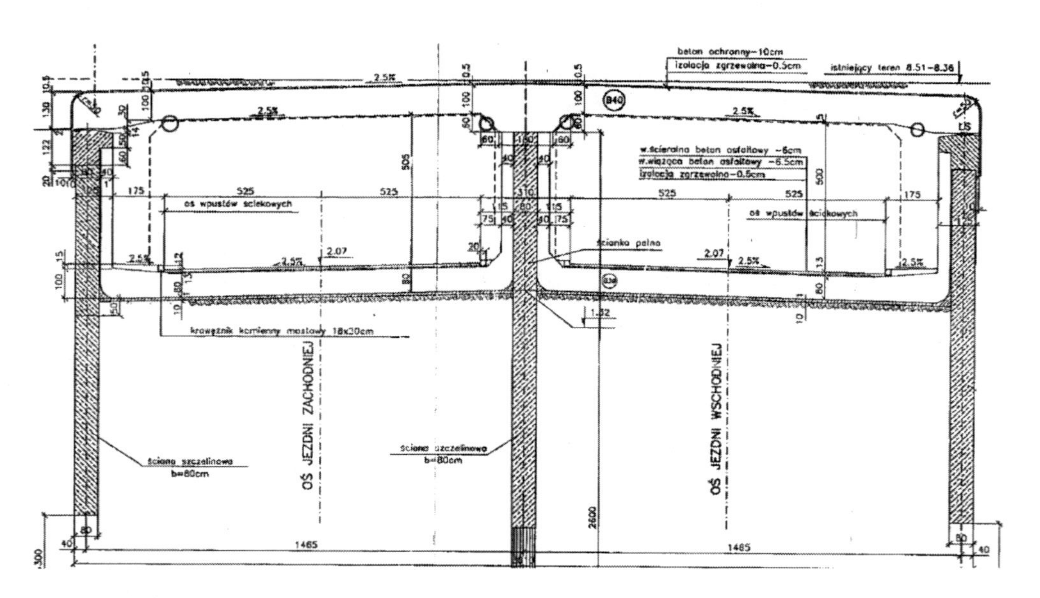

Figure 3. Cross section of road tunnel.

road tunnel exist. The tunnel was built with the use of the Milan method with deep external diaphragm walls and even deeper barrettes under the middle wall. Tunnel structure (Fig. 3) is a two-nave reinforced concrete frame comprising roof plate of the tunnel supported on the peripheral diaphragm walls and on the barrettes in the middle. In order to connect the

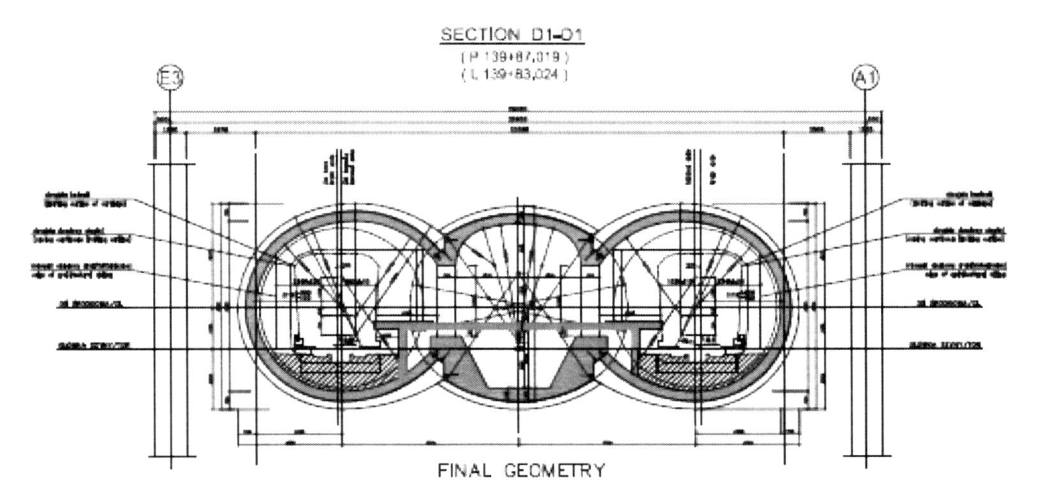

Figure 4. Cross section of connecting tunnels.

platforms and the tracks in west and east part of the station the connection consisting of three tunnels with a length of about 40 m is built (Fig. 4), (AGP 2010).

Due to the vicinity of river geotechnical conditions of the site are very complex and classified according to Eurocode 7 to the third geotechnical category. In general, soil conditions at the construction site are as follows:

- directly under the surface, 5–10 m thick anthropogenic soils occur (so called fills),
- below the layer of fills, 4–10 m thick fluvioglacial formations occur in the form of medium to coarse sands,
- under the sandy formations the Quaternary Pliocene clays are found, which occur as clays, silty clays or stiff silty clays. In this layer 1.20 to 5.70 m thick lenses of silty or fine sands and sandy silts are often found.

The main groundwater level stabilizes at the level +0.60 m above the Vistula river level. The surface of that water table depends largely on the river water level and the amount of rainfall.

2 DESCRIPTION OF THE CONSTRUCTION OF CONNECTING TUNNELS

Connecting tunnel consists of three tunnels—north (left on the Fig. 4), south (right on the Fig. 4) and central. It was decided that during the construction of the tunnels traffic in the road tunnel above will not be stopped. Some panels of barrettes of the road tunnel interfere with the excavation of the connecting tunnels. Their bottom level is lower than the connecting tunnel axis level. During excavation the concrete panels should be demolished at the bottom, up to a max. height of 5 m.

NATM construction method was selected. Temporary lining consisted of steel arcs HEB 240/75, the steel mesh and 0.30 m thick shotcrete reinforced additionally by steel fibres. Permanent lining was designed as reinforced concrete with a thickness of 0,70 m in the invert vault and 0.60 to 1.35 in walls. The cross section area of each of the tunnels was app. 70 m^2. The length of the connecting tunnels is 36 m, the total width –22.6 m The construction works were divided into three stages.

- First stage—construction from both sides of the road tunnel the east and west part of north tunnel, the last phase is the cutting of the baretts of road tunnel in the middle,

- Second stage—construction of south tunnel in the same way,
- Last stage—construction of central tunnel.
- The works started with the construction of the west part of the north connecting tunnel. The construction stages were designed as follows:
- drilling from the west part of the station of 15 horizontal bore holes with a diameter of 190 mm, spacing 0.40 m, grouting and installation of steel tubes with a diameter of 139.7 mm, thickness 12.5 mm and a length of 10 m and fiber glass tubes with a diameter of 110 mm, thickness 10 mm and a length of 10 m—forming the umbrella above the tunnel excavation,
- stabilization of the soil at the front of the tunnel with fiber glass bars, 18 m long with diameter of 60 mm.
- excavation of first 3.5 m, installation of temporary lining—steel profiles (HEB 240) spacing 0.75 m, steel mesh and shotcrete,
- in the upper part of excavation execution of jet-grouting columns 6 m, 8 m and 10 m long with diameter of 800 mm; columns were designed as not reinforced,
- stabilization of the soil at the front of the tunnel with fiber glass bars, 18 m long with diameter of 60 mm.
- excavation of the next 3.5 m, installation of temporary lining—steel profiles (HEB 240) spacing 0.75 m, steel mesh and shotcrete,
- in the upper part of excavation execution of an umbrella with 68 jet-grouting columns, 6 to 12 m long, not reinforced,
- excavation of the west part of the north tunnel until the baretts of the road tunnel in the middle, installation of temporary lining,

The stages of construction of the east side of the north tunnel were the same as the west side.

At the night from 13th and 14th August 2012, during the excavation of next 3.5 m of the west part of north connecting tunnel and installation of the fifth steel profile (HEB 240) an outflow of water under pressure from the crown of the tunnel occurred. As a consequence, the bottom slab of the road tunnel settled down about 60 mm and the traffic inside the road tunnel was suspended. In order to stabilize the flow of water with the soil from the area underneath the bottom slab of road tunnel to the inside of the western part of the "Powiśle" station, the station was filled with water up to level of the first underground slab (Fig. 5).

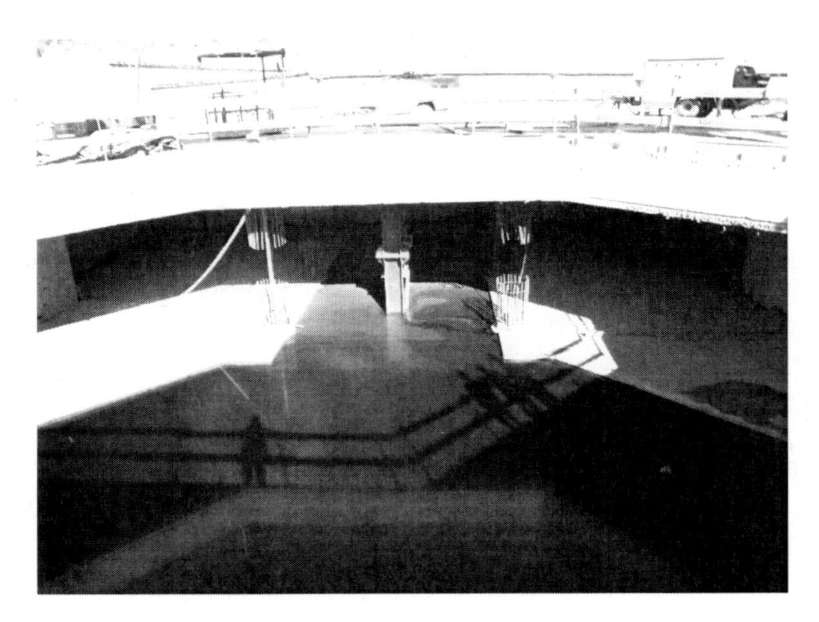

Figure 5. The west part of "Powisle" station filled with water with sand up to the level of first underground slab.

After that, the balance of the hydraulic loads between the inside and outside of the station was obtained and the structure displacements of the road tunnel brought to a stop. It was stated that station was filled with app. 6500 m^3 of soil (sand). In order to identify the areas affected by the loss of material the georadar and seismic surveys were conducted. The surveys confirmed the presence of one cavity under the bottom slab of the road tunnel, located between diaphragms walls and extending for a distance of approx. 60 m underneath the western road and 50 m underneath the eastern road. The cavity was filled with water up to approx. 2 m from the bottom of the slab.

3 SAFETY MEASURES

To provide the safety of the road tunnel and the works in the connecting tunnels as well as to stabilize the displacements of the bottom slab of the road tunnel, the following measures were proposed:

- execution of a plug, which will prevent further transport of the soil inside the station from the area underneath the tunnel,
- filling of the existing cavity underneath the bottom slab of the road tunnel with the rapid setting concrete with auto-levelling behavior through 250 mm boreholes in the bottom slab.
- after control surveys, consolidation of the voids in massive filling with cement mortars pumped at low pressure,
- consolidation of the soil underneath the massive filling with low pressure grouting of cement mixes and with the use of tube a manchette (Fig. 6); the area of consolidation to be limited by impermeable thin panels made by means of jet-grouting in the shape of Y. The depth of grouting was designed to reach the depth of the invert vault of connecting tunnels,
- after the consolidation is executed on the entire area and after completion of verification tests of the subsoil (CPT, core drilling) the traffic in the road tunnel could be restored.
- during all phases of reparation works, the structure of road tunnel, and the west part of the station should be monitored.

The grouting was supposed to be executed in non-cohesive and cohesive soil layers like fine sands, silty sands and Pliocene clays. During the complementary geological investigation

Figure 6. Low pressure grouting through the bottom slab of road tunnel.

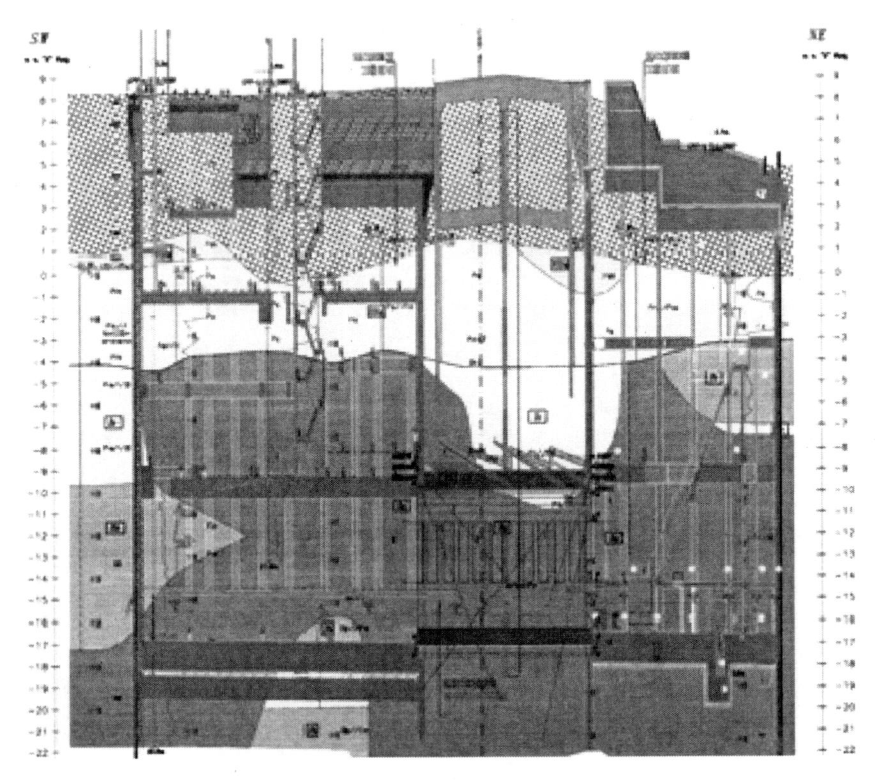

Figure 7. Geological cross section along the north connecting tunnel.

geotechnical soil conditions were verified (Fig.7) and the presence of silty sand layers filled with water in the upper part of west north connecting tunnel as well as a great difference in the levelling of the top boundary of the Pliocene clays layer were confirmed. In fact, the erosion notch occurred deeper than it was identified in the original geotechnical report. It is clearly shown on the geological cross section along the north west part of connecting tunnel (Fig. 7). That was the one of the reasons of the collapse in the north connecting tunnel.

During the next six months the grouting with cement and silica mixtures was carried out down to the invert level, in order to compact the soil in the area of construction of connecting tunnels and to restart the excavation works. The effect of grouting was not sufficient because of cracking (claquage, Fig.8), which occurred in the Pliocene clay layer. The monitoring of the displacements of the road tunnel confirmed the sensible movements of the bottom slab due to the grouting pressure. The area of collapse filled with the mixture of different soils and the water was very difficult to carry out the grouting. For all this reasons the grouting had to be stopped and it was not possible to restart the excavation. In consequence, the decision was made that the ground freezing around the excavation area is the only solution for the safe construction of connecting tunnels underneath the existing road tunnel.

4 THE PROCESS OF GROUND FREEZING

The artificial ground freezing was chosen in order to ensure consolidation and the waterproofing of the sandy and silty layers. In the sands the diffusion of the refrigeration was good; the silts had a lower refrigeration capacity and conductivity, so they needed more time to freeze. The lower Pliocene clay layers had a low permeability coefficient, which made it a good layer for sinking the freezing treatment.

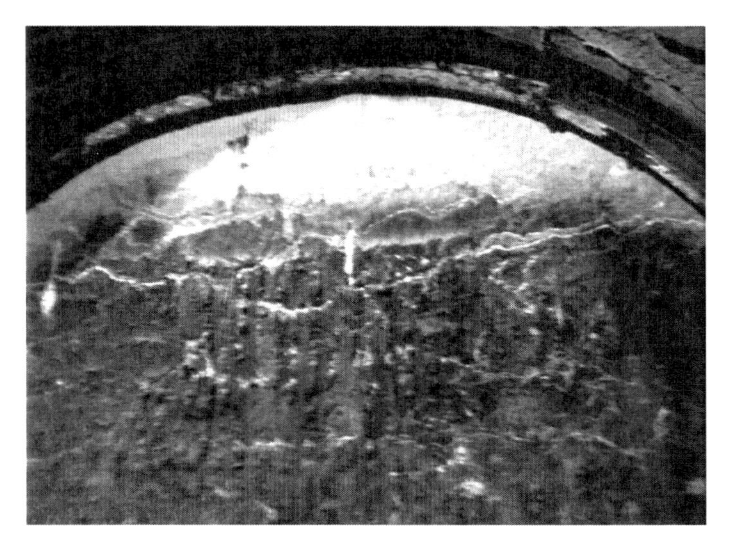

Figure 8. The effect of cracking—calquage in pliocene clay layers.

Figure 9. Excavation of north—west connecting tunnel in frozen ground.

The treatment based on the application of liquid nitrogen was adopted. The open circuit method was used in the first phase, as well as in the first maintenance stage. After that, the maintenance was switched to close circuit method with calcium chloride brine as a coolant. The treatment was carried out with 8 m high water table, measured from the upper part of connecting tunnels.

The stages of construction of the west and east parts of all three connecting tunnels remained the same as before the collapse. The freezing process around the arches of each connecting tunnel was carried out respectively from west and east parts of the station. The sub-horizontal tubes along a cylindrical surface surrounding the excavation profile were drilled above tunnel arches. It has been decided to start with the excavation of east—south part of connecting tunnel.

The final stage of each tunnel construction—cutting of central barettes—was executed in vertical frozen ground curtain made from the inside of the road tunnel. The curtain was made with the use of liquid nitrogen and after demolition of barettes, maintained with calcium chloride brine.

The construction of south and north tunnel was completed with final concrete lining. The lining was stopped 2 m from the central frozen wall, and the barettes were demolished. The construction of central connecting tunnel and cutting off the walls in between north and south tunnel is the final stage of construction of connecting tunnels underneath the existing road tunnel. In a few days the freezing maintenance will be turned off.

During all freezing and excavation process the settlements of the bottom slab of the road tunnel were monitored, especially in the stage of cutting of the central diaphragm wall. No excessive settlements were noticed.

5 CONCLUSIONS

The results of low pressure grouting and jet-grouting of the silty sandy and clayey layers, under the high water table, proved that proper consolidation and impermeability of those layers is very difficult to be obtained. The fine granulometry of the soil prevented the regular and homogeneous diffusion of the consolidating grout. The claquage phenomena occurred and the process of consolidation was stopped.

On the other hand, even such a low pressure grouting in combination with very high water pressure (water level) caused substantial and continuous movements of the road tunnel and led to the instability of the structure.

The ground freezing, finally adopted for the construction of the metro station underneath the existing road tunnel in service was the appropriate solution in very complex and difficult geotechnical conditions (silty and clayey layers with sand lenses carrying water under pressure as well as high water table due to the vicinity of the river).

REFERENCES

Analysis of the complementary soil investigation performed in order to precise the leveling of the top boundary of Pliocene clays layer in the area of designed C13 "Powiśle" station. 2012. Warsaw University of Technology. Geotechnics and Underground Structures Division.

Building Investment C13 "Powiśle" station, Section A-A, Section D1-D1, Design and construction of the underground line II from "Rondo Daszyńskiego" station to the "Dworzec Wileński" station in Warsaw, Consortium AGP, Astaldi S.p.A. 2010.

Geotechnical and structural design, C13—action needed to complete the excavation of tunnels, October 2012, Consortium AGP, Astaldi S.p.A. 2012.

Grodecki W. & Siemińska-Lewandowska A. 2012 Expertise concerning the determination of the causes of the accidents and the methods of its removal on the C13 station of the central section of the II metro line in Warsaw. *Internal Report*, Metro Warszawskie Sp. Z o.o.

Ground freezing process around the tunnels of C13 station under Wisłostrada tunnel. General Report, Consortium AGP, Astaldi S.p.A. 2013.

Mitew-Czajewska M. & Siemińska-Lewandowska A. 2014. Uncertainty of geotechnical conditions—a case study of metro tunnel construction underneath the existing road tunnel. In. *Proc. of the ITA-AITES World Tunnel Congress 2014. Tunnels for a better life, Fozz do Iguassu, Brasil.*

Siemińska-Lewandowska A. et all. 2009. Construction of the 2nd metro line in Warsaw – problems and execution methods. In: *Proc. of the ITA-AITES World Tunnel Congress 2009. Budapest.* Rotterdam: Balkema.

Underground Infrastructure of Urban Areas 3 – Madryas et al. (Eds)
© 2015 Taylor & Francis Group, London, ISBN 978-1-138-02652-0

Hydraulic analysis of functioning of the Przemyśl town sewage system

D. Słyś & J. Dziopak
Department of Infrastructure and Sustainable Development, Rzeszów University of Technology, Rzeszów, Poland

A. Raganowicz
Zweckverband zur Abwasserbeseitigung im Hachinger Tal, Taufkirchen, Germany

ABSTRACT: The article presents a results of modeling sewage system of the Przemyśl town and research of hydraulic capabilities of the wastewater transport in the Zasanie quarter. The hydrodynamic model of the urban area and sewage system was basis of the research. The objective of development was to determine the increase hydraulic loading system possibilities and indentification of the sections of the sewage system, which are particularly exposed to hydraulic overload.

1 INTRODUCTION

Together with development of urban areas and their infrastructures, an increase of watertight surfaces in the catchment area usually occurs, hindering significantly infiltration of rainwater to the ground. Consequences of such changes include intensification of surface rainwater runoff from the catchment area. Situations involving occurrence of a short but very intensive runoff of rainwater from the catchment area to rivers, in most cases through closed sewage systems, result in a number of adverse effects, including the drop of ground water level resulting from disturbed balance between rainfall, evaporation, soaking and surface runoff, more frequent and higher surges in rivers, disturbances in rivers biological life, erosion of river beds and other.

Activities undertaken in order to counteract such effects are focused mainly on three tasks aimed at:

- reduction of rainwater volumes that must be transported to rivers by reduction of impervious surface areas;
- development of calculation methods and technical solutions aimed at introduction of rainwater to the ground (Scholz and Yezdi 2009, Datry et al. 2003),
- utilization of rainfall water for household purposes (Sturm et al. 2009, Matsumora and Mierzwa 2008, Słyś 2009) and temporary retention of said waters in order to reduce the peak flow rate values in sewage systems and receiving bodies (Scott 1999, Dziopak and Słyś 2009).

Increase of rainwater amounts routed to sewage networks poses frequently a significant technical and financial issue and results in necessity to increase flow capacity of sewers, extend the existing sewerage system components and, in many cases, construct new facilities.

2 FORMULATION PROBLEM

The town of Przemyśl is located in south-eastern part of Poland in the vicinity of the Ukrainian border and is one of the most important urban centers in Podkarpackie province, inhabited by 67 thousand people.

The spatial layout of Przemyśl is a result of natural conditions creating and determining its boundaries, San and Wiar rivers, railway lines, roads and other local conditions. Three diverse areas are typically identified in the town, including:

- Stare Miasto (Old Town), a residential area located south on right bank of the San river;
- Zasanie residential quarter located north on left bank of the San river;
- town areas located east on both banks of the Wiar river.

The town of Przemyśl has a gravitational combined sewage system. Sewage from the town area are transported by means of main collector sewers, located on both sides of the San river, to the municipal wastewater treatment plant situated in the bifurcation of the San and Wiar rivers.

The fundamental problem that currently needs to be solved pertains to the way the sewerage system should operate in the course of intensive rainfall. Along with physical development of the city, surface of watertight areas within the catchment increases resulting in increase of rainwater volumes channeled from Zasanie quarter through the main sewer.

The article presents a results of modeling sewage system of the Przemyśl town and researches of hydraulic capabilities of the wastewater transport in the Zasanie quarter.

3 A HYDRODYNAMIC MODEL OF A SEWAGE SYSTEM

The Storm Water Management Model 5.0 software prepared by EPA (US Environmental Protection Agency) for modeling river-basins and sewerage systems is the application used in this hydraulic analysis.

A hydrodynamic model of the Zasanie catchment area was calibrated on the basis of data on quantities and numbers of storm water discharges into the San river.

In the SWMM program, a rainfall depth is transformed into an effective outflow of a linear reservoir whose filling equals the quantity of water that fell on a given surface, including all losses occurring due to water evaporation, infiltration, and retention in terrain low spots. Therefore, it is important to know parameters such as: the sealing ratio, Manning's coefficient n, terrain slope, the quantity of terrain retention, and for non-sealed surfaces the ground infiltration capability. Detailed information on these parameters is very important because they influence the total effective rainfall depth and consequently the quantity of rain water outflow from a given river basin. The outflow rate of water directed to the sewerage system is calculated by equation (1).

$$Q = W/n\,(d - d_p)^{5/3} s^{1/2} \tag{1}$$

where: Q—outflow from a river-basin (dm^3/s), W—outflow zone width (m), n—Manning's coefficient (-), d—water layer height (m), d_p—water layer height for water retained in terrain low spots (m), s—terrain slope (%).

In the case of the Zasanie catchment area, these parameters were selected on the basis of information concerning the sewer types, catchment management, terrain slopes, and also data from the literature.

In the calculation of water losses, Horton's infiltration model was used, which is described by the following relation (2).

$$f = f_c + (f_o - f_c)e^{-kt} \tag{2}$$

where: f—infiltration ability (mm/h), f_C—final infiltration ability (mm/h), f_0—initial infiltration ability (mm/h), k—time-constant of infiltration rate decline curve (1/h), t—time from an occurrence of precipitation (in days).

This model assumes that the terrain absorbing capacity declines asymptotically from an initial quantity f_0 to a final quantity f_C, while a time-constant k determines the shape of the infiltration rate decline curve, illustrated in Figure 1.

The hydrodynamic model of the sewerage system includes 81 main sections with a total length of 177.15 km. These are concrete sewers with circular, oval and rectangular cross sections. Due to the age of the sewers and their technical condition, Manning's coefficient n adopted for the calculations was between 0.013 and 0.018.

The simulations were carried out using real rainfall depth data from the years 2007 and 2008, which were based on real measurements from pluviometers located within the Zasanie quarter. Table 1 shows the parameters and their quantities used to calculate the quantity of rain water outflows from the analyzed catchment area, while Figure 2 presents the layout of the sewerage system.

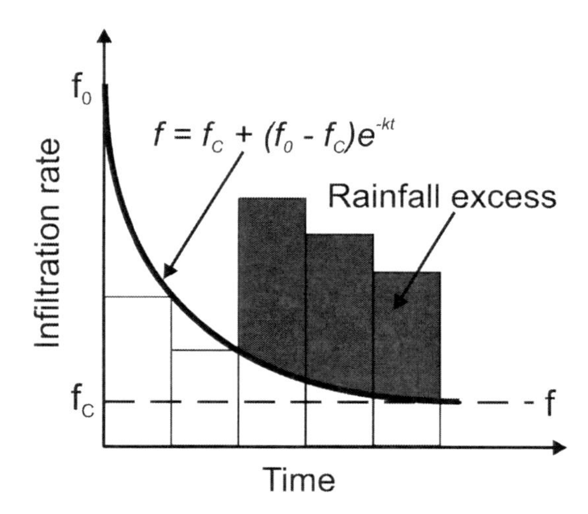

Figure 1. Infiltration curve as per the Horton's model.

Table 1. Characteristic input data for the Zasanie quarter adopted for the hydrodynamic model.

Parameter	Quantity
Catchment area sealing rate	10–60%
Manning's n coefficient (non-sealed surface)	0.25
Manning's n coefficient (sealed surface)	0.015
Retention water height (non-sealed surface)	7.0 mm
Retention water height (sealed surface)	1.5 mm
Minimum infiltration rate	20 mm/h
Maximum infiltration rate	90 mm/h

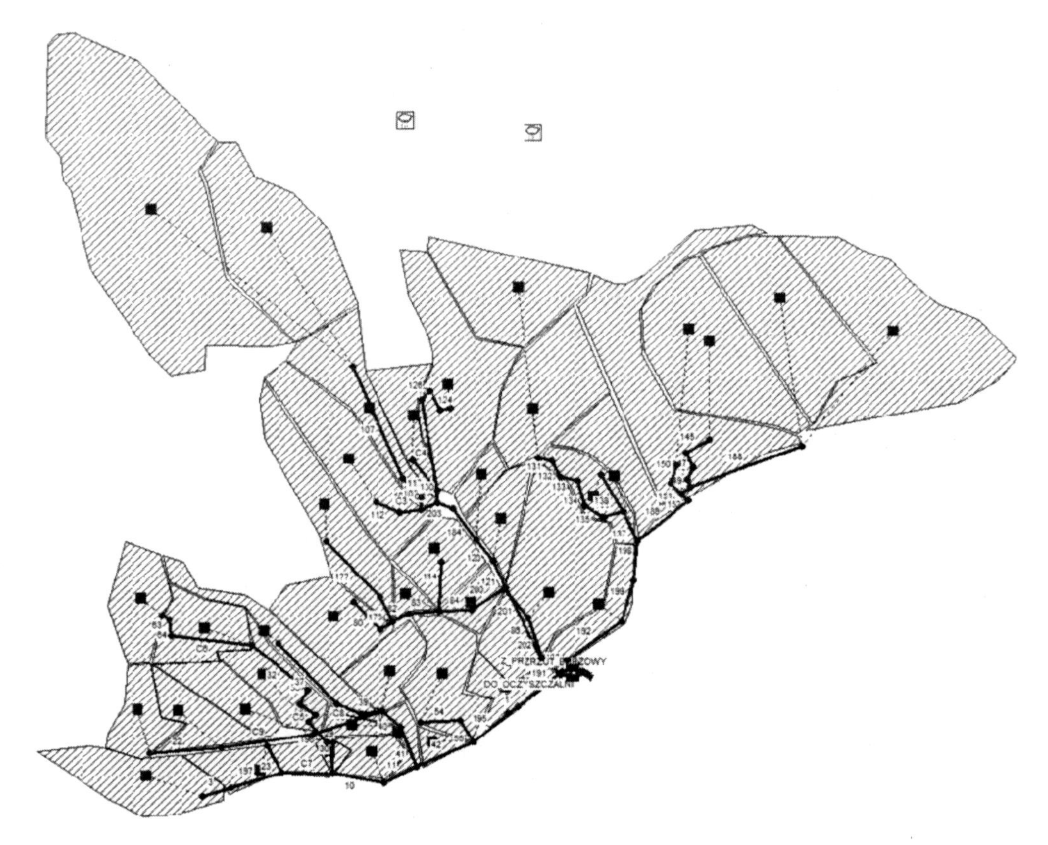

Figure 2. Layout of main trunk sewers in the Zasanie quarter of Przemyśl.

3 ANALYSIS OF THE HYDRAULIC CONDITIONS IN THE TRUNK SEWERS

As a consequence of this hydraulic analysis, the sewer sections particularly susceptible to hydraulic overloads were selected. As a criterion, the presence of at least two occurrences was used, which means a sewer link completely filled during rainfalls and a diameter bigger than 0.6 m.

The analysis demonstrated the existence of 29 trunk severs within the Zasanie quarter which were frequently overfilled. There are circular sewers of diameters between 0.8 and 1.6 m, as well as oval sewers with a height between 1.2 and 1.7 m. Figures 3 to 5 show the results of simulated wastewater fillings in the selected sections of the sewerage system most exposed to the risk of overfilling.

The analysis also included a hydraulic assessment of other sections of the sewerage system in terms of their flow capacity utilization and possible wastewater retention. Figure 6 shows the locations of sewer sections most exposed to overloads during rainfalls.

The analysis also provides a hydraulic assessment of other sections of the sewerage system in terms of their flow capacity utilization and possible wastewater retention.

This research allows us to ascertain that idle flow capacities exist in sewers with small diameters between 0.3 and 0.6 m operating as a sanitary sewage system. Within the combined sewage system characterized by considerable dimensions of cross-sections, with the exception of the sewer links marked 86, 191 and C4, there are no sewers with idle flow capacity. The results of the calculations for sewer links 86 and 191 are presented in Figures 35 and 36 respectively.

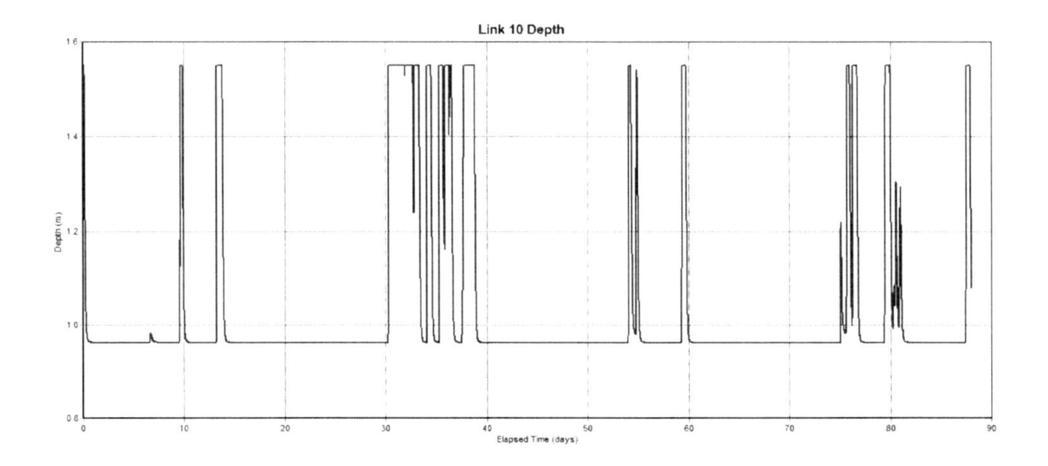

Figure 3. Fillings in sewer link 10 for archival rainfall depths from the year 2007.

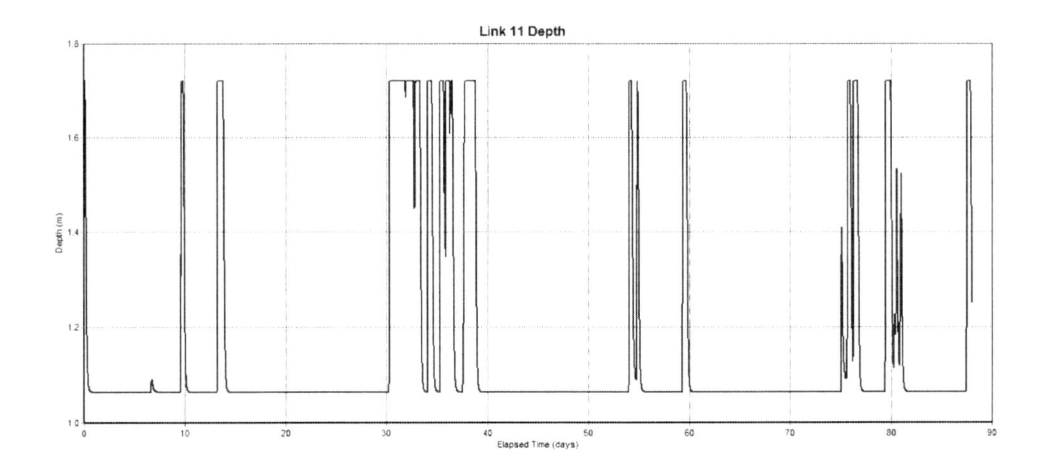

Figure 4. Fillings in sewer link 11 for archival rainfall depths from the year 2007.

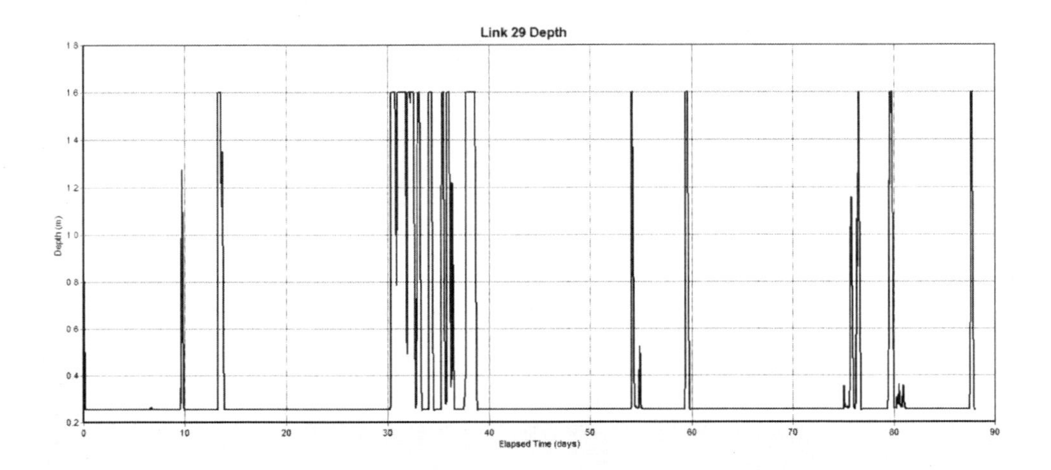

Figure 5. Fillings in sewer link 29 for archival rainfall depths from the year 2007.

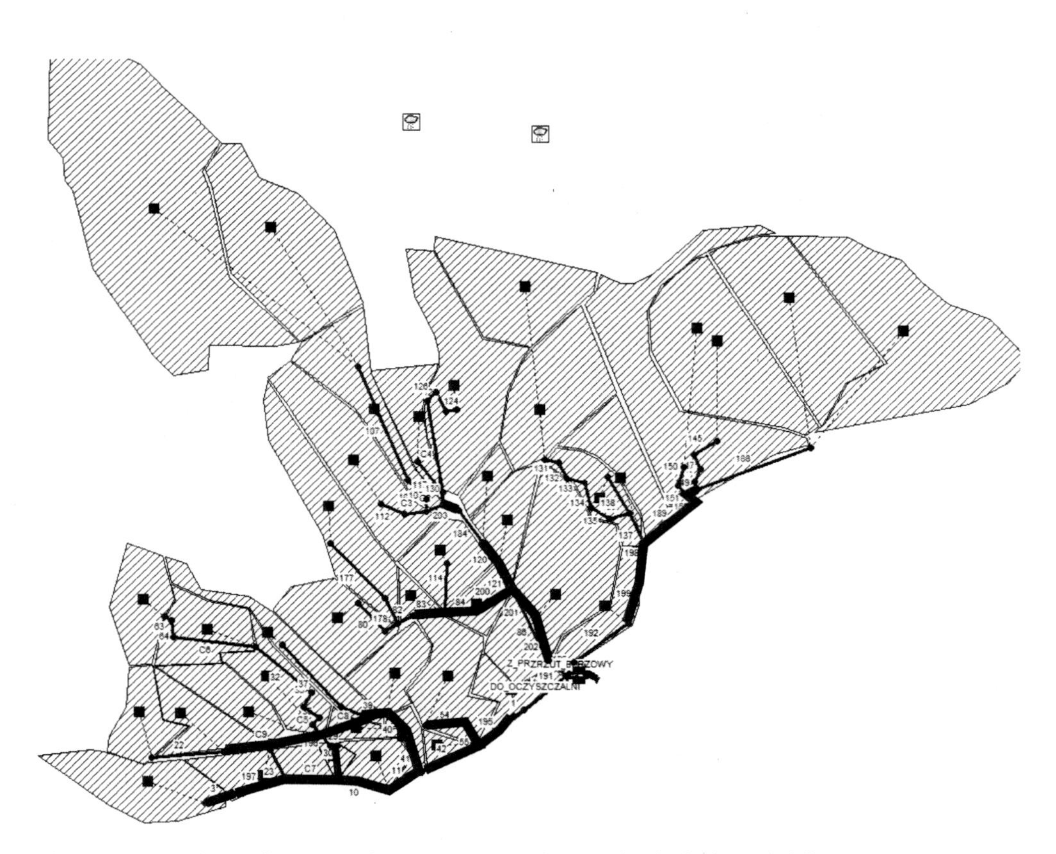

Figure 6. Locations of sewer sections most exposed to overloads during rainfalls.

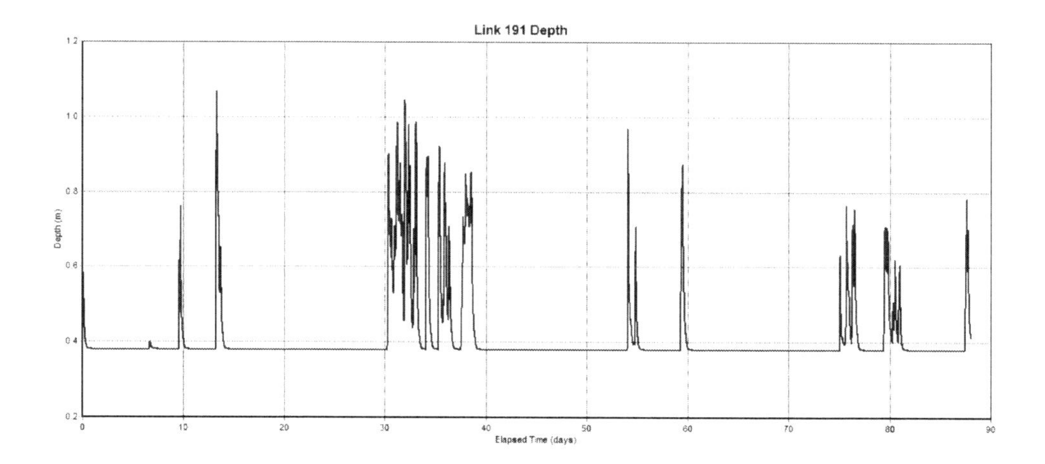

Figure 7. Fillings in sewer link 86 for archival rainfall depths from 2007.

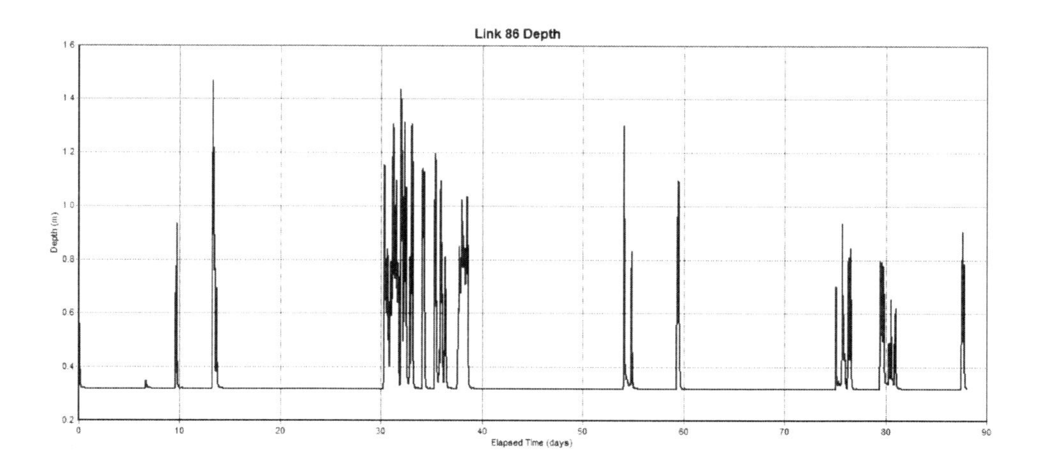

Figure 8. Fillings in sewer link 191 for archival rainfall depths from 2007.

4 FINDINGS

The hydrodynamic analysis of sewerage system performance in the Zasanie quarter enables us to draw several cognitive and applicable conclusions.

1. As seen from the analysis conducted using simulation modeling and archival meteorological data, most of the trunk sewers in Zasanie sewerage system have pressure flows. In many cases these flows are long-lasting and occur over periods of several hours.
2. The secondary sewer links are still the least hydraulically loaded; these sewers often have small diameters.

3. Because the sewers with sufficient hydraulic flow capacities have limited cross-section dimensions, there is no basic possibility to retain wastewater in the sewerage system, with a few exceptions for secondary sewer links.
4. The results of the computer-aided hydrodynamic analysis verify the results obtained by measuring flows and fillings in the sewerage system and the observed flood occurrences within the city area.

REFERENCES

Datry T., Malard F., Vitry L., Hervant F., Gibert J. 2003. Solute dynamics in the bed sediments of a stormwater infiltration basin. Journal of Hydrology. 273, 217.

Dziopak J., Słyś D. 2009. Simulation researches of pump-gravitational storage reservoir and its application in sewage systems. Underground Infrastructure of Urban Areas. CRC Press, London, UK. 75.

Matsumura E.M., Mierzwa J.C. 2008. Water conservation and reuse in poultry processing. Resources Conservation & Recycling. 52, 835.

Scholz M., Yezdi S.K. 2009. Treatment of Road Runoff by a Combined Storm Water Treatment, Detention and Infiltration System. Water Air Soil Pollut. 198, 55.

Scott P., Santos R., Argue J.R. 1999. Performance, environmental and cost comparisons of onsite detention (OSD) and onsite retention (OSR) in re-developed residential catchments. Wat. Sci. Tech. 2, 33.

Słyś D. 2009. Potential of rainwater utilization in residential housing in Poland. Water and Environment Journal. 23, 318.

Sturm M., Zimmermann M. Schutz K., Urban W., Hartung H. 2009. Rainwater harvesting as an alternative water resource in rural sites in central northern Namibia. Physics and Chemistry of the Earth. 34, 776.

Underground Infrastructure of Urban Areas 3 – Madryas et al. (Eds)
© 2015 Taylor & Francis Group, London, ISBN 978-1-138-02652-0

Buried steel structures in urban infrastructure—underpass on the Kapitulna street in Włocławek

P. Tomala, A. Marecki & M. Woch
ViaCon Polska Sp. z o.o., Rydzyna, Poland

K. Mikołajczak
Husar Budownictwo Inżynieryjne Sp z o.o., Włocławek, Poland

ABSTRACT: Paper presents an alternative solution to the classic ones—a flexible burried steel structure as the superstructure of pedestrian underpass at Kapitulna street in Wlo-clawek which provides crossing under the tracks of the railway line Number 18 Kutno-Piła. Underpass was built in 2012 by HUSAR BUDOWNICTWO INŻYNIERYJNE commissioned by Włocławek city.

The use of the corrugated steel structure allowed fast realization of the investment, without detriment to aesthetics and functionality of underpass.

1 INTRODUCTION

Modernization and expansion of urban infrastructure, due to nature of the project area, is associated with a number of technological problems, but also the need to ensure the traffic, at least in a limited scope, over the reconstructed structure. The total closure of traffic in this case means more funding and social costs.

The most critical point of the entire project was very tight work schedule. The reason for that was the fact that the proposed pedestrian crossing was located under electrified multi-lane railway line, which complete closure was impossible. In addition, the construction site was located close to the city center and prolongation closure of the communication would be burdensome for citizens.

Work was carried out in stages, and due to the use of corrugated steel plates as the superstructure, it was possible to limit the time of technological breaks, and execution time of the entire project to minimum. In conventional solution, in which concrete is used, it is necessary to make work brakes due to the maturation of the concrete. Thanks to possibility of assembly of steel superstructure in parts outside the designed location and the closure of part of the railway tracks, the task could be completed without any delay.

2 CHARACTERISTICS OF DESIGNED STRUCTURE

2.1 *Primary solution using precast concrete*

Main parameters of structure:

- Dimensions of cross section: 4.50 × 3.00 m
- Length: 49.85 m
- Length of precast elements: 2.49 m
- Concrete of precast elements: B50
- Concrete of headwalls: B40
- Reinforcing steel: A-IIIN

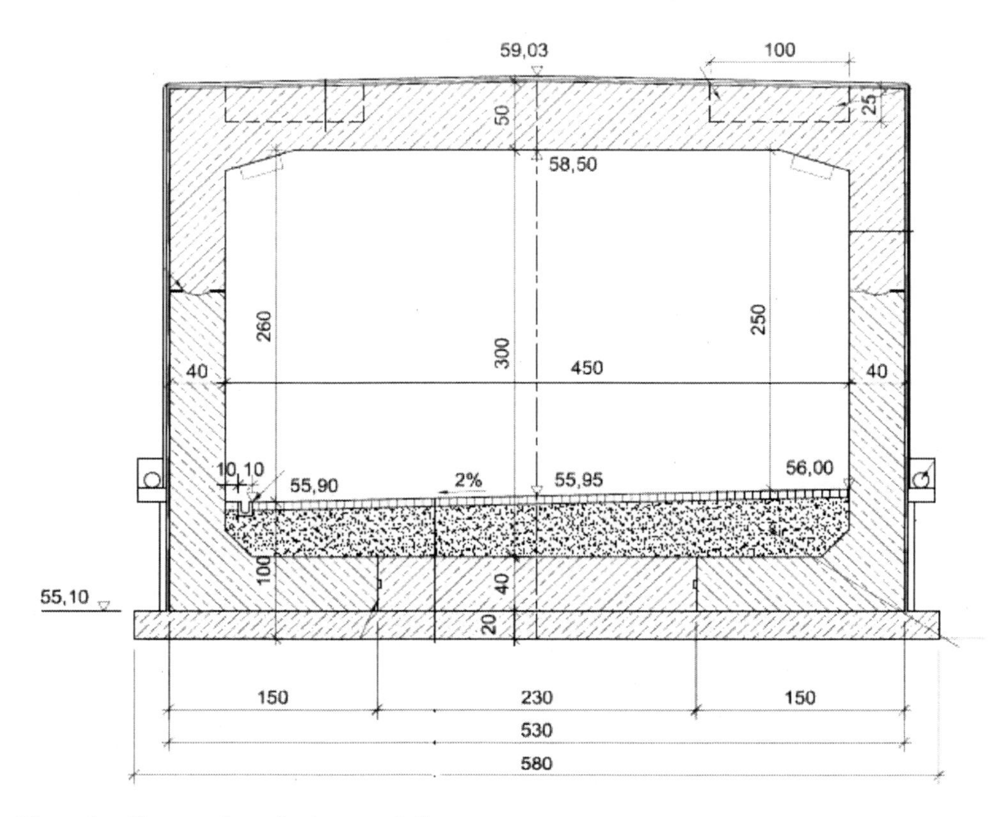

Figure 1. Cross section of primary solution.

Structure was designed as frame made of precast concrete elements with inlet and outlet as reinforced concrete retaining walls. Function of structure is pedestrian underpass which provides crossing under the tracks of the railway line Number 18 Kutno-Piła. Superstructure of underpass was calculated for rail load k = +2 according to PN-85/S-10030.

Because in primary solution, in which concrete is used, it is necessary to make work brakes due to the maturation of the concrete, contractor began to search alternative solution which would speed up works requiring partial closing of railway no 18.

2.2 *Alternative solution using soil-steel structure*

Main parameters of structure:

- Clearance box of underpass: 4.50 × 3.00 m
- Dimensions of culvert in cross section: 4.91 × 3.95 m
- Corrugation: 200 × 55 mm
- Plate thickness: 5.5 mm
- Length of structure: 49.85 m
- Length of precast elements: 6.00–9.60 m
- Concrete of headwalls: B40
- Reinforcing steel: A-IIIN
- Steel of structure: S235JR

Alternative solution was based on steel-soil structure made of corrugated steel plates joined by blots M20 class 8.8. Cover above superstructure of underpass measured from bottom of tie to neutral axis of corrugation was 1.15 m. Solution of inlet and outlet remained unchanged. It was only adapted to the shape of the steel structure.

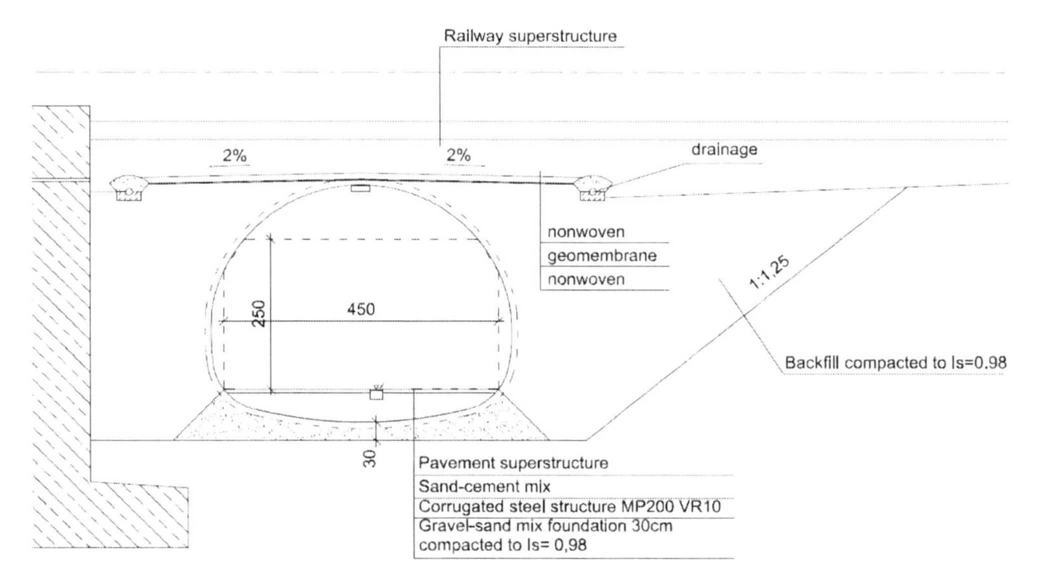

Figure 2. Cross section of alternative solution.

Due to the closed shape and relatively good soil conditions, foundation was proposed as 30 cm layer of sand-gravel mix compacted to Is = 0.98.

The applied thickness 5.5 mm and additional anti-corrosion coatings allowed to achieve capacity allowing rail load k = +2 according to. PN-85/S-10030 with maintaining calculation durability of over 100 years.

From the inside structure was protected with a zinc coating thickness acc. to PN-EN ISO 1461:2000 and epoxy paint coating thickness of 200 μm. From the outside, also was used zinc coating with additional coating based on bitumen with a minimum thickness of 4 mm.

To protect the object against rain water infiltrating through the backfill designed to "umbrella" protection in the form of geocomposite (nonwoven 500 g/m^2 + geomembrane HDPE + nonwoven 500 g/m^2).

Drainage was realized by the 2.0% two-sided decrease in cross-section that brings rainwater to drain on both sides.

Inherent in the case of steel-soil structure is compacted backfill to Is = 0.98, which should be built in symmetrically in 30 cm layers on both sides of steel structure. As the material had to be used to backfill of sand-gravel mix fraction of 0–45 mm, uniformity coefficient $Cu > 5.0$ and coefficient of curvature $1 < Cc < 3$.

3 CONSTRUCTION OF THE UNDERPASS

3.1 *Stage I*

The task could be divided into four key stages. Work began with the demolition of the existing embankment. Wall of excavation was secured by berliner wall. This allowed for the execution of works related to the partial demolition of the existing abutment of overpass above Kapitulna street.

3.2 *Stage II*

The next step was to close five of the seven railway tracks and partial execution of the object.

Figure 3. Excavation of existing embankment.

Figure 4. Assembly of steel structure.

Before earthworks started there was made retaining wall of steel sheet piles which was protection of the excavation necessary to perform the object. The schedule included a performance of 26 m sheet piling to a depth of 6 m in less than 6 h.

Parallel to the works related to the excavation, gravel-sand mix foundation and foundation of headwall, there was carried out a partial prefabrication of superstructure with the implementation of external paint coating based on bitumen. The use of the structure of corrugated steel plates, significantly lighter than originally designed precast reinforced concrete, allowed to use of longer segments without the use of heavy equipment. Precast 3 sections with a total length of about 35 m. Assembly of superstructure with the implementation of paint coating took totally 4 days.

Before backfill was built in it had been necessary to make reinforced concrete headwall. For the headwall was used concrete B40 reinforced with steel A-IIIN.

The quality and degree of compaction of backfill has a very significant impact on the soil-steel structure capacity therefore each 30 cm thick layer of backfill was controlled.

Figure 5. Merge between two phases of construction.

All works associated with the construction of the culvert under five railway tracks were carried out within 16 days while maintaining the technological regime.

3.3 *Stage III*

Work on the remaining part of the underpass began after the traffic was released on the reconstructed tracks. As in the earlier stage, superstructure was initially precast.

Merge between two stages was made after hole in sheet piling was cut.

Inlet also was made as reinforced concrete head wall. For aesthetic reasons, the joint of the headwall and steel structure was hidden under steel collar. Works related to building the second part of superstructure took a total of 6 days.

3.4 *Stage IV*

The final stage was mechanical raise the track in order to achieve the designed elevation of the rail head. Additional works which no longer required the closure of the railway tracks were finishing works and the construction of a MSE wall system ViaBlock, which aim was to secure the unstable slope along the sidewalk.

4 SUMMARY

Time of work execution was in this case a key aspect of the planning and optimization of a design. Contractual penalties associated with any delays that may have affected the traffic on the railway line No. 18 Kutno-Piła were 1 164 000 zł per day which was a third of the total value of the works.

Execution of tasks was conducted in a 24 hour, 7 days a week while the schedule has been planned to the hour. Dividing the works related to the superstructure of underpass in two stages, allowed the traffic of passenger and freight trains with access to the station platform Wloclawek without the need for substitute means of transport. Time to complete the whole task was scheduled from 11.07.2012 to 15.11.2012 while performing works related to the superstructure of underpass requiring the closure of the railway line were a total of 29 days.

Despite the short notice provided for works related to the occupation of the railway, works were completed on schedule, with full technological regime and with building craft.

Figure 6. Finished structure.

REFERENCES

Janusz, L., Madaj, A., 2007. *Obiekty inżynierskie z blach falistych. Projektowanie i wykonawstwo*, Warszawa: Wydawnictwa Komunikacji i Łączności.

PN-EN ISO 1461:2000. *Powłoki cynkowe nanoszone na wyroby stalowe i żeliwne metodą zanurzeniową. Wymagania i metody badań*, PKN.

PN-85/S-10030. *Obiekty mostowe. Obciążenia*, PKN.

Underground Infrastructure of Urban Areas 3 – Madryas et al. (Eds)
 ISBN 978-1-138-02652-0

Application of prestressed concrete pipes to deep pipelines

R. Wróblewski & A. Kmita
Wroclaw University of Technology, Wroclaw, Poland

ABSTRACT: Application of prestressed concrete pipes at great buried depths (exceeding app. 5.0 m) is discussed in the paper. Modification of these pressure pipes is considered which requires limitation of a crack width. Experimental investigation of cracking performance is presented in the paper. The experiments were carried out under line, static load with records of patterns of cracks. Since that type of loading is used only for the pipe testing, differences to more realistic models are also presented. The modification methods leads to material and construction proposal which fulfil requirements related to cracking (cracking load and crack width) and load bearing capacity.

1 INTRODUCTION

1.1 *Problem description*

Two-directionally prestressed concrete pipes are designed and used as elements of municipal pressure pipelines for water delivery. The design rules of these pipes are soundly stated (Stephenson 1989; Raju 2008; EN 642:1995) and evaluation of damage tolerance is also available (Hovland & Najafi 2009; Valiente 2001; Zarghamee et al. 2003), so they are recognised and competitive elements of infrastructure. Moreover the pipes were extensively tested and analysed (Kmita & Wróblewski 1998; Kmita & Wróblewski 2006). A manufacturer in search of new markets decided to use the pipes for pressureless (e.g. sewage) pipelines which require deeper and different bury conditions. According to EN 1992-1-1:2004 pressureless concrete pipes should fulfill requirements of both ultimate and serviceability limit states (ULS and SLS). In the last case the crack width is limited to 0,3 mm. In contrast to the last requirement, cracks are not allowed in the pressure pipes in SLS and that is equivalent to requirement of ULS. The less severe restriction made to the pressureless elements gives a certain reserve of load bearing capacity of pressure pipes. Nevertheless control of the crack width is required in the prestressed elements, but they do not include any reinforcement in a tension zone that appears after the change of bury conditions. The authors present methods that will enable modification and fulfillment of ULS and SLS by pressure pipes intended to use in the same way as pressureless elements.

1.2 *Elements requiring modification*

The prestressed pretensioned concrete pipes are designed and produced as pressure elements in three classes of nominal pressure (0.5, 1.0, 1.5 MPa). The elements are made of concrete—class C50/60. The pipes are two-directionally prestressed (Fig. 1). In circumferential direction the spiral strand of diameter ϕ 4, 5, 6, 8 mm is used. The strand diameter and its spacing are strictly determined by the pipe class. Meridional prestress reinforcement is made of ribbed bars. All the bars are tensioned simultaneously and the tension force is controlled through the elongation measurement.

Technology of prestressing is unique: the tension in spiral reinforcement is forced through the rubber shell of the cast. The shell is filled with water under pressure. The cast pushes

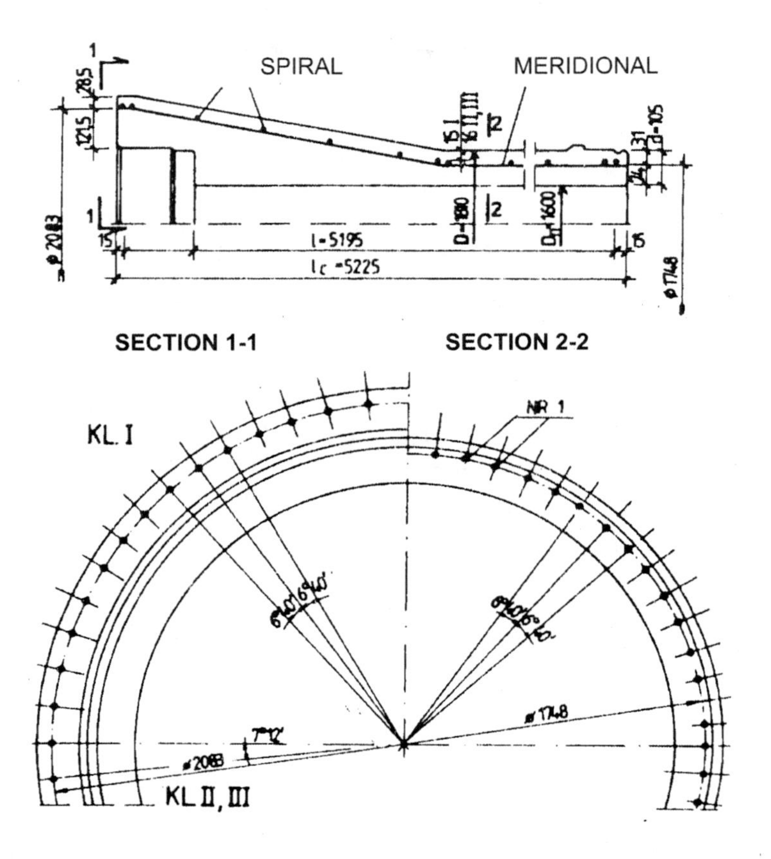

Figure 1. Typical pipe sections.

on consolidated concrete mix and then concrete pushes on spiral reinforcement. The prestress force in the spiral reinforcement is controlled via water pressure in the rubber shell and through the measurement of the cast deformation. The circumferential reinforcement (spiral) is moved close to the outer surface but minimum concrete cover of 15 mm is obtained.

Since the indirect methods used to control the prestress force in spiral reinforcement were considered uncertain, tests were carried out on the stress in strands. The results proved that the stress in this reinforcement $f_s \approx 650 \pm 75$ MPa. The variation of the stress is larger than 10%, but the tests proved that the indirect control methods are sufficiently effective.

2 BACKGROUND OF MODIFICATION

2.1 *Design problems*

Increased and modified loadings change distribution of stress and thus axial forces N and bending moments M in the section of the pipe. An absence of internal pressure increases influence of compression forces and increased depth magnify moments and so tension and compression zones appear instead of solely compression. Furthermore this leads to necessary verification of crack appearance or calculation of crack width. According to EN 1992-1-1:2004 the design crack width is calculated with:

$$w_k = s_{r,\max}\left(\varepsilon_{sm} - \varepsilon_{cm}\right) \tag{1}$$

where $s_{r,\max}$ = crack spacing and $\varepsilon_{sm} - \varepsilon_{cm}$ = difference in medium strains in steel and concrete between the cracks.

In elements reinforced only with prestressing steel (Fig. 1), without any reinforcement close to the internal surface, estimation of crack width can be made experimentally as presented in section 2.2 or can be calculated. In the last case prestressing reinforcement will play any role in controlling the crack width only if tension zone will appear near it. This can be achieved for large eccentricity of the force (bending moment is essential) with very narrow compression zone in concrete. In the pipes in soil or water such combination of stress can hardly be achieved because the loads cause simultaneously axial forces and bending moments (Fig. 2). But assuming appearance of cracks, effective control of crack width will essentially reduce the section thickness (to approximately two concrete covers of prestressing steel) in calculation.

As an alternative to ineffective crack width control it is possible to limit the loads in a way that the cracks will not occur, and so the following conditions have to be fulfilled:

a. Pipe in the ground

$$N_{Ed} \leq N_{cr} = \frac{f_{ctm} + \sigma_{cp}}{\frac{1}{A_{cs}} + \frac{e}{W_{cs}}} \tag{2}$$

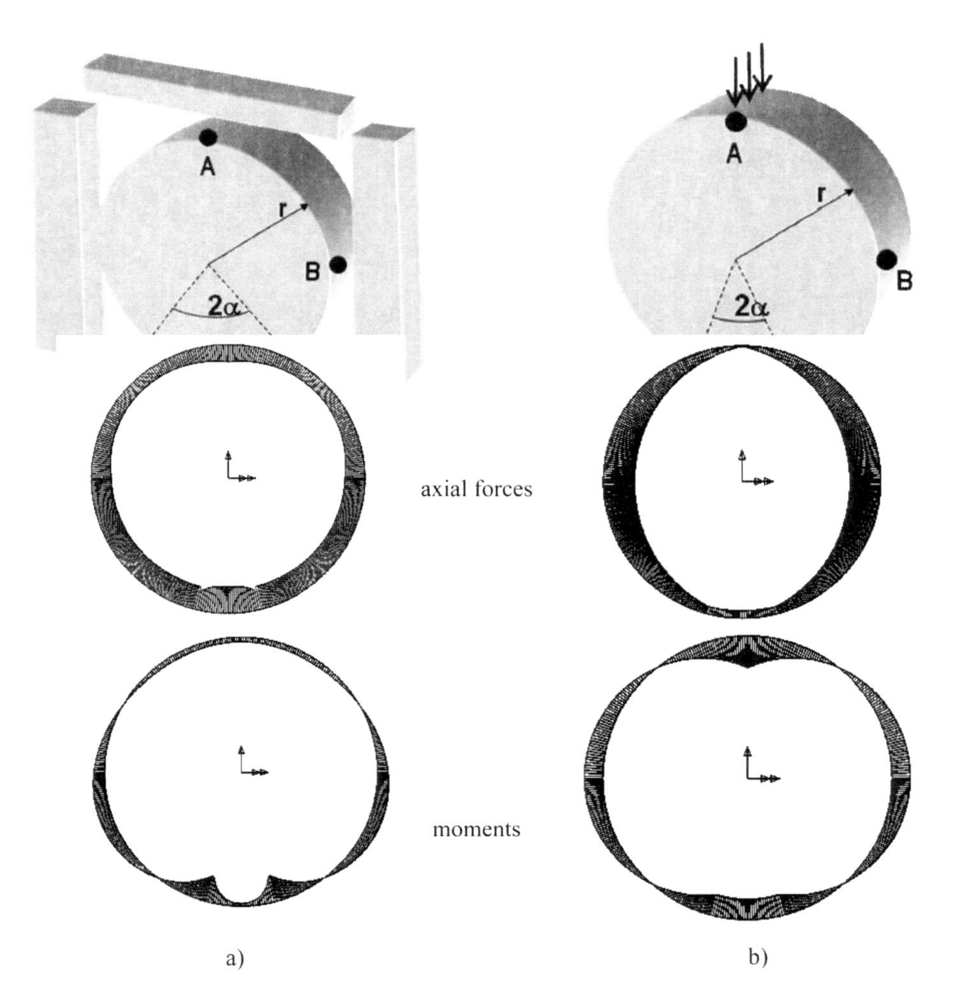

Figure 2. Schematic loads and support conditions in a pipe with distribution of axial forces and moments: a) in the ground, b) diametral line load used to verify the class of a pipe.

b. Verification of the pipe class (point A in Fig. 2b)

$$M_{Ed} \leq M_{cr} = W_{cs}\left(f_{ctm} + \sigma_{cp}\right) \tag{3}$$

where N_{Ed}, M_{Ed} = axial force and bending moment—results of loads; N_{cr}, M_{cr} = cracking force and moment; f_{ctm} = concrete tensile strength; σ_{cp} = compressive stress in concrete from prestress; A_{cs} = cross-section area of concrete including steel and W_{cs} = cross-section modulus of concrete including steel.

Different loading and support conditions are considered when a class of pipe is required according to EN 1916:2002. A diametral line load is applied in design and in experimental verification (Fig. 2). This type of load is very different and much more severe than real loads in the ground (Fig. 2, e.g. from ground, water and live loads from vehicles). Load bearing capacity and crack control have to be checked for these conditions in which bending moment in point A (Fig. 2) is dominant. A link between two types of conditions (a and b in Figure 2) is possible via force—moment interaction curves calculated for the pipe cross-section.

According to the reinforcement arrangement presented in Figure 1 there is not reinforcement close to an internal surface of the pipe. If the tension zone appears in this area of the section it is not possible to calculate crack spacing and medium strains in steel according to EN 1992-1-1:2004. With this in mind one should find a method for determination of crack width or apply additional reinforcement for crack control.

2.2 *Experimental verification*

Experiments were performed on several typically manufactured prestresed pipes. They were carried out on a stand (Fig. 3). The pipes were loaded with diametral line load as presented in Figure 3. Monotonic load was applied until first crack on an internal surface appeared and

Figure 3. Loading stand.

then the load was increased and appearance of subsequent cracks was recorded. As a practical application criteria the limit of internal crack width $w_{max} = 0.2$ mm was set. The failure load (Fig. 4) exceeded significantly the load accompanying the internal crack width 0.2 mm (Fig. 5). After overcoming w_{max} the load was increased until element failure (Fig. 4). Large deformations (50 mm change of diameter) were forced to obtain the pipe failure mechanism (Fig. 6).

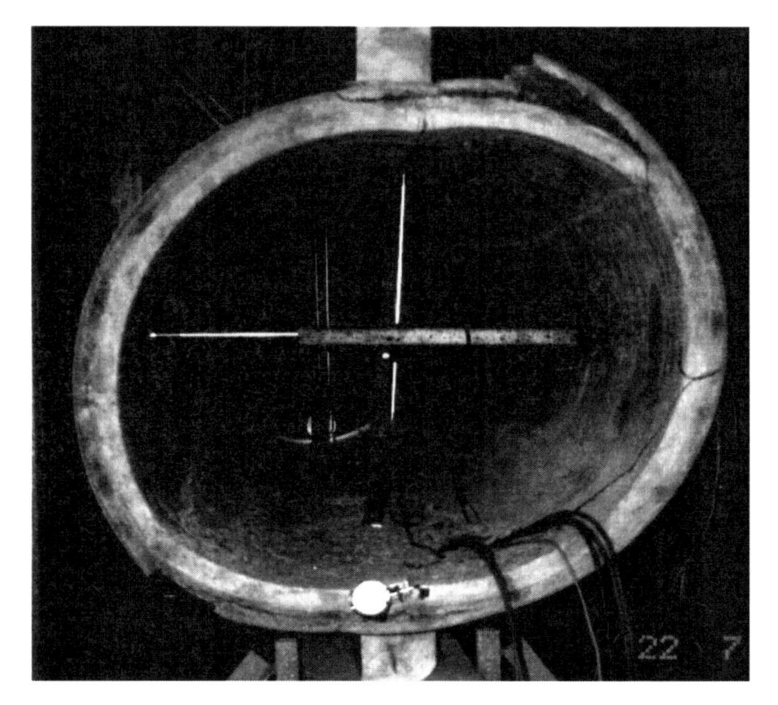

Figure 4. Failure mechanism.

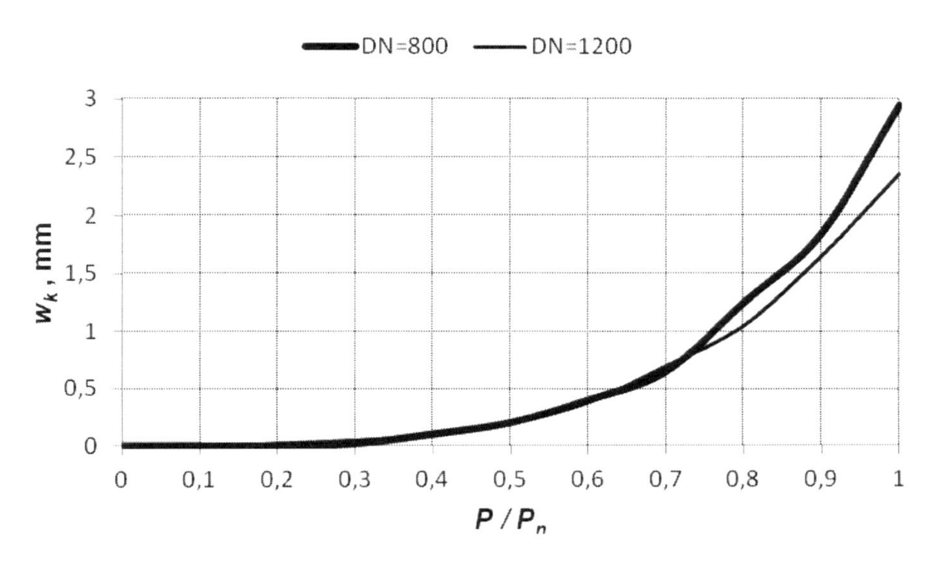

Figure 5. Experimental relationship between load level and crack width for two pipes: DN = 1200 mm and DN = 800 mm (P—load; P_n—failure load; DN—nominal, i.e. internal diameter).

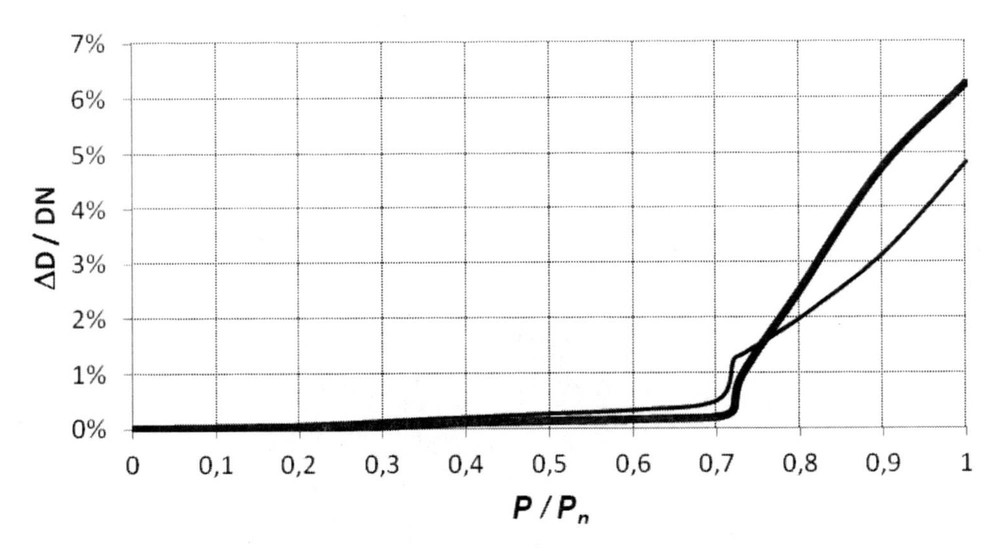

Figure 6. Experimental relationship between load level and diameter deformation (between points A and C in Fig. 2) for two pipes: DN = 1200 mm and DN = 800 mm (P—load; P_n—failure load; DN—nominal, i.e. internal diameter).

3 MODIFICATION METHODS

Various modification methods are possible but not all of them are can be applied in existing manufacturing process. The following ones are found to be promising:

a. According to formulas (2) and (3) increase of prestressing force will increase cracking resistance
b. According to formulas (2) and (3) increase of concrete tensile strength will increase cracking resistance.
c. Application of mild reinforcement in tensile zone (internal reinforcement) that will limit crack width
d. Mixed method of a), b), c).

Both increases of prestressing force and concrete tensile strength have technological restrictions. Moreover increase of prestressing force would require replacement of essential equipment in a plant, so it is practically ineffective. In case of concrete tensile strength it is limited to a cement based concrete (it would not require an application of a new material, e.g. polymer concrete), so only concrete class can be increased.

In Figure 7 calculated influence of concrete strength on the crack width and in Figure 8 on load bearing capacity is presented. Increase of cracking force N_{cr} or moment M_{cr} is linear according to (2) and (3) and is presented in Figure 9. The pipe of DN = 800 mm and wall thickness $t = 65$ mm made of concrete C50/60 is considrered. The dimetral line load was set constant (causing $w_k = 0.2$ mm) and than the crack width was recalculated with various values of f_{ck}. The crack width was calculated in point A of Figure 2b, because there exist only axial force from prestress and bending moment form the loads. But it would not be possible to calculate crack width in any section of a pipe in the ground, because in this case larger axial forces accompany bending moments.

Application of mild reinforcement near the internal surface enables calculation of the crack width and experimental verification is no longer necessary. Load bearing capacity can be also theoretically obtained. In this case maximum crack width for partially prestressed structures should be considered according to EN 1992-1-1:2004. Mild reinforcement enables effective use of a section under eccentric loading and control of crack width and limitation of forces and moments according to (2) and (3) is no longer necessary.

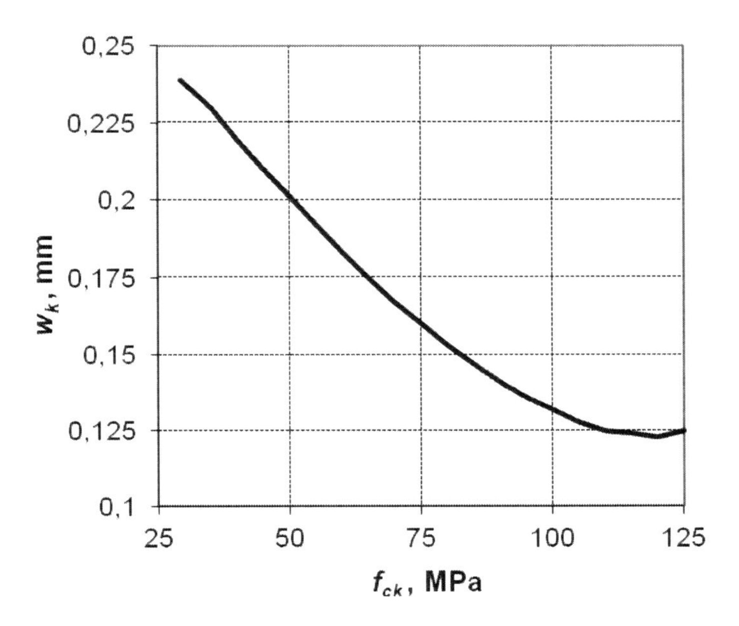

Figure 7. Influence of concrete strength on crack width w_k of the pipe DN = 800 mm.

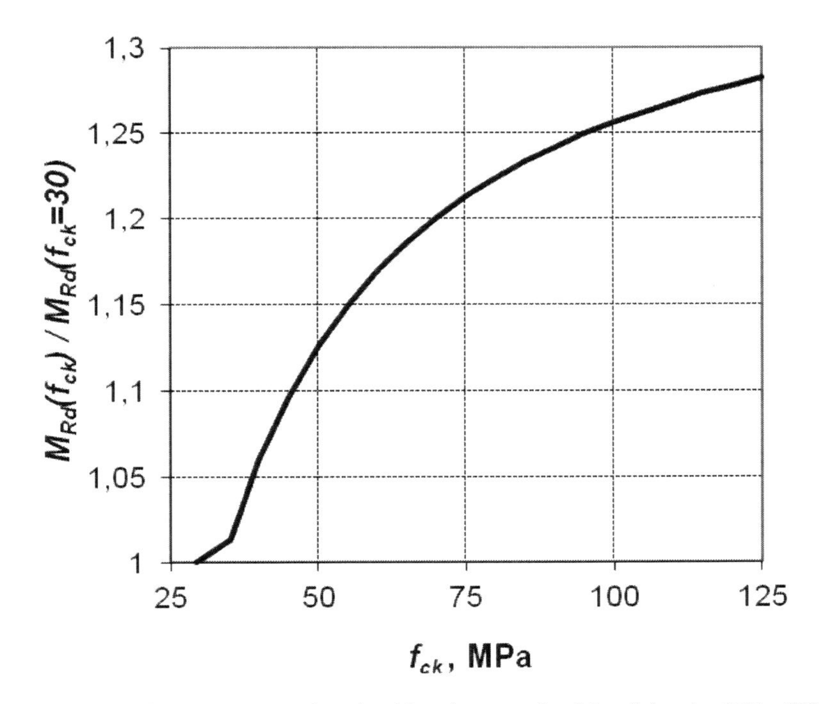

Figure 8. Influence of concrete strength on load bearing capacity M_{Rd} of the pipe DN = 800 mm.

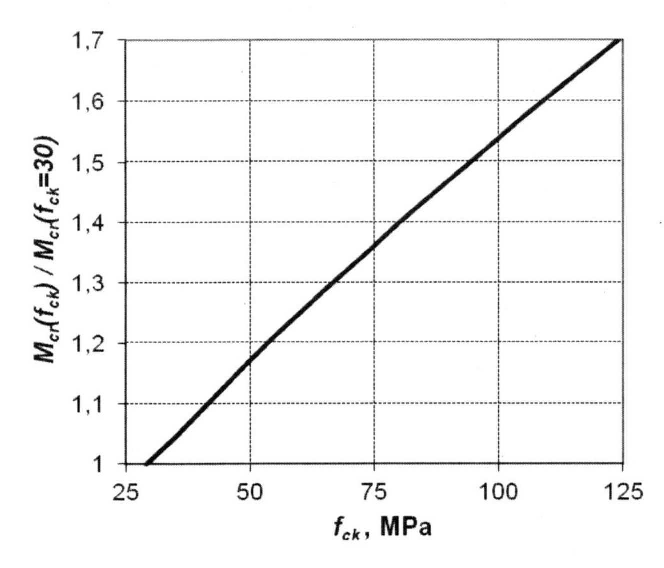

Figure 9. Influence of concrete strength on cracking moment M_{cr} of the pipe DN = 800 mm.

4 CONCLUSIONS

Presented results indicate that application of prestressed pressure pipes for construction of pressureless pipelines is possible and after some modifications they can be settled deeper than pressure pipes. There are two practically possible improvements to the pipes: increase of concrete strength and application of additional mild reinforcement close to the internal surface of the pipe. A combination of both methods is also possible and is considered as the most effective. Moreover application of high performance concrete is effective due to increase of durability (chemical and mechanical).

REFERENCES

EN 642:1995. *Prestressed concrete pressure pipes, cylinder and non-cylinder, including joints, fittings and specific requirement for prestressing steel for pipes.*

EN 1916:2002. *Concrete pipes and fittings, unreinforced, steel fibre and reinforced.*

EN 1992-1-1:2004. *Design of concrete structures. General rules and rules for buildings.*

Hovland, T. & Najafi, M. 2009. Prestressed Concrete Cylinder Pipe. Inspecting Pipeline Installtion: American Society of Civil Engineers: 143–155.

Kmita, A. & Wróblewski R. 1998. Experimental verification of plane stress concrete FE models—Beams and shells. *Proceedings of the Euro-C 1998 conference on computational modelling of concrete structures.* Rotterdam/Brookfield: A.A. Balkema: vol. 2, 79–985.

Kmita, A. & Wróblewski, R. 2006. High Perfomance Concrete in Pipes-Case studies. *European Symposium on Service Life and Servicability of Concrete Structures ESCS-2006 Espoo*, Finland.

Ojdrovic, R.P., Nardini, P.D., Zarghamee, M.S. 2011. Verification of pccp failure margin and risk curves. *Pipelines 2011: A Sound Conduit for Sharing Solutions.* American Society of Civil Engineers: 1413-1423.

Raju, N.K. 2008. *Prestressed concrete.* Tata McGraw-Hill Education.

Stephenson, D. 1976. *Pipeline design for water engineers.* Oxford: Elsevier Scientific.

Valiente, A. 2001. Stress corrosion failure of large diameter pressure pipelines of prestressed concrete, *Engineering Failure Analysis* 8(3): 245–261.

Zarghamee, M.S., Eggers, DW, Ojdrovic, R.P., Rose B. 2003. Risk analysis of prestressed concrete cylinder pipe with broken wires. *New Pipeline Technologies, Security and Safety.* American Society of Civil Engineers: 599–609.

Underground Infrastructure of Urban Areas 3 – Madryas et al. (Eds)
© 2015 Taylor & Francis Group, London, ISBN 978-1-138-02652-0

The specificity of construction of eco civil engineering objects in transport infrastructure using trenchless technology

A. Wysokowski
Road and Bridge Department, University of Zielona Góra, Zielona Góra, Poland

C. Madryas
Municipal Engineering Department, Wroclaw University of Technology, Wroclaw, Poland

ABSTRACT: With the sustainable development of transport infrastructure can not be avoided environmental issues. In view of the undeniable advantages of trenchless technology area of their application is multidirectional and applies to both the materials used, contracting technologies (machinery and equipment) as well as modern methods of calculation and design. Although Poland has significant experience in the application of this technology, and even what it is worth noting some of the national achievements in this field are a record, it is still too rarely these types of technologies are used in the transport infrastructure in particular, taking into account the above their undeniable advantages.

1 ECOLOGICAL STRUCTURES IN TRANSPORT INFRASTRUCTURE

Considering dynamic development of transport infrastructure in our country, on no account the ecology-related issues could be passed over. It refers to construction of new roads and railway lines and also to their modernizations. To this purpose, it becomes necessary to construct ecological facilities of transport infrastructure—crossings for animals and combined function culvers.

It is of special importance for motorways, dual carriage ways and high-speed rail networks which are continually increase their lengths, the animal crossings being their integral components (Kurek 2010) which is essential as these lines are provided with fencing.

Also of importance is a thorough analysis of the locations of such crossings, their number and technical dimensions in view of local conditions and needs. This bears exceptional economical significance when such facilities are constructed and, even more important, in their further maintenance (Wysokowski et al. 2008, Wysokowski et al. 2007).

The authors of the paper have brought up the need of preparing unified *recommendations for design, construction and maintenance of animal crossings*, which would explicitly straighten up all issues related to these structures. At present, such recommendations are being in ongoing preparation phase.

Such recommendations, drawn up and approved by interdisciplinary group including various specialists (design engineers, environmental biologists) and also investors, could contribute to further minimizing the problem of accidents on communication routes, both concerning animals and people. Furthermore, the rules specified in the recommendations could eliminate all irrational solutions as early as during the preliminary design stage.

In terms of communication projects, and especially when we consider their number planned, this issue is very demanding and of big importance, and—in authors' opinion—requiring urgent solution (Conference proceedings 2013).

An exemplary structural scheme of underground animal crossing under communication routes is shown in Figure 1.

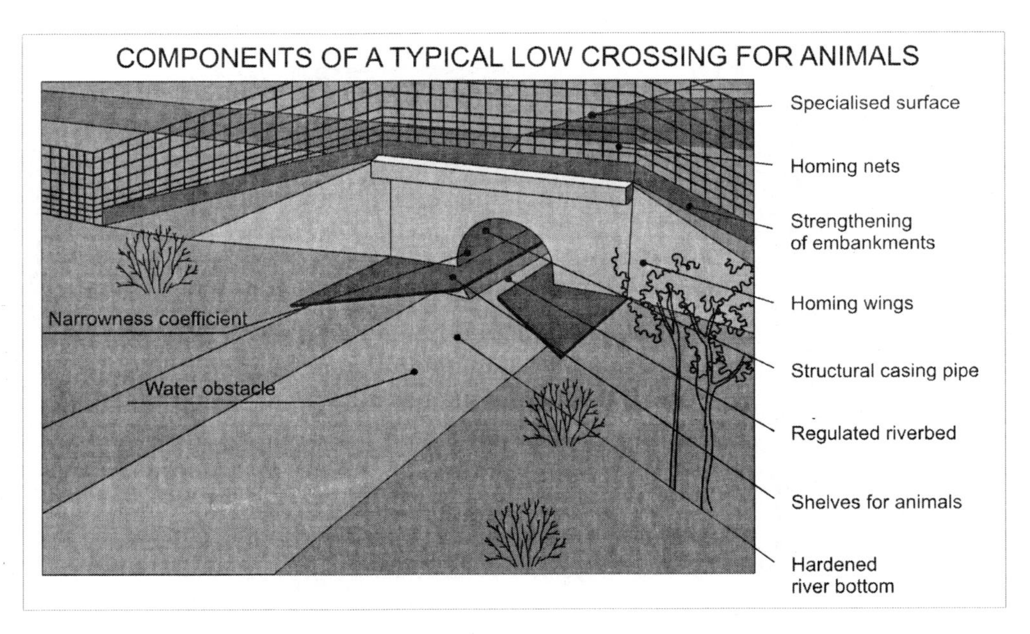

Figure 1. Structural scheme of animal crossing under transport routes including component elements (A. Wysokowski).

The animal crossing shown in Figure 1, or similar crossings, can be constructed in excavations or using trenchless technologies. This paper reveals potentials and advantages of trenchless technologies applied in constructing such types of structures.

2 UNIQUE ISSUES IN CONSTRUCTING TRANSPORT INFRASTRUCTURE COMPONENTS USING TRENCHLESS TECHNOLOGIES

2.1 *Trenchless technologies for proper structures*

Bearing in mind their undeniable advantages, widely described in relevant literature (Kuliczkowski et al. 2010, Madryas et al. 2006), trenchless technologies are wider and wider applicable in Poland. It can be even said that the number of these applications in municipal network structure, and lately, also in constructing the networks beyond developed area, makes Poland a one of leading countries in this field. The range of analyses on applicability of trenchless technologies is multi-directional and refers to materials, machines and other equipment (e.g. control systems) used and modern design methods as well.

The authors, upon reviewing the type of structures under consideration, are in opinion that classic micro-tunnelling, described in detail in relevant literature (Madryas et al. 2006), could be most often used for such projects. As this technology is highly developed, the top-advanced equipment is easily accessible as well as a wide range of materials, hereinafter outlined. Hence, a fast and safe execution of animal crossings is possible from 100 to 4,000 mm in diameter, which meets the scope of actual demands.

It should be also emphasized that the currently used equipment for micro-tunnelling enables considerable reduction of construction costs. Furthermore, this technology is much more friendly for natural environment. This is due, but not limited to, the total replacement of combustion engine driving units with hydraulic sets.

Among indisputable advantages of mentioned technologies are the following:

- animal crossings can be made without interrupting continuous traffic, for both road and railway lines,

- high accuracy of completed structures,
- fast execution time when comparing it with open excavation method,
- the works can be run for difficult soil and water conditions,
- high economical efficiency of the project while allowing for factors related to necessity to provide by-pass roads or traffic holding (in case of railway lines) in case of conventional method.

Despite the fact that a large experience is already gained in Poland with application of various trenchless technologies, and even, as it was emphasized above, some domestic achievements in this field are of record-breaking nature, still these technologies are too rarely used in transport infrastructure in spite their unquestionable merits. Hence, in authors' opinion, they should be more often used to construct animal crossings, but also culverts, pedestrian crossings, underpasses for cyclists, etc. (Wysokowski et al. 2010).

Wide application of these technologies would expedite construction of linear communication projects which are always related with crossing problems (especially for already existing lines). In case of new projects, acceleration and facilitation of works would consist in preparing standardized embankments of uniform compaction; then, a grade-separated animal crossing would be made using trenchless technology (which in practice does not disturb the soil structure). This way the embankments are of uniform parameters, so the problems with differential soil settlement existing in excavation technology would be reduced. Differential settlement causes appearance of *a hump* or *a trough* on embankment surface (crosswise to its axis). This is due to different settlement of embankment in the cross-section with the crossing and beyond it. The first case (i.e. the hump) will appear when settlement on both sides of the structure are larger than that in the structure of cross-section. The second case (i.e. the trough) will come into sight for opposite situation. In both cases, the reason of trouble lies in difficulty of compacting the soil so as its rigidity will correspond with that of the soil-structure system. The *hump* emerges when the structure stiffness is higher than that of soil volume of corresponding dimensions on both sides of crossing (this is usually the case when crossings are made with non-deformable materials, e.g. concrete or polymer concrete). The *trough* comes forth in opposite situation, when the crossing is made as a flexible structure with excessive flexibility, which could be the case when such materials as thermoplastics (e.g. PEHD) or hardening plastics (e.g. GRP) are used. Trenchless technologies facilitate elimination of this problem because, as aforementioned, they enable to standardize the soil parameters (mainly its compaction) over the full length of embankment (compaction is made for the embankment of the same height). It is obvious that for uniform-parameter embankment a pipe to be selected shall have the stiffness identical to that of the soil block extracted during installation of micro-tunnel. This is possible now thanks to a wide range of materials used for pipe fabrication (provided respective static/strength calculations are made). The problem outlined is well known for a long time and reported in relevant literature (Kuczyński 1975). However, at that time the trenchless technologies have not been sufficiently refined and open face pipe ramming was dangerous and leading to numerous failures (sink holes, etc.).

2.2 *Materials used in constructing ecological structures with trenchless technology*

The conditions of executing underground installation of pipes as animal crossing under road or railway embankment and operating conditions of such structure require the pipes with high strength and quality features. Pipes installed with trenchless technology must meet a series of rigorous conditions, both related to their strength and durability. Material used for pipe fabrication must be of high compression strength to withstand horizontal forces during pipe ramming, must be resistant both to external and internal corrosion (caused by chemical and biological aggression of soil and rain water). It is also favourable if the pipes are fire resistive. This is due to grassland burning during spring (obviously forbidden). Under such conditions, high temperatures could lead to plasticization of pipe material (this is especially relevant to thermoplastics) (Kuliczkowski 2001). Among a wide range of available materials for pipe fabrication, pipes for micro-tunnelling are as follows:

– reinforced concrete pipes (diameters up to 4,500 mm),
– stoneware pipes (diameters up to 1,400 mm),
– polymer concrete pipes (diameters up to 2,000 mm),
– pipes made of polymers reinforced with glass fibre (GRP) (diameters up to 4,000 mm).

The listed products are successfully used in trenchless technologies where high horizontal and vertical loads exist. The most often applicable materials used to construct underground animal crossing with trenchless technology are summarized in Table 1 (Wysokowski et al. 2008–2013).

As it was mentioned earlier, an important issue in micro-tunnelling technology is the workmanship of casing pipes. As opposed to the technology of laying pipes in open excavation, the external wall of casing pipes must be, as far as possible, of smooth surface (Advertising brochures and catalogues from the manufacturers of elements for trenchless technologies). Smooth external surface of pipes considerably reduces friction forces on side surface, hence diminishes also the costs related to application of additional lubricating agents, e.g. bentonite.

Casing pipes are joined with integrated system of connection fittings ensuring required uniformity of the pipeline structure and tightness. Modern types of connection fittings and availability of various-length pipes enable to route the casing pipes along an arch.

Table 1. Most popular materials used in constructing ecological structures of transport engineering with trenchless methods.

Materials used in underground animal crossings made with trenchless methods		
1	Concrete, prestressed concrete, reinforced concrete	
2	Polymer concrete	
3	Earthenware	
4	Glass Reinforced Plastic	

2.3 *Preparation of construction site*

In case of micro-tunnelling technology, it is necessary to construct appropriate starting and receiving chambers, also called the shafts. Their earth slopes are usually protected with sheet pilings, most often with steel sheet piles. Dimensions of chambers should be each time adapted to the specific project. Diagram of starting chamber is shown in Figure 2 (Madryas et al. 2006).

Like other trenchless technologies, micro-tunnelling requires small area of construction site. The components which determine the size of this area are, first and foremost, the lengths of sections between chambers, diameters of pipes and geological conditions. In case of animal crossings, these will be individual sections of several dozen meters, while geological conditions will be identical as those for embankment (except rare cases when crossing is made below the embankment in natural soil). The mentioned conditions impose equipment requirements and organization of two basic working stands, the starting and the final shafts.

Preparation of the starting shaft, its dimensions and the dimensions of adjoining area depend on configuration of drilling set, including the need or no need of drilling fluid. The total size of the stand on the starting shaft is usually within the dimensions 30 m × 20 m, and when the drilling fluid is necessary (for coarse grain soils)—30 m × 50 m (Madryas et al. 2006). The construction site must not be of regular (rectangular) shape, so the available land can be used more freely and, for instance, it allows to leave trees, etc.

The basic component of the construction site is the starting chamber. Typical dimensions of such chamber for pipes 6 m long are 4–5 m × 12 m, and for pipes 3 m long—4–5 m × 8 m (Madryas et al. 2006). A pipe storage site with crane manoeuvring area are located along the

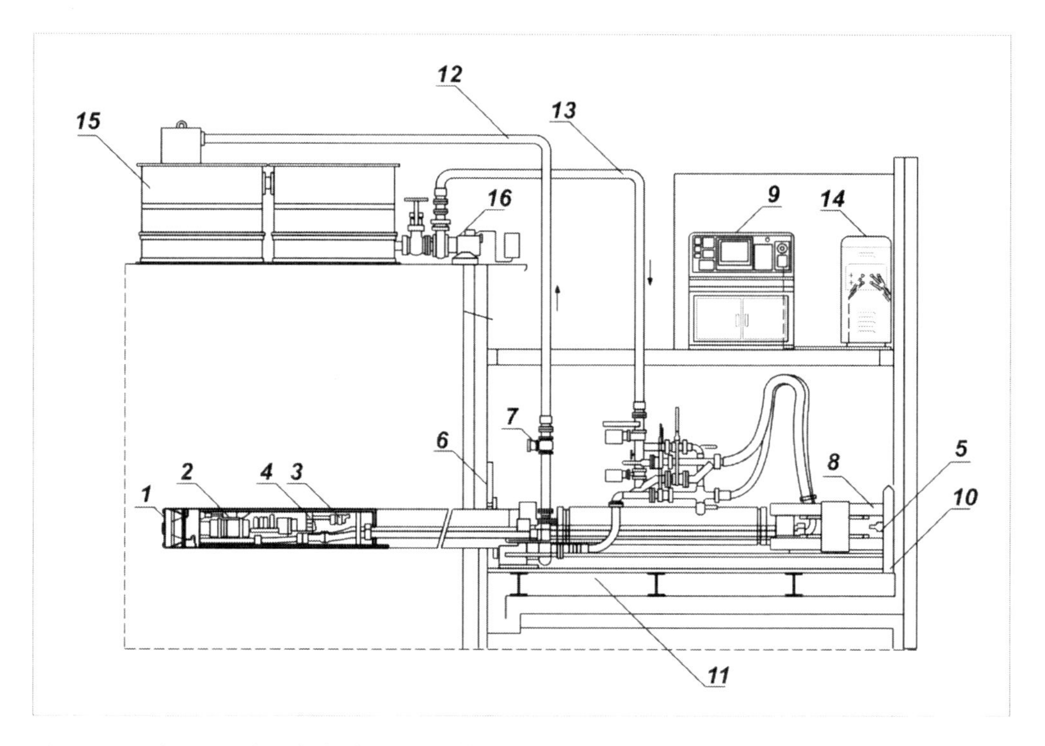

Figure 2. Diagram of typical micro-tunnelling system. 1—soil working head, 2—disc, 3—TV camera, 4—target disc, 5—laser, 6—input window (hole in excavation protection structure), 7—pump of flushing mixed with winning, 8—cylinder station, 9—control panel, 10—retaining block, 11—track and track subgrade, 12—pipe of flushing mixed with winning, 13—pipe of clean flushing, 14—power supply unit, 15—flushing separator and clean flushing tank, 16—clean flushing pump.

chamber. The remaining outfit elements like: container-control unit, flushing preparation and circulation unit, water tanks and possibly flushing tanks, workshop containers, amenity and office containers and power generators can be located nearby when no place is available directly at the starting shaft. In case when the micro-tunnel is drifted from the ground level through the embankment base, which is often true for animal crossings, there is no need to make deep starting and final chambers. Then, the depth of chambers results from the thickness of subgrade for bottom plate of the chamber, the thickness of this plate and the height of guides (track) for pipes. The difference refers also to the way of installing the retaining block which cannot transmit horizontal loads on excavation wall. It is then necessary to make independent construction to stabilize the position of the retaining blocks.

To ensure approach to the construction site, when it is not on hardened area, a preliminary access road (e.g. using MON plates) able to receive heavy construction machines (approx. 30 tons) need to be made.

On the final chamber side, the construction site area is limited to about 15 m × 20 m and the chamber with dimensions 3–4 × 5–6 m. The site must be accessible via a process road to allow extraction and transport of the head when tunnelling is completed. The construction of final shaft can be the same as that of the starting chamber.

To execute a micro-tunnel, the construction site need to be provided with power and water supply. In case of animal crossings located generally where no networks of such media are available, the water could be delivered by its transportation and high-efficiency power units need to be ensured. Then, fuel (usually diesel oil) need to be delivered. The technology is then less friendly for environment.

3 UNIQUE FEATURES OF ECOLOGICAL STRUCTURES IN TRANSPORT INFRASTRUCTURE MADE WITH TRENCHLESS TECHNOLOGY

3.1 *Outfit elements specific for ecological structures constructed with trenchless technology*

One of basic structural elements facilitating animal migration in ecological transport facilities—especially those with combined functions—are the shelves for animals. This is a very modern and economical solution making use the structure of micro-tunnels (or culverts made in an excavation) for both letting a water course through and for animal migration. The shelves in road animal crossings may be made on one or on both sides of the crossing. The double-sided shelves are used mainly in structures of diameters over 1.20 m (as it results from effective minimum width of single shelf). In general, shelves for animals can be classified due to their construction (geometry, location and assembly) and due to the material used for their fabrication. Distinguished are shelves located on the culvert bottom (within side walls) and shelves mounted directly to culvert structure using additional fasteners. Shelves located on the culvert bottom are most often made of natural aggregate, gabions protected with geotextiles, or made of structural concrete (usually as prefabricated elements). In such case, one should remember to cover the shelf with material "friendly" for animals. Most often cohesive impervious soils with appropriate durability are used for this purpose. Another type of shelves are those secured directly to side structural walls of the tunnel. Various solutions are adopted depending on the structural material of the tunnel. Among a wide range of materials, the basic ones used in construction of animal shelves are:

- reinforced concrete shelves,
- steel shelves—usually as profiled sheet metal,
- shelves out of polyester resins reinforced with glass fibre,
- mixed solutions.

In case of both types of construction, it is absolutely necessary to remember that the top surface of animal shelves should be above average water level in the culvert—as determined on the basis of hydrological calculations.

The newest material solutions include application of polyester resins reinforced with glass fibre. In this case the shelves are laminated to structural pipes of the culvert using mats with glass fibre and polyester resins. Such solution offers required mechanical resistance to damages and increased resistance to environmental factors. Furthermore, laminated shelves could be coloured, e.g. to be green to adapt them to surrounding environment.

In view of the fact that issues related to design of animal shelves in culverts are relatively new, there are still many aspects for their improving and optimizing.

For the sake of the characteristics of trenchless method, it is impossible to use elements preinstalled to the micro-tunnel structure. This is due to the fact that the structural pipe bore is used for process purposes, e.g. transport of winning. In such cases, main outfit elements of crossings (like mentioned shelves, elements of surfacing, dead lights, etc.) are installed following the crossing structure completion. Figures 3 and 4 show exemplary shelves for animal migration applied in underground crossings for herpetofauna and small animals under communication routes.

There is also a problem of dead lights necessary when ecological structures of transport infrastructure made with trenchless technologies are of considerable length (e.g. under motorways). Such elements create no problems for excavation technology. The problem is case-to-case resolved by constructing wells/dead lights over the completed tunnel. In authors' opinion, a system solution need to be developed in the future.

The problems mentioned for the outfit show hindrances existing in constructing underground animal crossings with trenchless technologies. However, this would not present an essential obstacle for application of these technologies as such works could be made in parallel to other necessary finishing operations.

3.2 *Finishing elements of underground animal crossings*

Finishing elements of animal crossings constructed with trenchless technologies can be made, in authors' opinion, in similar way as in case of structures made in open excavation. Such elements like, to name a few, heads and directing wings (being also of structural nature) (Wysokowski et al. 2008) require some preparations of the area directly adjoining the inlet and outlet of the crossing. This involves necessity of foundations for these structures and also formwork in case the elements are to be made in reinforced concrete technology (Jasiński et al. 2006).

For the sake of ecological aspect, in case of structures of heads and wings of animal crossings, now "heavy" reinforced concrete solutions are given up. Preferred are solutions more environmentally friendly, e.g. in the form of soil reinforced with geotextile, geograte with

Figure 3. Installation of animal crossing structure made in CC-GRP technology with visible shelf (HOBAS Advertising brochure).

Figure 4. View of completed small animal crossing made out of concrete pipes with single-sided shelf attached to them (photo by A. Wysokowski).

appropriate mineral filling or other solutions of earth structures. Such solutions create more "friendly" area within access area to the crossing.

4 ECONOMICAL ISSUES

Economical efficiency of engineering structures, including animal crossings under consideration, represents the ratio of expenditures incurred and materials used to the value of effects gained thanks to these expenditures. Trenchless technologies applied in constructing or reconstructing roads and railway lines give, at increased production expenditures, much higher effects than traditional open excavations. Especially notable are technical and economical effects in case of high road or railway embankments which is often the case.

As aforementioned, large losses, mainly hardly-measurable social and ecological costs, are generated when the works are carried out in open excavation, which necessitates setting up diversions and closing the main communication routes. Hence, endeavours shall be made to use solutions which in technical and economical aspects allow to get the best results while, at the same time, which generate the lowest costs (including the social costs as well). Unquestionably, the trenchless technologies belong to such solutions.

Despite the fact that when analysing the direct costs of constructing ecological transport structures with traditional technologies, they are often lower, when considering the pace of works, the chance to get uniform parameters of embankments (identical with those of the embankment-soil system), lack of hindrances in continuation of earth works, and also social costs, the combined costs appear to be more advantageous in trenchless technology (Wysokowski 2012).

5 CONCLUSIONS

Despite the hindrances outlined above and sometime higher project expenditures, trenchless technologies are more and more often used to construct underground animal crossings, culverts and structures combining these two functions.

This undoubtedly results from outlined advantages of these technologies and from more and more common trend to include hardly-measurable social costs and also pace and safety of works in technical/economical analyses of building projects.

In the face of continually improved trenchless technologies and successful experience with their application in other areas of civil engineering (mainly in constructing underground urban network infrastructure), it seems that the number of their application for animal crossings would be rising.

The problems depicted are relatively new and so far have no wider representation in theoretical studies. In authors' opinion, it would be appropriate to undertake research work on this subject which implementation effect should be algorithms enabling technical/economical analysis of appropriateness of trenchless technologies for defined local conditions.

REFERENCES

Advertising brochures and catalogues from the manufacturers of elements for trenchless technologies HOBAS Advertising brochure.

Conference proceedings 2013 XII Świąteczna Drogowo-mostowa Żmigrodzka Konferencja Naukowo-Techniczna pt. *Przepusty i przejścia dla zwierząt w infrastrukturze komunikacyjnej*. Żmigród, grudzień 2013 r. Infrastruktura Komunikacyjna Żmigród-Wydawnictwo Nowoczesne Budownictwo Inżynieryjne, Kraków.

Jasiński W., Łęgosz A., Nowak A., Pryga-Szulc A. 2006. Wysokowski A. *Zalecenia projektowe i technologiczne dla podatnych drogowych konstrukcji inżynierskich z tworzyw sztucznych*. GDDKiA-IBDiM Żmigród.

Kuczyński J., 1975. *Budowle sanitarne*, PWN, Wrocław-Warszawa.

Kuliczkowski A. 2001. *Rury kanalizacyjne. Własności materiałowe.* Monografie, Studia, Rozprawy nr 28. Wydawnictwo Politechniki Świętokrzyskiej, Kielce.

Kuliczkowski A., i inni. 2010. *Technologie bezwykopowe w inżynierii środowiska*. Wydawnictwo Seidel-Przywecki Wrocław.

Kurek Rafał T. 2010. *Poradnik projektowania przejść dla zwierząt i działań ograniczających śmiertelność fauny przy drogach*. Stowarzyszenie Pracownia na rzecz wszystkich istot. Bystra.

Madryas C., Kolonko A., Szot A., Wysocki L. 2006. *Mikrotunelowanie.* Dolnośląskie Wydawnictwo Edukacyjne. Wrocław.

Wysokowski A., 2012. *Technical and economic effectiveness of using trenchless technologies in Constructing roads and railway infrastructure*. Underground Infrastructure of Urban Areas 2 Balkema Book Taylor&Francis Group, London.

Wysokowski A., Howis J. 2008–2013. *Przepusty w infrastrukturze komunikacyjnej-cz.*

Wysokowski A., Howis J., 2008. *Stosowanie konstrukcji gruntowo-powłokowych jako przejść dla zwierząt w infrastrukturze komunikacyjnej*. Materiały Budowlane 04.

Wysokowski A., Madryas C., Howis J. 2008. *Stosowanie rurowych elementów betonowych jako przejść dla zwierząt w infrastrukturze komunikacyjnej.* Konferencja Dni Betonu 2008. Tradycja i Nowoczesność. Wisła, październik 2008. Polskie Stowarzyszenie Producentów Cementu Kraków.

Wysokowski A., Madryas C., Skomorowski L. 2010. *Development of the transport infrastructure* in *Poland with the application no-dig with CC-GRP materials.* International No-Dig 2010 28th International Conference and Exhibition 8–10 November. Singapore.

Wysokowski A., Staszczuk A., Bosak W. 2007. *Przejścia dla zwierząt w budownictwie komunikacyjnym.* Inżynier Budownictwa 12, s. 72–75.

Underground Infrastructure of Urban Areas 3 – Madryas et al. (Eds)
© 2015 Taylor & Francis Group, London, ISBN 978-1-138-02652-0

Subway in Wroclaw

Marek Żabiński
Deputy Director of the Wroclaw Development Office, Poland

Michał Gallas, Katarzyna Herma-Lempart, Monika Kozłowska-Święconek,
Mateusz Majka, Przemysław Matyja, Anna Rygała & Błażej Stopka
Municipal Office of Wroclaw, Wroclaw Development Office, Poland

ABSTRACT: Subway is associated with prestige and modernity, with reliable and fast means of commuting. It seems that a large city, such as Wroclaw, may have aspirations of having a subway. When considering the matter of subway in Wroclaw, a lot of questions were asked. First and foremost: Is it even worth to think about a subway in Wroclaw? Is subway in Wroclaw the best way to improve the standard of providing services by the public transport? In what time perspective do we see subway in Wroclaw? And finally—Can we afford the construction and maintenance? The document "Subway in Wroclaw" covers some of the issues and problems relating to the subway, in particular those closest to spatial planning and development, as well as the development of functional and spatial structure of the city.

1 UNDERSTANDING SUBWAY IN WROCLAW

This paper is a presentation of the document "Subway in Wroclaw" prepared by the Development Office of Wrocław, the objective of which was to explain the issues and determine the scope and level of the conditions that are associated with subway.

When considering and analyzing the matter of subway in Wrocław we cannot forgot asking fundamental questions that arise immediately as soon as we think about subway. First and foremost, it is it worth it even think about a subway in Wroclaw? The answer to this question is the very fact of creating the document by the Wroclaw Development Office. An affirmative answer to the first question allows us to inquire further. Therefore:

1. Does subway mean prestige for the city? Or does merely thinking about the subway mean prestige for the city?
2. Is subway the only and the best "cure" for changing the standard of public transport in Wroclaw?
3. In what time perspective do we envision subway in Wrocław?
4. Who might finance its construction? And finally?
5. Can we afford to maintain the subway?

2 DEMOGRAPHICAL AND SOCIAL ASPECTS

2.1 *Demographic forecast*

Wroclaw is affected by a deepening phenomenon of aging of the population. This is, above all, a consequence of a low fertility rate. The fertility rate, and thus the coefficient determining the number of children born per one woman of childbearing age (15–49 years) in Wrocław in 2005 amounted to 0.97, and then slowly began to grow. In 2009 it reached 1.19, but still it remains at a level far too low to ensure the simple replacement of generations. Subsequent generations are therefore becoming less numerous than their parents' generation.

Table 1. Population forecast in Wrocław according to CSO forecast.

Year	2011	2015	2020	2025	2030	2035
population	627 590	624 734	622 803	620 111	615 304	609 943

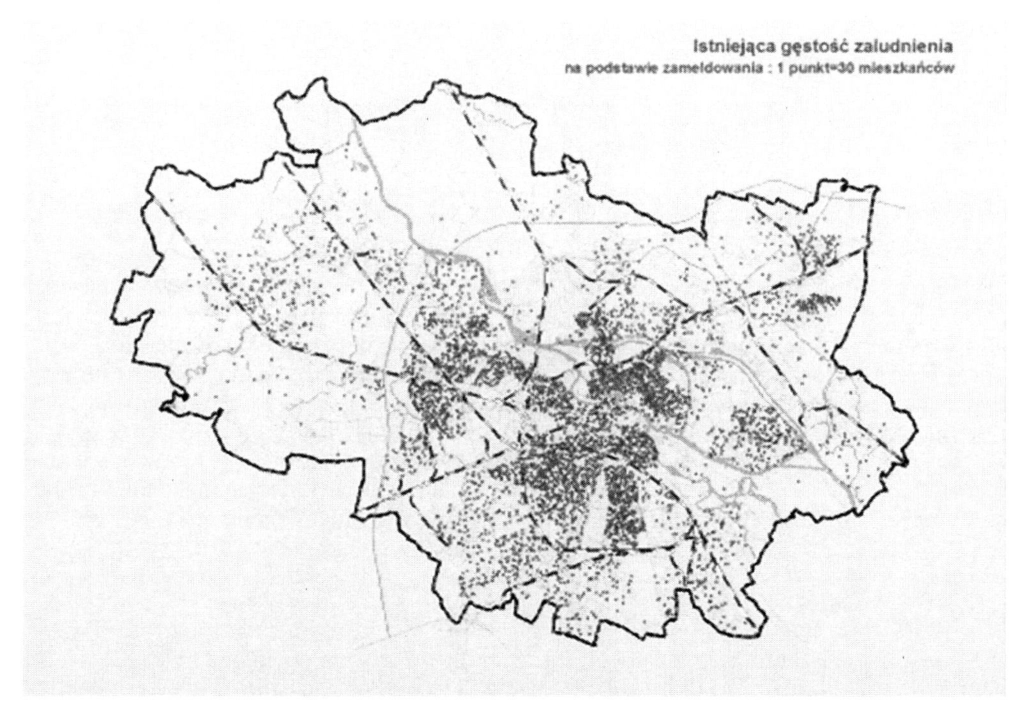

Figure 1. Existing distribution of residence based on the number of people registered (1 point—30 residents).

Aging of the population is not a problem in itself—more important are the economic and social consequences of the occurrence of such a phenomenon. The trend of population aging causes the need to change the approach to various aspects of the development of the city, including the development of public transport. Public transport should take into account the changing physical, mental, health, and material aspects of the society. The consequence of the aging population of the city is that a growing number of potential communication customers will be assigned to collective transport.

2.2 *The spatial distribution of population and employment in the city*

Population distribution was performed on the basis of the data on the registration of residence—as of 31.12.2010. The current state of population and employment is presented on the map below.

Larger population centers are concentrated in the regions of large housing estates such as: Kozanów, Gądów Mały, Nowy Dwór, Gaj and Różanka. Very high population density is also in the downtown area, in the northern part: the estates of Ołbin, Nadodrze, Plac Grunwaldzki, while in the southern part: the estates of Powstańców Śląskich, Huby, Grabiszynek and Przedmieście Oławskie.

The next map shows the spatial distribution of jobs, which shows a very strong concentration in city center and downtown. Most of the daily and most important commutes in the direction home-work and work-home occurs within those parts of the city.

Figure 2. Existing distribution of employment (1 point—30 workplaces).

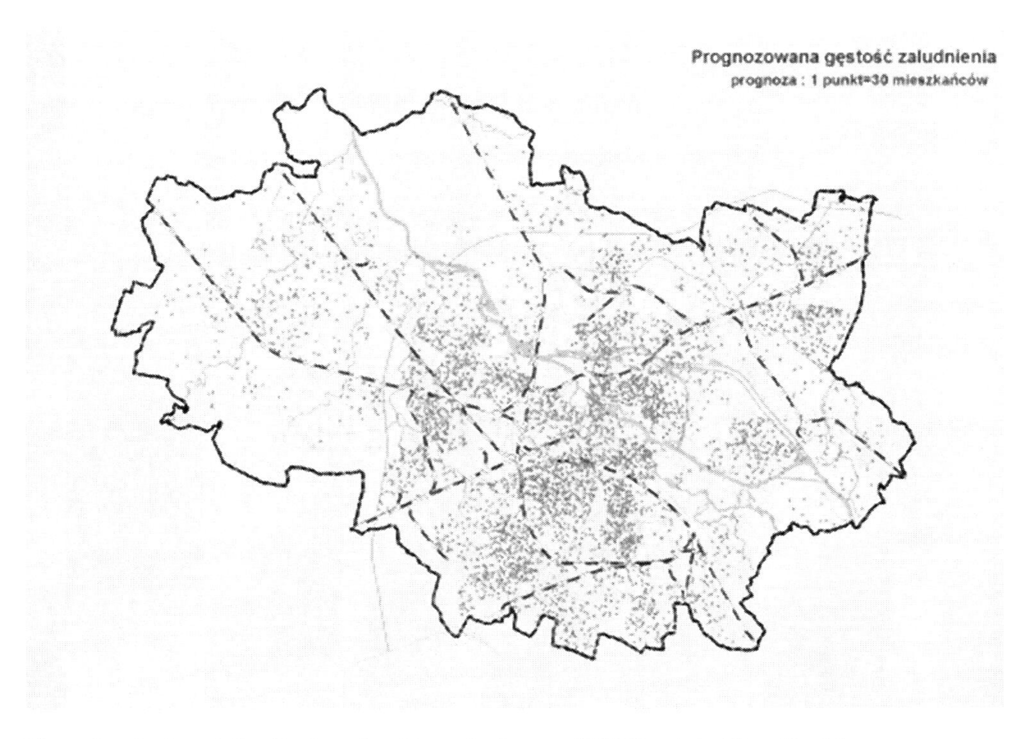

Figure 3. Projected distribution of residence in the year 2033 (1 point—30 residents).

2.3 *Forecast of population and employment*

Due to the fact that the demographic forecast does not introduce significant changes in the number of inhabitants of Wroclaw, and even a decline in the number of residents is predicted in the perspective of the year 2035, the population forecast is based primarily on the population movements within the city.

The emergence of new investment areas of residential character will result in thinning out the population and changes in the spatial distribution of the city. Despite this phenomenon, large housing estates still remain in the range of areas with a high concentration residence.

A significant influx of residents, according to the forecast, can be observed in the areas of settlements: Muchobór Wielki, Wojszyce—Jagodno, Grabiszynek, Krzyki and Psie Pole.

2.4 *Communication behaviors of the inhabitants of Wroclaw*

According to the results of the Comprehensive Research of Traffic of 2010/2011 (CRT), the primary destination for the inhabitants of Wroclaw is home (45%). The second most significant destination is work (21%). Most traveling is done by people aged 25–39 years. The share of traveling by collective transport in Wrocław amounts to 34,7%.

Travelling by different modes of transport, divided by age group, is as follows (the results refer to a standard weekday [in %]):

	Age				
Means of transport	6–15	16–25	26–39	40–60	60+
On foot	56,7	15,8	11,7	13,5	32,6
Car	21,6	23,1	59,4	49,4	23,2
Public transport	17,9	54,8	23,1	30,4	38,0

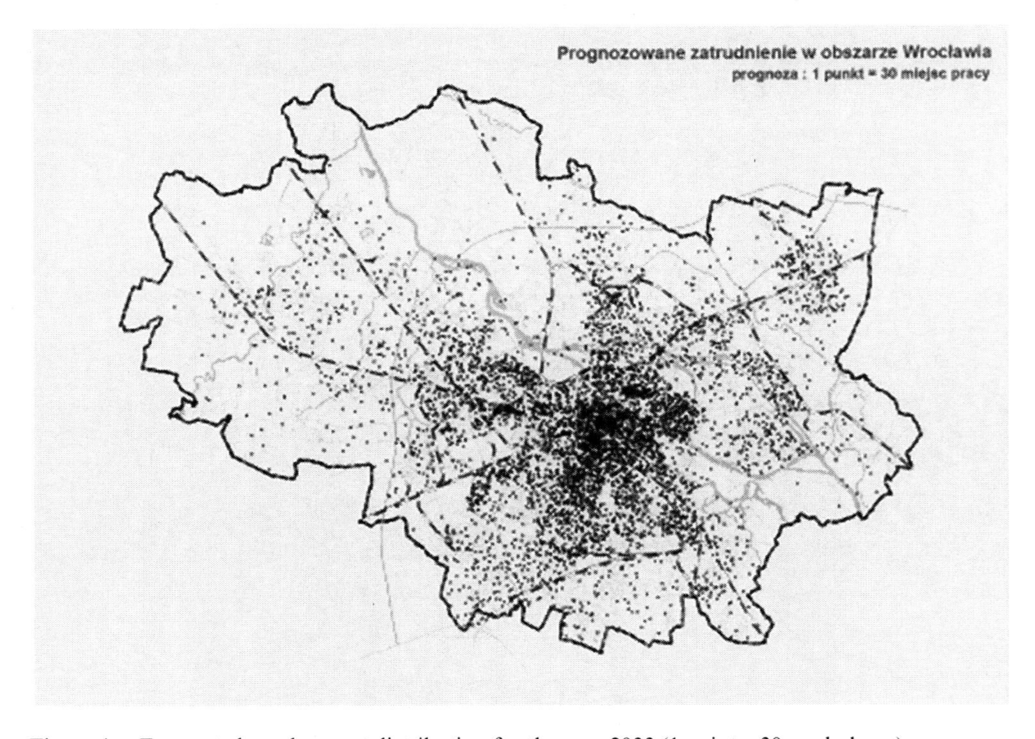

Figure 4. Forecasted employment distribution for the year 2033 (1 point—30 workplaces).

Motivation for traveling					
Home	49.6	46,5	43,8	45,0	46,4
Work	–	11,3	29,7	30,4	–
Minor shopping	–	–	–	–	11,4
School/university	36,4	23,2	–	–	–
Business matters	–	–	6,5	4,0	–
Other	8,2	–	6,0	8,6	20,7

School-age children (6–15 years old), on a weekday travel mostly on foot (57%), which merely confirms that the parents are guided chiefly by the distance from the house when choosing a school for their children. In the second place is traveling by car (21.6%), which is associated with parents transporting their children. The amount of traveling made by collective transport is optimistic, i.e. by bus, tram or bus + tram—17.9%. With a relatively considerable extensiveness of measures to promote by collective transport there is a chance to form ecological behavior in young people and thus reduce the desire to travel by car in the productive age.

The activity of the 16–25 age group occurs more or less evenly throughout the week, while on a weekday they perform more traveling with the use of collective transport (55%).

Pro-car behaviors are dominant in the group of 26–60 years, so of working age, and are mainly related to professional activity and, thereby, increase their social status.

Car activity falls after the age of 60 in favour of frequent travel by public transport. It may also be due to a change in activity and a decrease in income (pension, retirement).

2.5 *Evaluation of public transport by the inhabitants*

Within the framework of the Comprehensive Research of Traffic, as part of the survey of households, evaluation of public transport performed. Respondents were asked about the three most important characteristics of collective transport, which are important from the point of view of the user.

The most important characteristics associated with collective transport, indicated in addition to ticket prices, were in large majority those that are associated with time: time of travel, punctuality and time of waiting at bus stops.

According to the results contained in Comprehensive Research of Traffic, assessing the validity of the actions currently being undertaken by the city in relation to public transport, the residents indicated the following actions as the most important: the modernization of the existing tram routes and the extension of tram system to include new routes.

2.6 *Traveling preferences of young people*

Age group 16–25 years is the part of society, which usually uses collective transport, (55%—collective transport, 23%—a car). It is, therefore, also a target group in relation to which a broad campaign should be carried out, promoting ecological means of transport (collective transport, bicycle). As shown by the results of Comprehensive Research of Traffic, in the age-group 26–39 there is a turn towards the pro-car behaviors and the proportions of the car and public transport use radically change (23%—collective transport, 60%—car).

This phenomenon is also noticeable in cyclic testing conducted by the Centre for the Registry of Vehicles and Drivers, whose task is to record the issued licenses.

It is possible to notice a growing trend in the number of licenses issued over the last years, both to the first and to the other age group. This phenomenon may be disturbing, given the demographic changes and the gradual reduction of the group of young people. However, more important is the fact that of a vast difference between these two age groups. The group of school and university students is not as interested in having a driving license as a group that is beginning its professional activity. The reason for this phenomenon can certainly be raising the social status in the age of 25, associated with getting a better job as a result of completing the period of education and acquiring an education and preparation for the profession.

The main drawbacks of collective transport, according to the age group 16–25 included: time of traveling and congestion of vehicles, while more extreme judgments are presented by pupils rather than university students.

3 SUGGESTIONS FOR THE SUBWAY ROUTES IN WROCLAW

These socio-demographic aspects formed the basis of the analytical work concerning subway in Wrocław. Subsequently, a multilevel analysis of other aspects of the city was carried out. The method of achieving the goal, which is the designation of the proposals of routes, would be "expert" in character—the method has temporarily been called the method of successive steps.

3.1 *Step 1—analysis*

The first step was to analyze the situation of the spatial distribution of residence and workplaces in Wroclaw, at the current state and in the perspective of 2025. On the basis of the diagnosed situation and based on the general information about the situation of the city and the direction of its development, important points have been identified on the map, which are important from the point of view of ensuring the availability of high-quality communication. Among them were places, which already at the moment should have the highest level of communication availability, because of their function in the structure of the city, but there are also those that will be important in the future—near or long-term. Such places have been divided into two types:

- specific objects such as hospitals, schools, railway stations,
- areas of high residential density and a large number of workplaces, with the points of identifying these areas located in indicative centers of gravity, or in places where a subway station would be most desirable.

The location of the described points is shown on the following map.

3.2 *Step 2—rationalization*

The next step was to examine the selected points in terms of rationality and the need for a subway line leading to them in the context of railway or tram service capabilities. Therefore, a rail network with stops and a tram network in the perspective of the year 2020 was "superimposed" on those points. As a result, it was determined which from among the points designated on the map are already or in short perspective will be provided with very good accessibility through public transport and because of this can be ignored when considering the construction of subway.

3.3 *Step 3—connections*

The next step is to create strings from the resulting grid points that will indicate the possible potential connections. Within the structure of the city, two clear directions of the possible need for excellent availability of high-speed means of transport are emerging. The west-central direction, beginning in region of the built Regional Hospital and ending at the Grunwald Square or further are the Centennial Hall. The North-south direction beginning in the area of Marino and ending in the Medical University at ul. Borowska. Additional three sections are emerging in the southern direction: in the south-western region the settlements

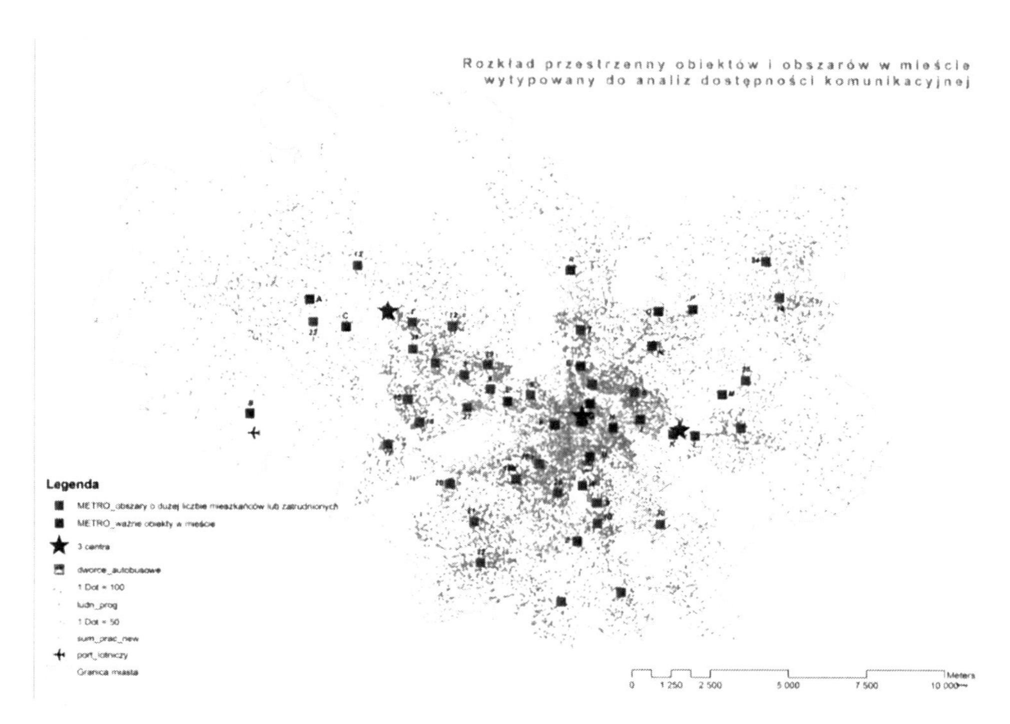

Figure 5. The spatial distribution of facilities and areas in the city, selected for the analysis of transport accessibility.

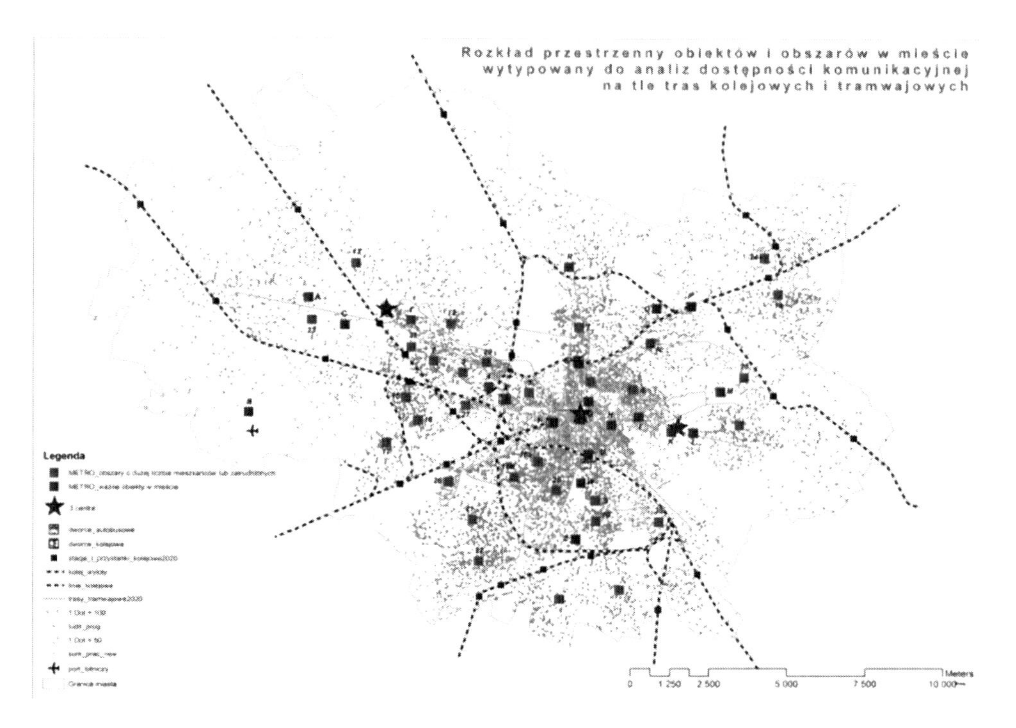

Figure 6. The spatial distribution of objects and areas selected for analysis concerning railway and tram routes.

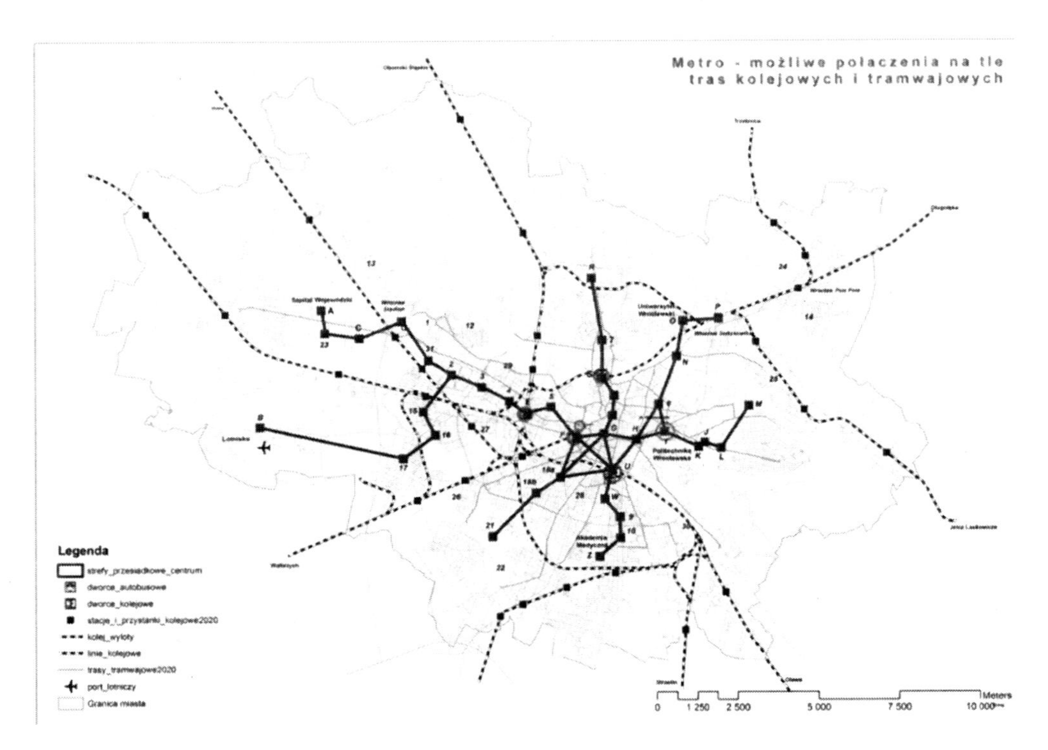

Figure 7. Possible connection on the background of railway and tram routes.

of Nowy Dwór and Muchobór, in the Southern Downtown area and Krzyki, as well as in the area Karłowice and Sołtysowice.

Therefore, the points on the map were first connected into two strings: one east-west, the other north-south. This experiment indicated that both lines intersect at a point in the vicinity of Market Square and the Old Town, within ul. Świdnicka. This just seems to be a key point in relation to both lines. There are also additional branch lines: towards Nowy Dwór and Krzyki, and possible connections, especially in the central part of the city. The resulting diagram of connections has been presented below.

In the context of the entire city, particularly important points, which should appear in the potential metro network, regardless of the option or investment variant, because they play a very important role in the transfer system of the city by integrating multiple modes of transport in one place, were clearly drawn. They include: Stadium (D), Wrocław Mikołajów Station (E), Wrocław Świebodzki Station (F), Świdnicka Square (G), plac Społeczny (H), Wrocław Główny Station (U), Wroclaw Nadodrze Station (S).

3.4 *Step 4—infrastructure*

One of the key technical elements of the subway, without which the construction of the station cannot be commenced is planning of the location of the technical and stop station. Such a station, in accordance with the technical terms, should be located in such a manner that it would have a railway siding, and so that its connection to the subway line was as short as possible and collision-free. Additionally, at least one technical-stop station should be connected to the subway line by at least two subway tracks.

The technical-stop station is a space-absorbent and not having any special architectural qualities facility. Therefore, it must be positioned in such a way that it is not exposed to the direct exposure of the important points of the city.

For the proposed subway routes, two variants of placement were indicated. The first is located in the south of the city, in the region of ul. Koszycka, by the southern rail freight bypass and may be connected with the rail line towards Sobótka. The area is not currently maintained, and the neighborhood is not sensitive. The second location is the area of the Wrocław Gądów freight station. The station is not currently fully used, has great spatial capability for the location of the technical-stop station and a direct possibility of connecting to the rail.

3.5 *Step 5—variants*

The next step of the analyzes is to enter subway routes corresponding to the requirements of their formation resulting from technical conditions into the above diagram. In particular, this applies to the horizontal curves' radii, inclusions into the railway stations and technical-stop stations, and the necessity of locating the metro stops on the straight sections. Due to the various options of connections, three variants of shaping routes and stops including different variants of connections in the city center were proposed. Sections from the airport and to the regional hospital were proposed as having growth potential, due to the current lack of transport potential and alternative capabilities of handling. For each of the variants, the number of inhabitants in the range of availability of the subway stops, that is in the zone of 500 m from its location, was calculated.

In addition, the analysis was extended with the information on the number of residents located in the zone of influence of each of the proposed stops. Counts were made for all the stations from all variants, in order to determine which one is most preferred from the standpoint of potential. For each of the stops a number of residents located in the isochrone of access with a radius of 500 m was calculated. It results from the analysis that the stops of the greatest residents potential include: the stops in the South Downtown are, in the vicinity of the Nadodrze station and Gaj.

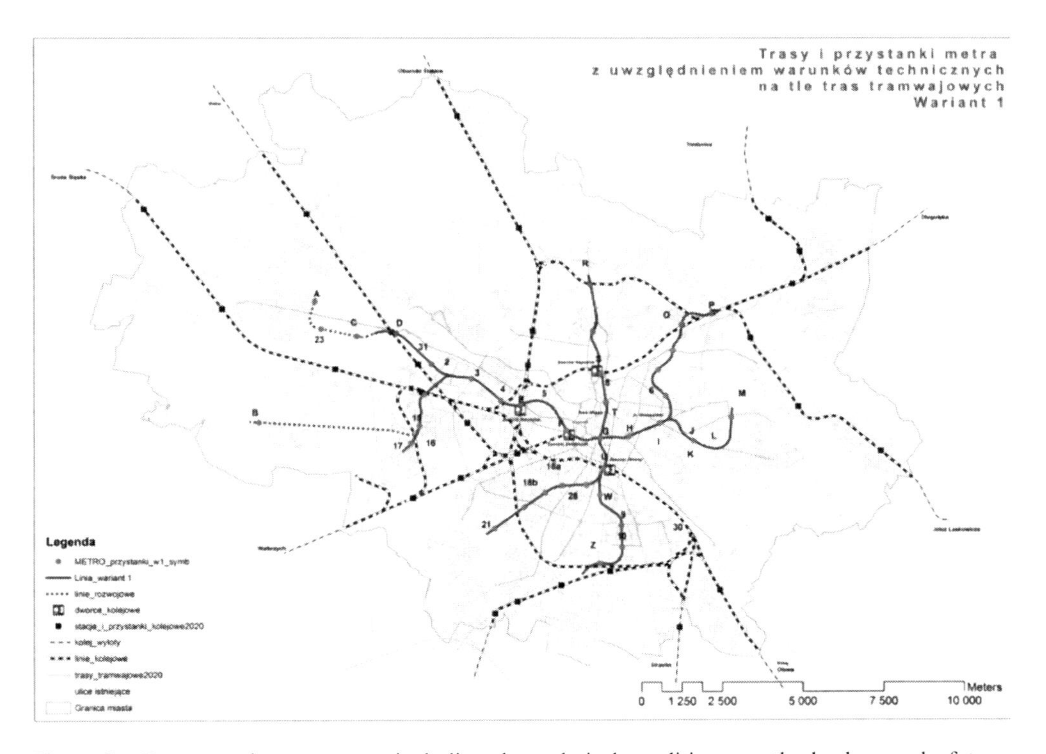

Figure 8. Routes and metro stops, including the technical conditions on the background of tram routes (Option 1).

Figure 9. The number of inhabitants within the range of subway for all stations.

Of the three proposed variants, for functional reasons Option 1 was recognized as the most favorable and it was recommended for further consideration. The following figure shows the preferred Option 1 on the background of many important communication determinants of the city. The planned Park&Ride parking lots, important railway stops and integrated transfer nodes in the vicinity of the city center were indicated. In the resulting system, the metro stations, which are connected to the railway stops and transfer parking lots are of the greatest importance. They will be most attractive to travelers from the external settlements and the area of Wroclaw agglomeration. These stations should include the Wroclaw Sołtysowice, Marino, Wrocław Stadion, Akademia Medyczna stations. In the city center the most important stations are those which allow for a smooth transfer to other means of transport. The most important of them are: Dworzec Nadodrze, Dworzec Mikołajów, Dworzec Główny, Dworzec Świebodzki and plac Grunwaldzki, and The Old Town Square station, which connects the two metro lines.

3.6 *Step 6—evaluation*

The final element of the analysis is the evaluation of availability of the public rail transport with new sections of the subway. The evaluation was carried out for the recommended option and it consists in identification of the number of inhabitants within the range of acceptable isochrones of access to stops of individual means of transport. The range of availability was shown in the drawings.

The above summary shows that the greatest improvement in the availability of rail transport is observed after the stage of development of the railway infrastructure, consisting in the enrichment of the new tram line with new stops and the introduction of an effectively working suburban railway network, also for handling the inhabitants of Wroclaw. With a well-functioning system of tram-rail, introduction of subway does not result in a spectacular strengthening of the coverage of the availability. This does not mean, however, that this

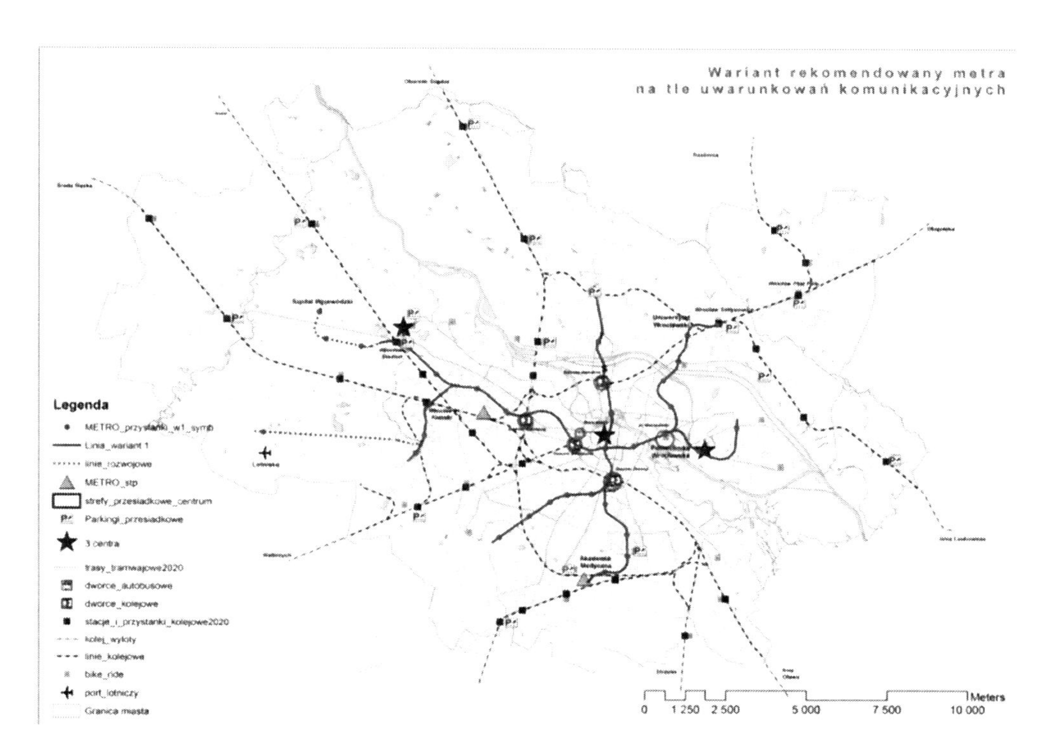

Figure 10. The recommended option of subway with respect to the communication conditions.

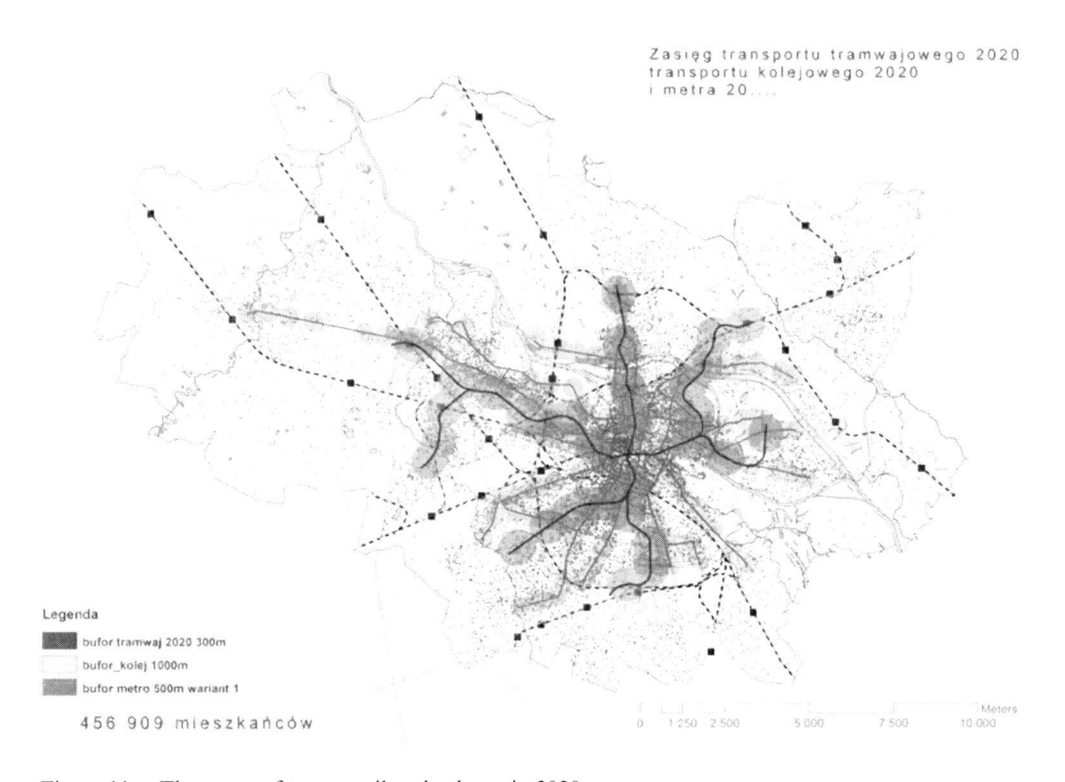

Figure 11. The range of trams, rail and subway in 2020.

argument lowers the position of the subway—it only shows that the existing, but developed rail system can be an alternative to the subway.

REFERENCES

Comprehensive Research of Traffic 2010.
GUS.
http://lubczasopismo.salon24.pl/transport/post/228849,przywrocmy-kolej-s-bahn-we-wroclawiu.
http://www.gazetawroclawska.pl/artykul/26237,jak-czekista-budowal-metro-2,2,id,t,sa.html.
http://e-sancti.net/forum/showthread.php?tid=2897.
http://businesstraveller.pl/przewodnik/prezentujemy-najlepsze-metra-swiata.
http://mic-ro.com/metro/categories.html.
http://pl.wikipedia.org/wiki/Transport_kolejowy.
http://pl.wikipedia.org/wiki/Lista_system%C3%B3w_metra_na_%C5%9Bwiecie.
Ostaszewicz J., Rataj M. 1979. *Szybka Komunikacja Miejska*. Warszawa.
SIP Wrocławia.
Studium uwarunkowań i kierunków zagospodarowania przestrzennego Wrocławia 2010.
www.metro.waw.pl.
www.prgmetro.pl.
www.prw.wroc.pl.

Underground Infrastructure of Urban Areas 3 – Madryas et al. (Eds)
© 2015 Taylor & Francis Group, London, ISBN 978-1-138-02652-0

The Wrocław metro: Geology, hydrogeology and tunneling perspective

A. Żelaźniewicz & PAN-METRO Group*
Wrocław Division, Polish Academy of Sciences, Institute of Geological Sciences, Wrocław, Poland

ABSTRACT: Discussions about building an underground public transport system in Wrocław started 80 years ago but with a negative outcome. Then the subject was revived from time to time. In late 2012, a cooperation was initiated between the Commune of Wrocław, Bureau for Development of Wrocław, and the Wrocław Division of the Polish Academy of Sciences, the PAN-METRO Group. The Bureau came out with a concept of two metro lines operating along the N–S (I) and W–E (II) routes. Based on this concept, the PAN-METRO Group identified and summarized the present knowledge of local geology, in particular on lithological, geotechnical and hydrogeological characteristics along the two lines. These are highly diversified: 5–16 m thick sandy-gravel fluvial/glaciofluvial Quaternary deposits overlie 5–30 m thick glacial tills, generally thinning westwards, which are in turn underlain by Tertiary clays up to 150 m thick. The uppermost aquifer has its water table at a depth of 1–7 m and hydraulic connection with river waters. The underground infrastructure does not exceed a depth of 12 m. All such data are of primary importance for planning and then constructing both the metro routes and stations. In the historical center of the city, where trenchless excavations seem an only rationale approach, tunnel bored in tills and clays with TBM at a depth of 20–25 m would be the most reasonable solution. Locally ground freezing and NATM may prove more suitable. This is to be decided upon recognition of details of the geological conditions, which requires 100–150 new boreholes and more detailed geotechnical studies. In Wrocław, however, these conditions do not differ significantly from those met in other European cities from Toulouse and Paris via Warsaw to Moscow.

1 INTRODUCTION

An underground public transport system for ~600 000 citizens in the city of Wrocław was planned some 80 years ago. Unfortunately, the idea was abandoned because local geological conditions confronted with techniques then available made such a venture too risky and difficult. Heavy damages of the city during the 2-nd World War and hard post-war times prevented people from thinking about other means of transportation than just tram or bus. Some 40 years ago, plans were made to construct a tunnel between the Central Railway Station and the Świebodzki Station but they failed too. At those times, city-planning predicted rather westward development of Wrocław, which was to be apparently attracted by the Lubin-Głogów copper district. These plans did not come into reality and political changes capitalism in the early 90-ties changed the urban planning substantially with massive housing projects developed on all sides of the city. Such new situation required new approach to the question of public transport in Wrocław.

*The PAN-METRO Group: K. Choma-Moryl, W. Grodecki, M. Kosmalski, C. Madryas, J. Malewski, L. Wysocki, A. Siemińska-Lewandowska & S. Staśko.

In the Wrocław Division of the Polish Academy of Sciences, four scientific commissions initiated theme workshops "The metro" and the City Council organized two debates "The metro in metropolis", which was further explored in a few MSc theses at Wrocław Technical University. In 2012, a letter of intention was signed by the President of Wrocław and the President of the Wrocław Division of PAS initiating studies to explore the conditions for the metro system in the city. On behalf of the Commune of Wrocław, the Bureau for Development of Wrocław was designated to cooperate with the PAN-METRO Group that was formed in the Wrocław Division of PAS. The metro is understood as an independent, collision-free urban railway system operating both underground and on surface.

Having analyzed population density, urban centers, existing city transport network, etc., in the year 2013 the Bureau came out with a concept of two metro lines operating along the N–S (I) and W–E (II) routes over a total distance of some 36 km. As the next step, the PAN-METRO Group analyzed: (1) recent state of knowledge on geological, hydrogeological and hydrological conditions along the preliminary recommended metro routes, (2) characteristics of shield machines, examples of their utilization for tunneling elsewhere and feasibility of using them in Wrocław; (3) impact caused by the tunneling on the existing infrastructure and environment; (4) feasibilities of constructing the metro stations using underground techniques; (5) feasibilities of constructing the metro stations using open-pit techniques, (6) risks connected with constructing tunnels and open trenches, and finally (7) reviewed some examples of the projects successfully realized elsewhere. The results were assembled in an upublished work and accompanying documents prepared for the Commune of Wrocław by Żelaźniewicz et al. (2013). This paper is a synopsis of that work.

2 GEOLOGY AND PHYSIOGRAPHY OF WROCŁAW

The city of Wrocław is situated on the Fore-Sudetic Monocline built of Permo-Triassic strata and on the Odra Fault Zone composed of crystalline basement. These tectonic units are concealed under a lithologically diversified cover of Cenozoic sedimentary deposits, the thicknesses of which vary over the area, in generally they are lower on the west and increase eastward.

In the western parts of Wrocław, Permian clastic rocks are overlain by Neogene deposits, the oldest members of which are Middle Miocene sands, silts, lignitic clays and clays with brown coal intercalations. Higher in the section, there are Upper Miocene 'flame' clays of the Poznań Clays (Dyjor 1970) that locally occur directly beneath the soil layer. Isolated patches of the Upper Pliocene Gozdnica Series are also in evidence (Łabno 1991). Such Neogene rocks were then covered with glacial deposits. Most widespread are those that accumulated during the Maximum Stadial of the Middle Polish Glaciation. Fluvioglacial sands and gravels, 3–6 m thick, dominate at the surface in the western parts of Wrocław over slightly less frequent glacial tills 3–5 thick. Holocene deposits are mainly fluvial sands, gravels, sandy silts and muds of flood terraces that occur 3–4 m above the river level (Łabno 1991).

In the central and eastern parts of the city of Wrocław, Neogene deposits directly overlie Triassic rocks of the Fore-Sudetic Monocline. They are formed by Upper Miocene grey and light grey, locally mottled clays with sandy and silty intercalations. Pleistocene is there represented by deposits of the South Polish and Middle Polish glaciations. The former left over silts, clays, glaciolacustrine sands, and fluvioglacial sands and gravels. Two layers of the South Polish glacial tills (Lindner 1988) are dark grey rocks with sand and fine gravel intercalations. In the center of the city, the tills form a continuous cover, the top of which occurs at different depths owing to erosional activity of the Odra river that drained at those times waters from the melted ice-sheet retreating after the glaciations. Tills of the Middle Polish glaciation are grey-brown, bluish, predominantly sandy with abundant pebbles of Scandinavian rocks (70–80%). They are separated from the South Polish tills by glaciolacustrine and fluvioglacial deposits. Locally the latter are missing and then all tills form a single thick layer. The North Polish Glaciation accumulated sands and gravels in over-flood terraces, 4–6 m high above the river level. Holocene deposits have developed as sands, gravels and muds in two flood terraces, of which the lower terrace is 2–3 m high and the upper one 3–4 m high above the Odra river level (Winnicka 1998).

In physiographic classification, the city of Wrocław is situated in the Silesian Lowlands, largely in the Wrocław Glacial Valley (Pradolina Wrocławska). Actually the city was founded and developed in the Odra river valley, the shape and dimensions of which were mainly controlled by the Middle Polish Glaciation and which is filled with relatively thin layer of Pleistocene and Holocene fluvial deposits. The 3–10 km wide valley was incised in the ground moraine left by that glaciation.

A drainage pattern in Wrocław is quite dense and embraces the Odra river system and its confluences: (1) on the south the rivers of Bystrzyca, Ślęza and Oława with oxbow lakes and creeks, (2) on the north-west the river of Widawa. The hydrologic pattern is further complicated by numerous old meanders of both the Odra river and the confluent rivers.

2.1 *Engineering-geologic conditions*

A recent state of knowledge of engineering-geologic conditions was summed up by the PAN-METRO Group (Żelaźniewicz et al. 2013). The proposed metro routes in Wrocław (Żabiński et al. 2012), was subdivided into several sectors. Line N–S, labeled I with sectors IA, IB, IC; line W–E is labeled II with sectors IIA, IIB, IIC, IID, IIE, IIF (Fig. 1).

Using all available data from boreholes located directly or close (<200–300 m) to the two recommended metro lines, geological cross-sections, general and auxiliary, were constructed for each sector. For the auxiliary sections, data from the deepest wells, drilled for various purposes in the city center (sectors IC and IID), were used. It is to be stressed that about 60% of all wells performed so far in Wrocław are merely 4–10 m deep, whereas those reaching depths of 20–50 m constitute only 0.65% of the database (Goldsztejn 2009). Therefore, before the

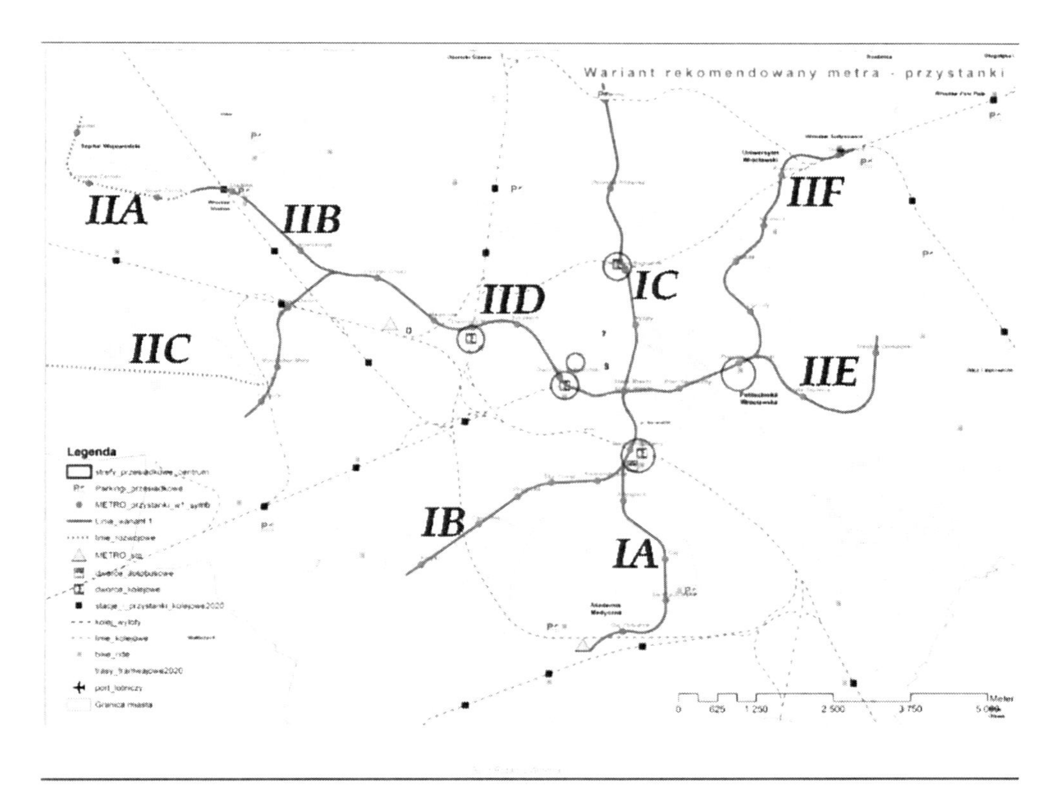

Figure 1. Sketch of the city of Wrocław to show the preliminary recommended metro lines I and II (blue) after Żabiński et al. (2012) subdivided into sectors for more detailed geological, hydrogeological and engineering geological characteristics (Żelaźniewicz et al., 2013). Note built-up areas and river pattern at the background.

final planning of the metro routes and stations, it will be necessary to drill 100–150 additional research wells to a depth of 40–60 m.

For the metro constructors, the geological characteristics and knowledge on spatial distribution of the uppermost Quaternary and Neogene deposits is of primary importance. In the city center, where the underground excavations have to be predicted, the top of glacial tills occurs at a depth of 5–20 m below the surface and dips gently in the NNE direction. The inclination is compensated by overlying sands, the thickness of which increases toward the Odra river, e.g. 5 m at pl. Legionów and 20 m under Most Pokoju (Goldsztejn & Skrzypek 2004).

The geology of Line I is variable. Along the 4 km long sector IA, at a depth of 12 to 25 m, laterally occur from the south to the north: tills with clays, loose deposits with lenses of clays, sandy tills. Along the 4 km long, sector IB, at depths of 12–25 m, from the SW to the NE, there is a lateral alternation of loose and cohesive deposits, and then sandy tills close to the junction with sector IA. In the area of *Sky Tower*, the top of the Poznań Clays was found at a depth of 36 m. Sector IC continues to the north from the IA/IB junction and covers the Old Town area. At depths of 10–20 m, sandy tills that occur on the south are rapidly replaced by cohesive sand/gravel deposits which continue to the end of the sector (Fig. 2). In its middle part, the sector runs underneath the Odra river.

The geology along Line II (W–E) is also variable. Sector IIA is built of the Poznań Clays in which there are deep erosional depressions filled with incohesive deposits, thus their palaeorelief is highly diversified and at depths of 10–20 m, one can encounter, while moving from the NW, laterally alternating loose deposits and clays, whereas at the SE termination of the sector, there is a highly diversified lithology of tills, sands, silts, and gravelly tills. Similar diversity is met at sector IIB, where generally incohesive deposits dominate over tills and clays. Sector IIC departs significantly form the above pattern as directly under the soil layer there are the Poznań Clays with local shallow erosional depressions filled with loose sands and the lithology gets more complex toward the junction with sector IIB. Along 7 km long sector IID, at depths of 10–20 m sandy tills occur, which are underlain by the Poznań Clays on the west and overlain by incohesive deposits on the east. Glacial tills occur at different depths, often being covered with moraine boulders. At sectors IIE and IIF, there are mainly incohesive deposits

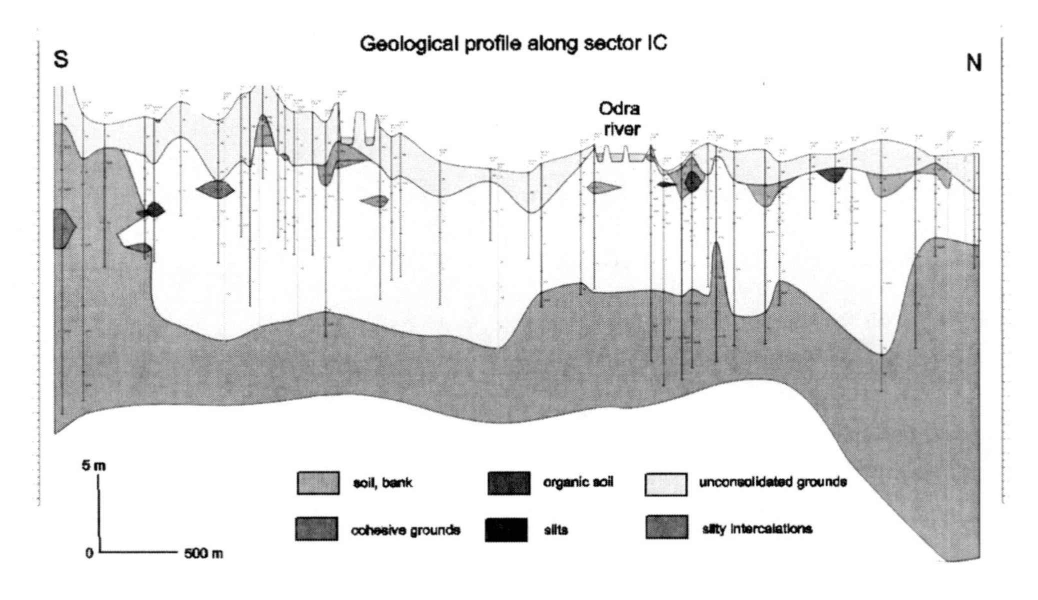

Figure 2. An exemplary geological profile along sector IC. To expose the geological structure and variability of engineering-geologic conditions the vertical scale in not equal to the horizontal scale, after K. Choma-Moryl (Żelaźniewicz et al. 2013).

few to several meters thick which overlie glacial till. The above synoptic review shows that the most diversified engineering-geologic conditions will be met along sectors IA, IB and IIB.

2.2 *Underground and surface waters in the city of Wrocław*

Both lines of the Wrocław metro will have to cross beneath the Odra river, its channels and old town fossa. Besides the surface waters, constructors will have to cope with variable hydrogeologic conditions in fluvial deposits and in local lenses of retained waters. In glacial tills, one may ecounter buried oxbow lakes and broad river channels filled with alluvial deposits, where significant influx of water can be expected, especially in the NW parts of Wrocław. To avoid troubles, such traps are to be identified already in the planning stage.

2.2.1 *Hydrologic conditions*

In the city of Wrocław, the present engineering of surface waters has been achieved owing to a series of hydrotechnical structures, building of channels, lockages, and inundating meanders and oxbow lakes, in particular in the center of the city and Old Town. The Odra river flowing through the city is a highly engineered, embanked, draining river, which due to the increase of water table at power plant lockages also have an infiltrating nature. An average water flow in the Odra river amounts to 141 m^3/s, in the Bystrzyca river ~9,3 m^3/s, in the Widawa river around 6,6 m^3/s and in the Oława river only 3,6 m^3/s. During the 1997 flooding, water flows in Odra, Widawa and Bystrzyca were 25, 36 and 50 times higher, respectively (Dubicki 2012). The Oława river is a main provider of drinking water to the city by supplying water to infiltrating trenches, channels and ponds, and directly to the water treatment plant. In the hydrologic system of the city, an important role is also played by the Odra–Widawa flood channel which allows to redirect high flooding waves to the north out of the metropolis. The latter are also reduced over three polders located in the southeastern part of Wrocław.

Another problem important for the underground railway system comes from the extremely high surface waters in times of flooding events. Heavy rainfall in July 1997, that reached in the Sudetes Mts. 415 to 619 mm/m^2, produced an enormous inflow and caused water level to raise up to 724 cm in Wrocław, which eventually resulted in the disastrous flooding over the densely built up areas. Such an experience forced modification of the flooding hazard project and launching a specially dedicated program which is still implemented.

2.2.2 *Hydrogeological conditions*

In the city of Wrocław and its suburbs, there are four underground water levels: Quaternary, Tertiary, Triassic and Permian. The shallowest and richest aquifer is confined to sand-gravel fluvioglacial and fluvial deposits of Quaternary age, and to sandy layers within Tertiary clays at depths of 60–120 m below surface. A characteristic feature of the first aquifer is its susceptibility to surface pollution and hydraulic connection with the Odra river waters (Kryza et al. 1989, Staśko 2005). Depth levels of the water table of the highest aquifer can be found on hydrogeological maps. A practical problem for the metro constructors is connected with fluctuations of the water table over the transformed urban areas (Kowalski 1977, Worsa-Kozak 2004). In the reference period 2003–2006, ninety eight wells monitored the water table that proved to occur at a depth of 1 m to 7 m below surface, being deepest (5–7 m) in the center of the city, with yearly fluctuations in a range of 1–2 m. Therefore in the center of the city also high water inflows (18–24 $m^3/h/1$ mS) may be expected and lower in the western and southern parts of Wrocław, where inflows of ~13 $m^3/h/1/mS$ occur.

The second aquifer, which is spread beneath the entire city area, is confined to lenses and layers of sandy-gravelly deposits within Tertiary clays at depths of 40–100 m (Mroczkowska 1995). Its waters are mineralized thus only locally can be used as drinking water. The water table of this aquifer is stabilized at a depth of 6,5–7,5 m below surface under sub-Artesian and Artesian pressure. Therefore, one may expect additional inflows of waters at the metro construction sites. The Tertiary aquifer locally supplies water to the overlying Quaternary aquifer and is drained to the Odra valley (Korwin-Piotrowska et al. 2013). Location of the aquifers and water-carrying rocks are shown on the cross-sections parallel to the proposed metro lines (Fig. 3).

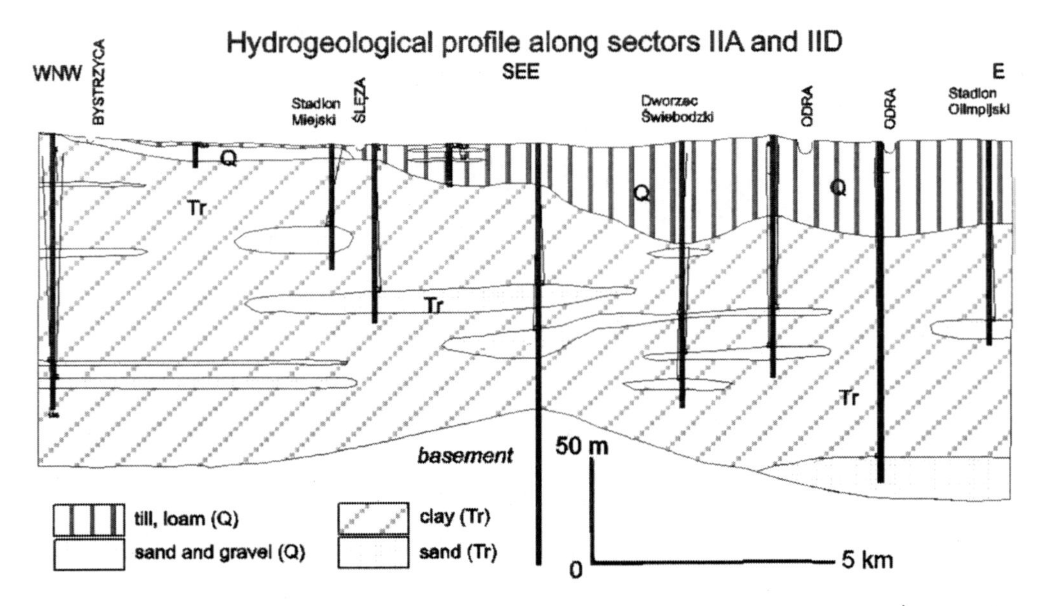

Figure 3. An exemplary hydrogeological profile along line II (W–E), after S. Staśko (Żelaźniewicz et al. 2013).

In the city of Wrocław, ground waters do not have natural chemical composition due to over 1000 years of human activities (Roszak 1991). The Odra river water are characterized by pH 6,5–8,1 and electrolytic conductivity in a range of 291–1220 μS/cm. Waters of Quaternary levels show mineralization 231–974 mg/dm³, average 723 mg/dm³, being highest in the center of Wrocław and decreasing with depth. Tertiary waters are more mineralized—do 2,6 g/dm³.

To sum up, before planning the metro, it is necessary to gain more detailed knowledge to complete the picture of already known parameters about: (1) unconsolidated deposits at depth of few meters down to 40 m, (2) water table at a depth of 1–7 m below surface, (3) usable aquifers 1–20 m thick, (4) water inflows, depending on local thicknesses and filtration coefficient fluctuating between 65 m³/h to 1–2 m³/h, (5) mineralization of ground waters, particularly important in case of possible hydraulic contacts or insufficient isolation of the deep-seated Tertiary aquifers in the western part of the city.

3 EXCAVATIONS FOR METROPOLITAN TRANSPORT SYSTEM

According to the system specification assumed by the Bureau for Development of Wrocław (Żabiński et al. 2012), construction of the Wrocław metro will require tunneling at least for the sectors which will be located beneath the densely built-up city center. As in numerous other cities, tunnels can be constructed by open pit or by underground excavations. Although apparently simple, an optimal choice depends on a number of factors and material parameters that are to be considered already during a planning stage in order to avoid troubles, delays and unpredicted events in the course of investment realization.

Trenchless technologies for building underground excavation with stabilized walls have been refined by men for ages. In London, in early XIX century a primitive shield boring machine was utilized for the first time to build a tunnel that crossed beneath the Thames river. Advancement of an excavation face was then very slow, and it was not much faster 150 years later when the first line of the Warsaw metro started to be constructed at a speed ~2 m a day. Fortunately since that time technologies have been much improved. The new, 10-km long tunnel "Słowacki Route" in Gdańsk was excavated 10 times faster, ~20 m per day. Ground water hazard was significantly reduced by freezing, utilizing compressed air or ground/soil sealing by various injections. Techniques for wall stabilization have also been greatly advanced and get more and more sophisticated.

Actually, in urban agglomerations, tunnels for more efficient public transportation can be constructed by means of several methods: classic mining, pit excavation, shield boring and also some minor techniques can be applied for special targets (Furtak & Kędracki 2005). In Wrocław, local geology generally excludes classic drill-and-blast method. Therefore either an open or closed trench excavation or underground shield boring can be under consideration.

In the 60-ties and 70-ties of XX c., newly introduced Tunnel Boring Machines (TBM) were a breakthrough in underground constructions and gave impetus for dynamic advancement of trenchless technologies for constructing public projects. TBM ensured support of the tunnel face, stability of walls and full safety for the staff owing to the separation of a workspace, where the ground is excavated, from the protected section inside the shield. High level of mechanization of drilling and pre-cast concrete elements to sustain the tunnel walls allow for excavation speed of some tens of meters per day. This is mainly due to the rapid advance in electronics and remote steering of drilling machines. Presently, TBMs can operates in all types of grounds, from crystalline rocks to watered disperse soils, and ensure fast progress of excavations. Delays in building large projects underground may be due to insufficiently known geology, therefore it extremely important to recognize the geological and hydrogeological conditions along the planned metro routes in every possible detail.

3.1 *Tunnel boring machines—TBM*

In Wrocław, subsurface infrastructure network is installed not deeper than 12,5 m, which means that most underground constructions should be safely located below this level. Local geological conditions do not allow to operate with TBMs without shield because excavations made in unconsolidated grounds require an immediate support, which is usually difficult and costly. Therefore only TBMs with shield can be rationally contracted for the underground projects in Wrocław. Not going into details, there will be a choice between full-face shield machines and open face shield machines.

3.1.1 *Types of TBMs*

The full-face shield machines can operate without support or with mechanical support of the excavation face and the choice largely depends on the type of ground in which tunneling is to be performed. It may happen that utilizing machines with mechanical support may be disadvantageous in glacial tills in which big boulders occasionally occur or in watered ground that can be subjected to uncontrolled subsidence. In aquifer grounds, TBMs with (1) compressed air application may be applied, which prevents collapse or damage of a ground layer above the tunnel, or (2) with slurry/fluid support, which not only helps to sustain the tunnel walls and but also to effectively remove the excavated soils out of the construction site. In the latter cases, the appropriate composition of the slurry is of great importance and needs to be monitored continuously.

Machines with Earth Pressure Balance System (EPBS) may prove to be especially economic because the excavated face receives then support from the ground which is dispersed during drilling. They are generally cheaper, assure lower drilling costs at higher speed of excavation and lower demands for surface area necessary for the project logistics. However, the choice between the EPBS and slurry system always depends on the soil parameters such as grain size/granulation or water pressure but also on the accessibility and dimensions of the construction site, costs, constructor experience, etc. The EPBS system can be advantageous for building the Wrocław metro, provided the underground routes will be positioned generally at a depth 20–25 m, that is mainly in cohesive grounds beneath the first aquifer. Multi-shield systems of boring can also be considered although it is expensive and the shields need to be reassembled during drilling operation to meet given ground conditions.

The open face shield machines can be used for excavations made in grounds above aquifers, or in grounds in which the water table can be temporarily yet safely lowered. Such machines can prove particularly useful in roughly uniform grounds of glacial till types that are rich in erratic boulders. However, the main disadvantage lies in a limited applicability in grounds characterized by laterally changing geological conditions, and this is the very feature that characterizes the underground Wrocław. Moreover, they require additional measures like compressed air, freezing, special injections, or anchors to stabilize the face.

Owing to varying geological and hydrogeological conditions along the metro routes, it is often necessary or desirable to combine various shield systems in a single boring machine. Combined or simply universal shields are also manufactured ready to use. For instance, constructors of the Duisburg metro used a shield working commutatively in a slurry and balancing ground pressure systems. Open face shields and EPB shields were employed for excavating the French sector of the Channel Tunnel. Such sophisticated shields are naturally more expensive than simple ones which by themselves are costly enough.

3.1.2 *Features to be considered*

Among advantages of shield machines are: (1) mechanization and high working speed, (2) precise profile of the excavation, (3) least possible impact on surface urban structure, (4) high safety for the staff, (5) environment protection including ground waters, (6) sealed and economic tube casing. Disadvantages include: (1) long duration of planning stage and assembling of shields, (2) needs for large areas to dump the drilling waste and to arrange pond for slurry monitoring and improving, which is however avoided with using EPBS machines (3) needs for individual planning of tubing.

A right decision which shield(s) will be appropriate to excavate a given tunnel is determined by the proper recognition of local geological and hydrogeological conditions along with engineering-geological and geotechnical ones projected onto local topographic and demographic characteristics. Urban infrastructure must withstand some possible ground deformation and minimization of ground subsidence, which closely depends on a chosen tunneling method. Further elements that must be considered and decided is cost optimization in relation to the assumed advance speed of excavation and time allocated for completing the project.

Another decision which is to be taken already in the planning phase has to determine whether a single track or double track tunnels will be constructed. In most cases, the existing metro systems favors the first option because of safety traffic reasons when the metro is in use but also because smaller shield diameter, shallower depth of tunneling and ~twice less drilling waste produced during the tunnel construction. Comparative studies conducted in several North American cities showed that single track tunnels offer more effective solutions as they are built faster, stations are simpler in design and cause less impact on the surface while being constructed, and the costs of organization and exploitation of the whole system are lower.

However recent advances in underground building make more and more popular double track tunnels drilled at a time by two parallel operating shields. Again the right choice depends on many parameters, even such like the type and width of the platforms at the underground stations.

Using shield machines, there is no need for temporary stabilization of the excavation walls, which is necessary when a cut and cover tunneling method is employed. TBMs allow for immediate installation of permanent tubing and sealing the space between the ground and tubes by grouting. In contrast, freezing is a high energy-consuming technique, while chemical means may be not neutral for the environment, even if they are susceptible to biodegradation.

As indicated above, the proper choice of tunneling method is determined by a number of factors which all have to be considered, of which geological and hydrogeological conditions are among the most important. Therefore the highest possible level of their detailed recognitions is absolutely essential for the project. However, in some instances when a tunnel is going to be short and constructed beneath densely built up areas, it is reasonable to employ classic mining methods which can be challenging to TBMs. As an example may serve the metro tunnel in Limburg (Wawrzyniak 2011). On the other hand, the Moscow metro is being extended by 150 km (now 300 km) owing to the FIFA World Cup scheduled for 2018. Local geological conditions—glacial tills and loamy sands—are ideal for EPB technique. Ten machines were purchased from Herrenknecht and their weekly progress amounts up to 125 m.

3.2 *Stations*

A task more complex than tunneling is the construction of metro stations. It is noteworthy that the metro stations will be built by mainly using cut-and-cover or classic mining methods, aided wherever necessary with temporary freezing and chemical/polymer injections into the ground. However such operations produce for time being serious problems in smooth functioning of the city and require costly and time-consuming relocations of the infrastructure networks, even when the Milan method is employed that allows for the simultaneous building of the underground and surface parts of the facilities. The choice of the right method(s) again greatly depends on the exact knowledge of local geological and hydrogeological conditions.

In Wrocław, detailed research should include rocks to depths 20–30 m below the base of future tunnels. To acquire such data one needs to drill 100–150 wells to a depth of ~50–60 m. These problems are limited only to the stations located in the center of the city, sectors IB, IC, IID (Fig. 1), and the choice would be between a trench method and one of the trenchless methods (in more cohesive ground). In these parts of the city, open pit excavations to a depth exceeding 20 m would be difficult but not technologically impossible. Examples coming from big European or Asian cities which have similar or even more challenging geological conditions show that metro routes can be constructed using both trench techniques and shield machines, thus from the technological point of view, the construction of the Wrocław metro is obviously possible. In Wrocław, the natural geological conditions speak in favor of trenchless technology which would logistically simpler and less inconvenient for people leaving there. Of course, budget is another question but this has to be considered in connection with the former issues.

In Poland, natural references are offered by both the lines of the Warsaw metro, which however was built almost entirely by a cut-and-cover method on the pre-assumption that tunneling would be shallow. The first line has stations ca. 150 m long and 20 m wide, located at 12–14 m beneath the surface. The second line is situated more deeply, 15 m to 26 m, with stations 22–25 wide which are distant by 400–900 m from one another.

3.3 *Tunneling and environment*

An open pit excavation technology requires deep trenches. The shallowly located highest aquifer requires temporary dewatering at the time of construction. The deep trench and large cone of depression cause the change in the state of stresses in the adjacent ground, which leads to either subsiding or upwelling of the ground and deformation next to the trench walls. Therefore not only geology but also geotechnical parameters of the ground along the metro routes have to be known already at the planning phase and possible distortions assessed in advance, especially hazard to which the existing buildings will be exposed. Tunneling or open pits may affect natural environment by lowering the water table and other changes in the aquifer systems, or by noise and vibrations affecting humans and birds and other animals. A detailed plan how to protect trees, in particular valuable ones, has also to be prepared in advance, perhaps resulting even in some corrections in the proposed metro routes.

3.4 *Other specific conditions related to metro*

The metro is an independent, collision-free mean of rail transport, a metropolitan railway, operating beneath the surface in the overcrowded city center and on the surface further off the center. A proportion of tracks under and over the surface may be even 1:1, or else, as it is in Cologne, Rotterdam or Vienna. As such, the metro is a specific and important but complementary elements in the composite urban transportation system (Fig. 4). Nowhere metro is meant or capable to entirely replace other means of public transport. This fact is often not well understood, in particular by inhabitants in the cities that have not developed any metro

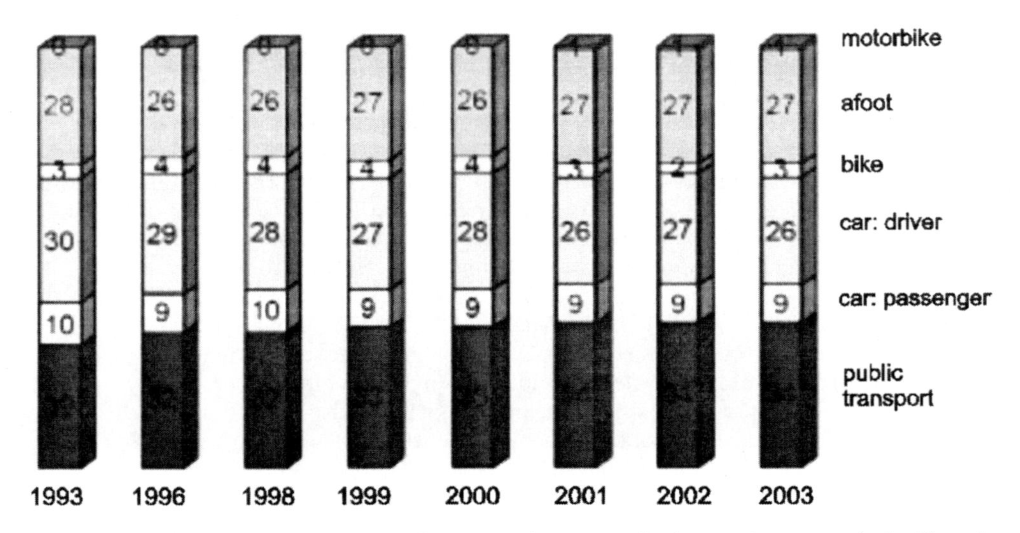

Figure 4. Demands for various means of transport in metropolis do not change much significantly over years. The Vienna example shows slight decrease of usage of private cars in favor of the public transport which provides service for only ~1/3 of the people who need to move in the Austrian capital city (Andreischitz 2005).

system but only think about it. Sometimes mass media publicity may intently or unintently deepen such misunderstandings. Nevertheless, the right message is that the metro is nothing else than the integration of various means of transport into one effective metropolitan system. Indeed, it is difficult to imagine that in any city the world over in which the metro system is operating, a public opinion polling would be in favor of closing the metro by saying that it is just a useless vagary and unjustified costly nonsense.

Particular emotions may sometimes be raised when one notices the numbers referring to the costs of the metro constructions in the city. None made a calculation how much the costs vary between projects depending on all the above mentioned parameters that have been taken into account during planning and budgeting. A very rough estimation fluctuates between US$ 90M to US$ 200M per kilometer. In general, the better planned and managed investment the lower costs of the whole venture. Otherwise, there is a risk of undesired delays and unjustified increase in costs.

4 CONCLUSIONS

- Geological, hydrogeological and engineering-geological conditions in Wrocław do not depart significantly from those that are known from other European cities which possess the metro facilities—from France to Russia.
- In Wrocław, excavations at depths of 10–20 m would often place very unfavorably the metro lines directly in Quaternary aquifers. Generally, unconsolidated deposits are disadvantageous whereas tills and clays are far more friendly for the underground construction. Therefore a bit deeper tunneling, at 20–25 m level, would advantageously locate the underground railway constructions in cohesive tills and Tertiary clays, and then most problems connected with the necessity of excavating unconsolidated water conducting rock layers will be easily avoided.
- The choice of the most appropriate method(s) of excavation must depend on the total economic calculation of gains and losses in the course of construction of tunnels of a given length in densely built-up urban areas. Potential hazards are easy to predict and eliminated already at the planning stage. At least in the city center, the TBM methods will be most

justified. An aiding technology could be NATM (New Austrian Tunneling Method) and locally ground freezing. With the usage of TBMs, the excavations for tunnels pose the least problem in the whole metro project from both technical and financial point of view.

- Before commencing any planning stage, some 100–150 additional research boreholes must be drilled down to 50–60 m below surface to ensure a satisfactory level of recognition of the geological and hydrogeological conditions. This knowledge is indispensable for selecting the most advantageous methods of excavations and reconciled with urban and architecture infrastructure for the final positioning of the metro routes and stations.
- The metro initiative is important not only for creating an efficiently and effectively integrated system of public transportation in the city but also for numerous positive side effects. The Warsaw example, with its first line terminating almost in the forest show that this line acted as the axis around which the city has built up and developed rapidly with urban housing and cultural infrastructure. The metro line has been functioning there as a strong magnet that attracts inhabitants ready to settle down in the vicinity of the easy transport route. In more central quarters, the metro infrastructure and stations can be seeds for growing shopping and cultural centers.
- In Wrocław, the metro should be also treated as a future investment which will promote the growth of the city. It is however necessary to start to think positively about it right now, starting from the revision of the strategic plans of the development of Wrocław during the nearest decades to come.
- The short review above is just a contribution to these plans. Its intention is to draw attention of the decision makers to the minimum actions that should be undertaken before any planning stage can be commenced. Prior to that also other important aspects should be considered: urbanistic, architectural, economic, sociological related to the long-term vision of the advancement of Wrocław.

REFERENCES

Andreischitz, P. 2005. System szynowego transportu publicznego w Wiedniu. In *Transport publiczny w Warszawie*. Warszawa.

Dubicki, A. 2012. Minutowy przebieg powodzi—stany i opady. In J. Sobota (ed.), *15 lat po powodzi na Dolnym* Śląsku: 7–32. Wrocław: Uniwersytet Przyrodniczy we Wrocławiu.

Dyjor, S. 1970. Seria poznańska w Polsce zachodniej. *Kwartalnik Geologiczny* 4.

Furtak, K. & Kędracki, M. 2005. *Podstawy budowy tuneli*, Kraków. Wydawnictwo Politechniki Krakowskiej.

Goldsztejn, J. (ed.) 2009. *Atlas Geologiczno-Inżynierski Aglomeracji Wrocławskiej. Baza danych geologiczno-inżynierskich* http://www.mos.gov.pl/artykul/5352_geologia_inzynierska/9910. PIG, Wrocław-Warszawa: Proxima SA.

Goldsztejn, P. & Skrzypek, G. 2004. Ukształtowanie powierzchni stropowej gliny zwałowe zlodowacenia sanu2 w centrum Wrocławia. *Przegląd Geologiczny* 2.

Korwin-Piotrowska, A.A., Krawczyk, J.I. & Russ, D. 2013. Wahania zwierciadła neogeńskiego piętra wodonośnego w obszarze Wrocławia. *Biuletyn PIG*.

Kowalski, S. 1977. Dynamika stanów pierwszego poziomu wód podziemnych terenu miasta Wrocławia. Zeszyty Naukowe Akademii Rolniczej we Wrocławiu. *Wydawnictwo Uczelniane Akademii Rolniczej we Wrocławiu, Rozprawy* 8.

Kryza, J., Poprawski, L. & Staśko, S. 1989. Główne zbiorniki wód podziemnych w rejonie wrocławskim—koncepcja optymalnego zagospodarowania i aktywnej ochrony. In *Materiały Konferencji Naukowo—Technicznej, Alternatywne źródła zaopatrzenia Wrocławia w wodę*: 33–38. Wrocław.

Łabno, A. 1991. *Objaśnienia do Szczegółowej Mapy Geologicznej Polski w skali 1:50 000 arkusz Leśnica*. Warszawa. Wydawnictwa Geologiczne.

Lindner, L. (ed.). 1988. Czwartorzęd. Osady. *Metody badań. Stratygrafia*. Warszawa. Wyd. PAE.

Mroczkowska, B. 1995. Zmiany jakości wód kenozoiku w zachodniej części Wrocławia i okolicach. In *Materiały LXVI Zjazdu PTG Wrocław*.

Roszak, W. 1991. Kształtowanie się składu chemicznego płytkich wód podziemnych w pradolinie Odry w rejonie Wrocławia. *Acta Universitatis Wratislaviensis*. 1981.

Staśko, S. 2005. Wody powierzchniowe i podziemne. In J. Fabiszewski (ed.) *Przyroda Dolnego Śląska*: 185–206. Wrocław. Oddział PAN.

Wawrzyniak, Ch. 2011. Technical challenges of high-speed rail tunnels in Germany. *Proc. of SMME Conference*.
Winnicka, G. 1988. *Objaśnienia do Szczegółowej Mapy Geologicznej Polski w skali 1:50 000 arkusz Wrocław*. Warszawa. Wydawnictwa Geologiczne.
Worsa-Kozak, M. 2004. Hydrogeologiczny model numeryczny w terenie zurbanizowanym (miasto Wrocław), *Acta Universitatis Wratislaviensis*. 2729.
Żabiński, M. 2013. *Metro we Wrocławiu—pierwsze spotkanie*, Wrocław, Biuro Rozwoju Wrocławia.
Żelaźniewicz A. & PAN-METRO. 2013. *Analiza nowoczesnych metod drążenia tuneli metra w odniesieniu do warunków Wrocławia*: 1–85. Wrocław. Polska Akademia Nauk, Oddział PAN we Wrocławiu.

Underground Infrastructure of Urban Areas 3 – Madryas et al. (Eds)
© 2015 Taylor & Francis Group, London, ISBN 978-1-138-02652-0

Analysis and evaluation of progress in the implementation of a road tunnel under the Martwa Wisła River in Gdańsk, Poland

I. Żygowska, P. Czech, S. Gapiński, A. Łosiński, K. Smolibowska & A. Templin
Gdańskie Inwestycje Komunalne Sp. z o.o., Poland

ABSTRACT: The road tunnel under the Martwa Wisła River is the part of the Slowacki Route, which provides the convenient connection between the Lech Walesa Airport in Gdansk and the Port of Gdansk. The tunnel together with Marynarki Polskiej junction is the engineering object with the total length of 2,155 meters. Length of the tunnel section is 1,377.5 meters including the TBM tunnel length of 1,072.5 meters. The dimension of the two tunnel tubes is 12.5 meters each. The study "Analysis and evaluation of progress In the implementation of to road tunnel under the Martwa Wisła River In Gdansk, Poland" shows the importance of the tunnel for the improvement of traffic condition in Gdansk and in the development of the city. This article presents an economic analysis of the project as well as technical analysis concerning the choice of TBM method of the crossing the river. It also widely presents the geological and hydrogeological conditions, the methodology of the surveys of these conditions. It is also describe the installation of the TBM, separation plant to clear slurry, principle of operations, boring advance of TBM, the assessment of the cutter head and replacement of the damaged tools. There is also described the fabrication of the tubing precast elements in the study as well as the process of building the cross-passages between the tubes of the tunnel. Finally the study provides the summary of the effects achieved during tunnelling under the Martwa Wisła River.

1 INTRODUCTION

The construction of a road tunnel under the Martwa Wisła River is the most important and also the most difficult part of the task IV of the investment project *The connection of the Airport and the Port of Gdańsk—Słowacki Route*, implemented since 2011.

When speaking about the construction of the tunnel, it should be mentioned that the construction of the Słowacki Route is of strategic importance for the improvement of traffic conditions in Gdańsk and the further, harmonious development of the city: it will provide a convenient connection between the Lech Walesa Airport and the Port of Gdańsk, as well as a network of regional, national and international routes, improve communication between the districts of Gdańsk, facilitate tourist traffic in the Kashubian Lake District, and create a new link road through the Martwa Wisła River being of a paramount importance to improve the competitiveness of the Port of Gdańsk as well as of the whole region.

As a part of implementation of the 4 tasks that make up the Słowacki Route, in parallel with the construction of the new roadways (approximately 10 km of dual carriageway and dual carriageway sections of the roadway), a thorough reconstruction of the existing crossings, construction of new access roads, sidewalks and bike paths have been carried out, new engineering objects have been created, including those being the subject of this study—a road tunnel under the Martwa Wisła River.

The project would not have been possible without financial support from the European Union. The project is co-funded (level of funding is 85% of eligible costs) from the Cohesion Fund, within the Operational Programme Infrastructure and Environment, Priority VII Environmentally-Friendly Transportation, Measure 7.2 Development of Maritime Transport.

As the engineering task, the construction of the tunnel under the Martwa Wisła breaks the technical threshold of entering the great engineering facilities, on which traffic takes place, underground. It is the development direction of all major cities in Europe and around the world. Limited space of cities still collides with unlimited development of road and rail traffic. Therefore, the elimination of thresholds, mitigation barriers and improving accessibility should be the essence of all engineering activities.

The selected, as a result of many studies (as well technical, as social and economic), method of implementation of the project, i.e. the tunnelling method, ensured the tunnel in Gdańsk status of Poland's first road tunnel performed with the use of a Tunnel Boring Machine (TBM). However, implementation of the project through tunnelling was not obvious from the beginning of the investment process. The first design studies have considered design solutions in the form of high or moving bridges, and in the form of an immersed tube tunnel, in the open trench and tunnelling. In 2008–2010, the execution of the bridge or tunnel in every possible variant was considered in details.

An economic analysis of the project to build a road tunnel under the Martwa Wisła River, because of its infrastructural character, differs significantly from the economic analysis conducted for 'productive' investments. Therefore, the analysis of such an investment must take into account the costs and drawn benefits—that is not only directly related to the implementation of the project, but also external effects for people and businesses not directly involved in the investment process. First, road transport planners and specialists assessed the relevance of the implementation of the Martwa Wisła crossing. Their analysis left no doubt that the benefits of such a crossing would not only make life easier for local residents and reduce the operating costs for the companies located along the new route, but would also streamlining the entire communication system of the Tricity, reduce traffic jams, fuel consumption, accidents, etc.

The next stage of the analysis concerned the choice of technical solution for the crossings, and the economic and social arguments of the investment were in this regard as important as technical. When analyzing, all the conditions of the crossing implementation were taken into account, including:

- the hydrogeology, engineering, geological and geotechnical conditions occurring for the location of crossing
- allowing the uninterrupted use of the Martwa Wisła River as the only navigable waterway, connecting the Motława and the Martwa Wisła rivers with the Gulf of Gdańsk, and then with the high seas, including the parameters of vessels using the navigable channel of the Martwa Wisła River, and the results of consultation with key users (the Port of Gdańsk SA, Gdańsk Repair Shipyard SA, Gdańsk Shipyard SA, etc.)
- necessity to allow using of power and water supply and sewerage systems, running below the bottom of the Martwa Wisła River in the vicinity of the crossing and the preservation of the existing port infrastructure and enclosed building structure being still used in the Port of Gdańsk, thus allowing full use of the area and the port aquarium together with the existing port quays, located on both sides of the Martwa Wisła and Motława rivers
- a combination of smoothness of land and water transport.

The conclusions of these analyzes have led to the decision to implement the crossing in the form of road tunnel, which would be implemented using the tunnelling method. Although the realization of the tunnel using a TBM machine is more expensive at the stage of construction work, it has been found that from the construction and operational point of view, the tunnel crossing in the form of a bored tunnel would be the most advantageous. The choice of this method was supported by the following arguments, inter alia:

- bored tunnel does not interfere in the waterway in any way. There is therefore no obstacles in the operation of the waterway through the Port of Gdańsk and entities operating there
- when implementing the tunnel by tunnelling, there are no work aimed to reconstruct the existing quays on both sides of the Martwa Wisła River

- from the durability and tightness points of view, the bored tunnel, due to the circular cross section, increases the guarantee of tightness in comparison with the tunnel of rectangular cross section
- tunnelling is beneficial to groundwater, since the stripping of the aquifer is less than for tunnels made in the open pit and submerged. This means that there is less risk of direct introduction of technical contaminants connected with the construction into the aquifer
- tunnelling does not affect the natural aquatic environment negatively
- duration of tunnel boring is similar to the duration of tunnel construction using immersed or open trench methods.

Implementation of the project *The connection of the Airport and the Port of Gdańsk—Słowacki Route*, total cost of which amounted to PLN 1 420 000 000 began in 2011. On October 14, 2011, a contract to perform the last step was signed—*task IV, Słowacki Route*, which included the construction of a tunnel with the three-level road Polish Navy junction. The General Contractor for the job is a consortium led by Spanish company Obrascon Huarte Lain SA. The size of the signed contract amounts to PLN 885 600 000.

Road tunnel under the Martwa Wisła River in Gdańsk, together with the *Polish Navy* junction, is the engineering object with a total length of 2,155 meters. It will connect the eastern and western port areas in the area of Dworzec Drzewny and Nabrzeże Wiślane quays. The length of the route, which consists of the tunnel and the access sections, is 1,700 m. Length of the tunnel section is 1377.5 m, including the TBM-excavated tunnel length of 1072.5 m and a diameter of 12.5 m. At the site of the tunnel crossing, current width of the Martwa Wisła is approximately 210 m and a depth is about 12.5 m. The tunnel will consist of two tunnel tubes, one for each traffic direction. Spacing of the tunnel tubes is 25 m. The tunnel at its deepest point will be 34.25 m below the water level of the Martwa Wisła, while the minimum recess of the tunnel beneath the Martwa Wisła will be about 9.0 m. The entrance section to the tunnel, in the form of an open trench, covers diaphragm walls, the bottom seal with the jet injection method and execution of anchor piles. The trench at the deepest point reaches about 22.0 m below the ground surface.

2 GEOTECHNICAL CONDITIONS

Engineering, geological and hydrogeological surveys related to the design of the structure were performed on the basis of the approved projects of geological works in two main stages. The first concerned the proposed immersed tunnel, and the second—the implemented bored tunnel. In addition, supplementary and monitoring surveys were performed related to the implementation of the tunnel. In total, surveys of the following types were performed: exploratory boring, boring for the hydrogeological purpose (wells and piezometers research in hydrogeological nodes), DPSH, CPT, CPTU and BDP probing, dilatometer tests (DMT), pressuremeter tests (PMT) and model testing of water flow.

Among these surveys, for the needs of two main steps, 136 lining borings were performed from the land and water surfaces, and 75 CPT and CPTU probing operations. Other tests were performed to a lesser extent.

The deepest borehole reached 70.0 m below the ground surface. During boring, measurements of ground water level and soil sampling were performed according to standard.

Laboratory tests included the determination of physical and mechanical properties of soils, with particular emphasis on grain size, because it determines the permeability of the soil. In addition, adhesion and abrasiveness surveys and mineralogical studies were performed. The waters were subjected to chemical analysis. Surveys of land for the content of gaseous hydrocarbons were also performed.

2.1 *Geological structure and hydrogeological conditions*

The tunnel under the Martwa Wisła River is located within the Żuławy Gdańskie Quaternary aquifer. Holocene sediments of this part of the delta of the Vistula River are composed mainly of fine-grained river sands, silts (dust) and clay with an admixture of organic debris, organic and mineral sediments and peat. Within the Pleistocene, sedimentary cycles of glaciation of the Oder River are the best preserved, represented by glacial till and series of glaciofluvial deposits. Quaternary aquifer is therefore built with gravel and sandy glaciofluvial and river series and permeable Holocene sediments belonging to the delta series, and to the Holocene marine sands closer to the bay. The average thickness of the sandy aquifer complex is 40 m. Groundwater can also be found in the Holocene aggradate muds and peat deposited above. Under natural conditions, the water of Pleistocene layer fed groundwater of the Holocene delta series. Such a system of pressures can be observed even today. It should be emphasized that under natural conditions, Żuławy Wiślane area was a marshy terrain, partially flooded with water, and different current situation is related to the exploitation of groundwater and land reclamation of this deltaic area.

On Żuławy Gdańskie, Quaternary sediments are deposited on Miocene sediments or directly on limestone-marl sediments of the Upper Cretaceous sediments. The oldest Pleistocene series, associated with the accumulation of San glaciation, do not have continuous spread and are represented by gray boulder clay, silt and marginal lake silts. Paleogene and Neogene sediments are represented here by the Oligocene sands and sandy silts with glauconite of small thickness, reaching up to 5 m, and the Miocene sands, silts and clays with inserts of brown coal. The ceiling of the carbonate series of sediments of the upper Cretaceous sediments (Campanian and Maastrichtian) is on Żuławy Gdańskie at elevations of about 100 m below sea level, while at the elevation of 150 m above sea level, glauconitic sands of cognac and Santonian level are deposited, forming artesian reservoir Gdańsk

In natural conditions, before the intensive exploitation of groundwater in the area of Gdańsk, the water of the Cretaceous level fed, due to its nature and stability of artesian water table several meters above ground level, Pleistocene level by ascending. The Quaternary groundwater level is lowered following from the edge of the plateau towards the sea. It is important that, in the shoreline, originally observed position of the water table was about 2 m above sea level, indicating that the drainage of the water took place just a few kilometres from the shore into the bay.

2.2 *Hydrogeological hazards related to the tunnel implementation*

The Martwa Wisła is an important factor influencing groundwater flow regime in the analyzed area. Results of observations of water table fluctuations in piezometers, presented in the geological and engineering documentation, demonstrated a direct effect of water table levels in the river on the location of the water table in the Quaternary aquifer. The strong influence of the river, stabilizing the water table during test pumping, was observed even in the piezometers located at a distance of few hundred meters from the shore. This almost direct contact between the aquifer and the Martwa Wisła contributed in the eighties of the twentieth century to a significant degradation of the groundwater resources. As a result of intensive use, there has been an intrusion of saline waters of the Martwa Wisła River and the Gulf of Gdańsk to the Quaternary aquifer. Until the sixties of the last century, water consumption from the Quaternary level was not exceeding 1,000 m^3/h, while in the eighties, it increased to more than 4500 m^3/h. As a result, the regional depression funnel has deepened and spread.

Lowering of the water table caused a change in the direction of flow and the aquifer received saline waters of the Martwa Wisła. The content of chloride ion in the Old Town area exceeded 300 mg/dm^3, and at Grodza Kamienna intake, it reached over 2000 mg/dm^3. In the late nineties, when the exploitation has stabilized at 1,500 m^3/h, a gradual reduction of the chloride ion concentration was observed. The intensity of the water sweetening off proved to be faster than would be expected. According to a survey conducted in 2010, salinity concerns a several hundred meter broad belt along the shore of the Martwa Wisła. In the piezometers located closest to the river bank, chloride concentration even above 3000 mg/dm^3 was observed. It is a proof of the almost direct contact of the saline groundwater with the waters of the Martwa Wisła. Storm phenomena and a backwater effect created by them cause that the water level in the Martwa Wisła ranges from 0.93 to 1.37 m ASL.

The resource renewal process continues, but can be inhibited by re-increased exploitation or possible dehydration. According to the design assumptions, tunnelling will allow to avoid dehydration. There are therefore no quantitative risks in relation to groundwater resources in the vicinity of the investment. The risks of water quality are the only risks arising from the construction of the tunnel. They may be related to the introduction of pollutants into the aquifer or with any unplanned dehydration, and hence the intrusion of saline waters of the Martwa Wisła. Because the tunnel is located within the main aquifer No. 112 (Żuławy Gdańskie reservoir), the water quality has been monitored from the start of earthworks (in addition to the observations of the water table level). It allowed the conclusion that the salinity zone would not move in the direction of the operated urban water intakes, and that no undesired impurities have entered the aquifer. Implementation of tunnelling is preferred due to the groundwater, since stripping of the aquiferous level is smaller than in the case of tunnel construction carried out in the open trench or immersed. Accordingly, the risk of direct introduction of engineering impurities related to the construction (e.g., construction machinery fuel) to the aquifer is lower.

In the natural conditions, ascending seepage was occurring from the Cretaceous to Quaternary stages, and also from the Holocene to Pleistocene layers. As a result of intensive use, artesian conditions disappeared in the Cretaceous stage. This resulted in a reversal of the direction of the vertical seepage to descending, from younger to older stages. Currently, due to the significant reduction in consumption, lift in piezometric surface in the Cretaceous stage is observed, as well as renewal of the resources in the Quaternary stage.

The measurement results presented in the geological and engineering documentation indicate that the water tables of the Cretaceous stage stabilize from 2 to 8 m above the ground level, while in the Quaternary level, the water tables stabilize at datums from 0.0 to 0.6 m. Similar values, i.e. within the range from 0.1 to 0.6 m, are recorded in weekly observations of the piezometers located in the vicinity of the tunnel in the Quaternary aquifer conducted since November 2011. Hydrodynamic conditions approach therefore natural ascensive vertical seepage of water from the older to younger stages. Speed of this seepage is small (<1 m/year). The currently observed resource renewal process will likely proceed. It is therefore expected that the level of the water table in the Quaternary sediments and the piezometric pressure in the Cretaceous stage will rise. During construction of the tunnel, encountering some unidentified but liquidated Cretaceous holes, many of which were made in the mid-twentieth century, mainly for industrial purposes, could not be excluded. These holes, after considerable lowering of piezometric pressures of the Cretaceous level were eliminated not always in the correct manner. If the tunnel would be executed in the open trench, encountering of such incorrectly liquidated hole could cause serious construction site hazard. Artesian water pressure would result in an outflow of Cretaceous stage waters, and due to the small grain of Cretaceous sands, this could even lead to the *quicksand* phenomenon.

Model calculations of groundwater flow showed that the tunnel does not appear in the balance of water flow at all, and the layout of hydroisohips is affected very slightly, causing deflection of hydroizohips of just 0.03 m. Similar results were obtained in other model studies performed at the assumption that groundwater flow direction is parallel to the Martwa Wisła. The aquifer filtration parameters are therefore sufficient to eliminate changes in the level of the groundwater table, which could arise as a result of the construction of the tunnel. Filtration coefficient values are ranging from $1.1 \cdot 10^{-4}$ to $0.3 \cdot 10^{-3}$ m/s, depending on the calculation method used. Aquifer formations can be defined as good, and very well-permeable. The best filtration parameters relate to interbeddings of the Pleistocene glaciofluvial gravels occurring at a depth of about 30 m, i.e. in a place where the tunnel runs directly under the Martwa Wisła.

2.3 *Geological and engineering conditions and geotechnical parameters*

Several geotechnical layers have been separated in the geotechnical documentation. Their classification and description is given below, with the proviso that I_D and I_L are the characteristic values:

I—peat and organic silts, highly plastic
II—humus and clay loam humus, highly plastic and plastic, $I_L = 0.51$
III—fine and medium-sized sands, loose, $I_D = 0.27$
IVa—fine sands and locally dusty, medium dense, $I_D = 0.49$
IVb—fine sands and locally dusty, dense and medium dense, $I_D = 0.68$
IVc—fine and very dense sands, $I_D = 0.89$
Va—medium and coarse sands, medium dense, $I_D = 0.51$
Vb—medium and coarse sands, concentrated, $I_D = 0.69$
Vc—medium and coarse sands, very dense, $I_D = 0.85$
VIa—medium and coarse sands, very dense, $I_D = 0.85$
VIb—clays interbedded with dust, of low plasticity, $I_L = 0.11$
VII—sand-gravel mix and gravel, very dense, $I_D = 0.82$
VIII—stiff dust, $I_L = 0.00$
IX—sandy clay, medium stiff, $I_L = 0.00$
Xa—clays, locally silty clays, dense, of low plasticity, $I_L = 0.1$
Xb—clays and silty clays, stiff, medium dense, $I_L = 0.00$.

Layers from I to VIb are Holocene soils, from VII to IX—Pleistocene, and Xa ro Xb—Miocene soils.

Ground adhesion and abrasion play an important role at tunnelling. Adhesiveness of ground depends on the plasticity index and the consistency index. The performed surveys showed that the clays were characterized by high, while silts by moderate adhesiveness. LCPC abrasiveness test showed that the sand was not characterized by abrasiveness or its abrasiveness was small, but very high abrasiveness characterized sand-gravel mix and gravels deposited in the area of the largest tunnel overdepth dredging under the Martwa Wisła River. It is related to their high density, grain size and quartz content.

Soil density below the bottom of the river was examined with a BDP probe. Geotechnical differentiation within the whole documented area, particularly in the open excavation area, is best illustrated by the selected CPT probing.

In the interpretation of the CPT results, rich regional experiences were also used, as Gdańsk was the first centre in Poland, which began commonly use the CPT probing back in the seventies of the last century. You should also pay attention to the results of the organic sediment checks, which, next to the water conditions, are critical in the design of the part of the tunnel located in the open trench. Previous surveys performed for the documentation purposes confirm these results.

2.4 *Conclusions resulting from the geological and hydrogeological recognition*

With regard to the geotechnical conditions, it should be emphasized that the results of all geotechnical surveys have been confirmed during tunnelling both in relation to the ground types, as well as the layout of layers. Significant differences were observed only for the presence of stones and boulders. In addition, there was no abrasiveness and adhesiveness in excess of the values specified on the basis of the completed geotechnical surveys.

Analysis of hydrological conditions showed that the filtration parameters of the aquifer proved to be adequate to eliminate changes in the level of the water table, which came as a result of tunnelling. Resignation from traps equalizing groundwater pressure on both sides of the tunnel structure found its full justification during the first tube boring. In the implementation of the transverse passages with ground freezing, there is an absolute need for continuous monitoring of groundwater quality, and in particular testing of chlorides content and mineralization. The case is extremely important due to the fact that the salinity of groundwater can significantly impede the freezing of saturated soil.

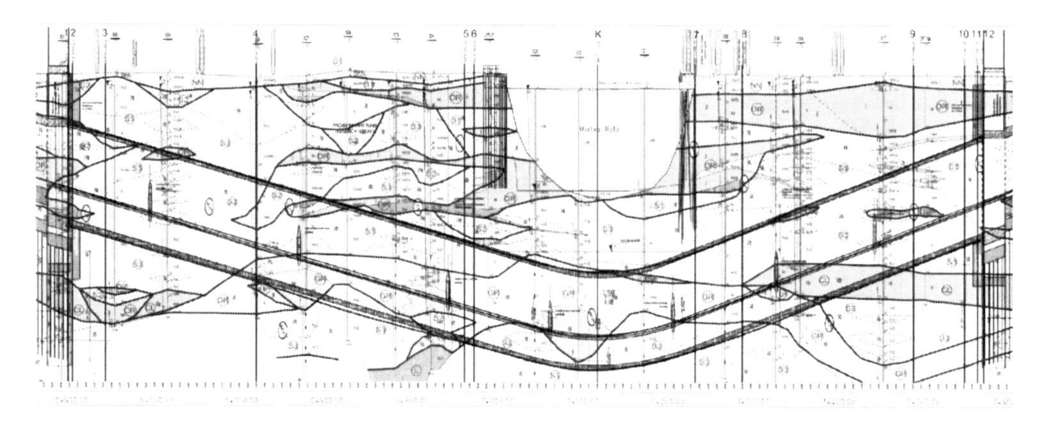

Figure 1. Geotechnical layers/*Report of Geotechnical Parameters*—OHL.

Already after construction of the first tunnel line, it was concluded that the choice of a cutter head with slurry system of transportation of excavated material was correct. This solution allowed for a smooth execution also of the second the tunnel tube, especially in places where there were rocks and boulders.

3 IMPLEMENTATION OF THE TUNNEL IN THE OPEN TRENCH, THE TBM STARTING AND EXIT CHAMBERS

Originally, the excavation of the foundation was planned with the use of the *wet* method that is under water. This solution was comprised of both winning soil from the excavation under water, as well as the performance of the bottom seal with the concrete plug, made by underwater grouting. In addition, the concrete plug should also be structurally connected under water with anchor micropiles protecting it against displacement under the action of buoyancy. It should be emphasized that work under water, in particular deep and a on a large area of the bottom, is difficult to be performed and is characterized, in comparison to the work carried out in a dry trench, by poorer quality and lower levels of control. The most important impediments of the *wet* work production technology may include:

- uncontrolled accumulation of dusty bottom sediments at the bottom of the trench while digging underwater (sedimentation), hindering the performance of the underwater grouting (risk for the quality and breaking up of the layers of underwater concrete)
- Difficulty in sealing the contact area between the concrete plug and the diaphragm wall (high risk of leakage through the joint)
- technological problems to ensure continuity of grouting over a large area of the trench bottom and the quality of underwater concrete
- problems associated with the fabrication of micropiles and anchoring under water
- the necessity, in case of the deepest sections of the excavation, to install steel struts under water (installation of struts under such conditions is very risky)
- the need to install a complex system of water management in the trench while digging under water, including filling and emptying the excavation and water discharge to ensure a constant hydrostatic overpressure of water in the excavation as well as construction of sedimentation tanks with the area of at least 10 000 m^2 and the 3 m high embankment.

Finally, *dry* work production technology was implemented, based on the execution of perimeter walls and horizontal Soilcrete sealing screens in areas without natural bottom seal, and which has the following technical advantages:

- Dry digging of the trench provides current control of section joints of the diaphragm wall being gradually exposed and allows immediate location and elimination of leaks.

- It ensures controlled conditions for the implementation of tight connection of the tunnel floor with the diaphragm wall, including the protection of the reinforcement against corrosion, which is important due to the safety and proper functioning of the tunnel.
- Thanks to the horizontal sealing below the bottom of the trench and additional vertical anti-filtration partitions, the opportunity for early check of separate sections of the trench for tightness at the time of water pumping has been achieved, which increases the security of the execution of works and enables faster location of leaks.
- In the deepest sections of the trench, massive Soilcrete screens, about 3.5 m thick (increased in relation to the proposed thickness of the underwater plug, i.e. 1.5 m), act as an additional horizontal struts and significantly reduce the displacement of the diaphragm wall in the construction phase, which significantly reduces the risk of wall fractures and leaks in the section connections.
- Fabrication of anchor micropiles in a dry trench, from the bottom of the baseplate significantly increases the quality of anchoring elements compared to the works under water. Fabrication of micropiles is controlled, which allows for ongoing correction of engineering parameters of boring and grout administration. This reduces the risk of deviations in the level of anchoring and deviations from the vertical, which can dangerously increase the tension in the anchor rods as a result of the additional bending when removing.
- Installation and commissioning of the anchors on the heads of micropiles is done in a controlled way, in more favourable conditions than under water. Quality of joints increases and positively affects security of the works.
- The ability to verify the load capacity of micropiles on production micropiles. The wet work process eliminated such a possibility of pile verification, since due to the work sequence of work micropiles had to be anchored in underwater concrete. Thus, the application of the dry method significantly increases the level of control over the implementation of the anchors, which are essential for the stability of the excavation and tunnel. In addition, in *dry* conditions, the micropiles can be inspected in accordance with the relevant standards.
- Method of spragging the diaphragm walls, which was not solved in detail in the original projects, has been adapted to the entire system of securing the foundation excavation.

The geometry and stiffness of the individual segments of the tunnel and its lining as well as of relevant sections of the diaphragm wall has also been taken into account.

3.1 *Tunnel in the open trench with road junction*

The experience gained during the construction of the tunnel in an open excavation permits the conclusion that by using steel sheet piling, diaphragm walls with a thickness from 80 to 120 cm, sprigging of diaphragm walls also under the ceiling, anchorage with reinforced concrete micropiles and the tunnel steel baseplate while using the impervious layers or anti-filtration screens to reduce the buoyancy acting on the baseplate, a tight lining of this tunnel part has been achieved, which will finally be closely connected with the bored part of the tunnel.

A very detailed analysis of the implementation of this part of the tunnel also allows conclusion that the success has been decided by the use of the latest technologies both in the execution of the diaphragm walls and fabrication of various kinds of piles and anti-filtration screens. Apart from the production technology throughout the construction site, requiring, for example, the implementation of 11 process partitions, attention should be drawn to the extensive quality control consisting of, inter alia, conducting a series of studies on experimental plots and checking of all structural elements of the tunnel in the course of and after completion of the individual works. Therefore, an important conclusion may be drawn that during the implementation of important structures in the hydrated surface, the issue of effective control, regardless of the production technology used, comes to the first place. Today, we can say that these facilities of the tunnel made in an open trench, were made in an exemplary way and should therefore be subject to further detailed consideration and analysis, particularly for training in the implementation of such facilities.

In discussing the construction of a tunnel in an open trench, attention should be paid to the need for very close cooperation between the contractor and designer, especially with the need to optimize the technology solutions and adapt them to the material and equipment possibilities. In relation to the discussed tunnel, we can confidently say the that designers' supervision really professionally dealt with work patterns of the structure in its current and the target conditions, and they made decisions in order to make full use of the potential of technology for secure construction and achieving the goal of the final tunnelling. It should be assumed that the effect of the designers' supervision will continue to be effective, but will require a very large step up, especially during the implementation of the transverse passages. In any case, it must be concluded, based also on the experience gained during the tunnelling, that ensuring the proper functioning of the designers' supervision is the primary task of the further stages of the investment in question.

3.2 *TBM starting and exit chambers*

Considering the progress of execution chambers to allow entry and exit of the TBM, we can say that very detailed analysis of the stratigraphy of the substrate, in which the excavation was carried out, was a very important and decisive for the adapted solution of the lining and bottom of the deep trenches. It is mainly the identification of layers of low permeability soils, including the determination of their thickness.

In order to determine the method of sealing the soil layer lying beneath the trench, it was important to determine very precisely the level of the tensioned water table and the thickness and permeability of the tensioning layer. A very thorough additional geotechnical surveys and comprehensive analysis were conducted, which allowed for the safe execution of the diaphragm walls of the trench lining, and the horizontal anchored anti-filtration screen in the jet injection technology.

Flawless execution of the starting and exit chambers in the selected production technology, including supplemented and thus more dense exploratory boring grid, became a proof of the need to allocate the number and types of exploratory boring to a particular construction facility and the considered process of its implementation.

With the current development of computer calculation methods, safe and cost-effective design of complex structures must take into account the spatial models of the considered structure and the latest and proven calculation methods, for example, the finite element method in a flat and spatial deformation conditions.

Construction of the TBM starting and exit chambers has shown that it is impossible to exercise important building structures without ensuring monitoring the behaviour of the individual components, which determine the safety of the entire project.

4 TUNNELLING

4.1 *Installation of TBM*

Machine number S-745 was manufactured in a factory Herrenknecht AG in Schwanau, Germany. This work lasted about 9 months. The machine had been disassembled and transported to Poland in many parts. Most of the items were transported by road to Gdańsk and 19 largest and heaviest parts were brought to Gdańsk by a Dutch freighter *Deo Volente*. On the ship, there were, among others: 245-ton engine, power unit, segmented shield of the cutter head. Due to the dimensions of the parts to be transported, their unloading was performed with the largest floating crane in Poland named *Maja*, which had brought them directly to the previously prepared units for transportation of oversized goods.

Making of the starting chamber mentioned above was a condition for the commencement of installation of the TBM on the construction site. The assembly process was divided into several stages. After completion of the preparatory works, the operation of merging and assembling the TBM in a position to start boring had begun. Cutting head and shield elements, the heaviest of which weighed 245 tons, were lowered into the shaft. The whole machine was placed on the previously assembled crib. Then, through an operating hole in the ceiling, the crawler crane with a lifting capacity of 600 t inserted previously prepared three gates. Weight of one assembled gate was 270 t. The machine was assembled into about 90 meter long system interconnected with the slurry separation plant.

4.2 *Mix face shield TBM—principle of operation*

The rotating cutter head detach soil over the entire surface of the tunnel face. The detached soil together with drilling fluid goes to the outlet in the partition. The space in front of the cutter head, is the mining chamber or excavation chamber, which is separated from the operating chamber by a partition. Excavated soil together with the slurry from the operating chamber is sucked by the pump to the inlet of the discharge line, and then pumped by pipeline to a separation plant located outside the tunnel. The removed volume of the suspension is supplemented with a bentonite mixture. Air cushion regulates the pressure in the operating chamber. Air cushion pressure balances the pressure of soil and water, preventing uncontrolled rock slides and loss of stability of the tunnel face.

In this machine, the pressure in the excavation chamber is not controlled directly by the suspension pressure but by a compressible air cushion, which keeps the suspension at the exact target pressure value through a compressed-air control system. In order to economically dispose the slurry, the bentonite suspension and soil mixture must be separated out of the tunnel into the solids and liquid. This allows return of the purified bentonite suspension back to the circuit to a relatively high degree.

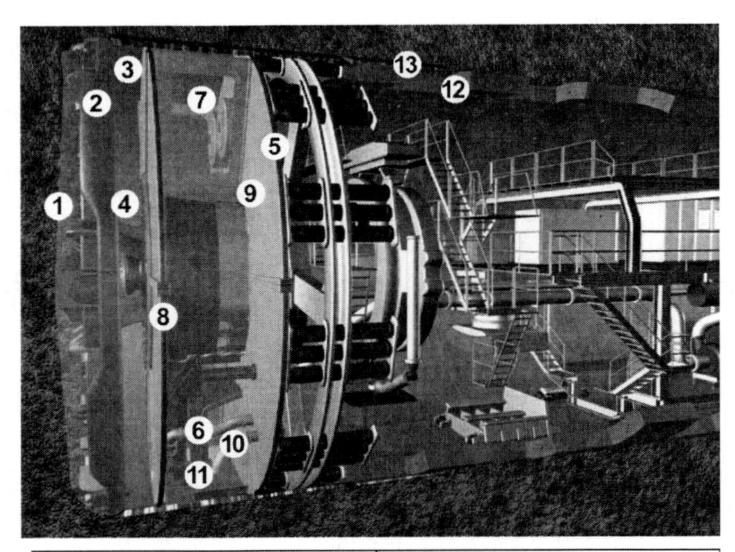

[1] Tunnel face	[8] Partition
[2] Cutter head	[9] Operating chamber
[3] Boring fluid	[10] Discharge line
[4] Excavation chamber	[11] Grid
[5] Pressure wall	[12] Segments
[6] Supply line	[13] Tailskin
[7] Air cushion	

Figure 2. Diagram of the TBM cutter head/*Work Quality and Organization Assurance*—OHL.

4.3 *TBM advance*

The cutter head advance is accomplished by actuators which are repelled from the tunnel prefabricated lining (segmental rings) built in the previous cycle. While boring runs in the face, segments that form a ring are unloaded on the tray. The space between the previously built rings and rock is filled with grout.

In the TBM type used, namely MIXSHIELD TBM, in each cutter head advance cycle, the following measures were carried out:

- boring of 2 m long section and simultaneous filling of voids outside the lining
- installation of the lining ring
- extension of track and plant (in 6 m increments)
- maintenance and repair work, if required.

4.4 *Structure of the ring*

After each two-meter advance of the cutter head, a new segmental ring is installed using the erector (feeder), which serves as a support for enabling forward motion of the cutter head. It creates a watertight tunnel lining. The geometry of the ring allows for the implementation of arcs, in both the horizontal and vertical directions. This is possible due to rotation of the next ring in relation to the previous one. Depending on the location of the key tubing, the required arc of the tunnel can be obtained. Rings fed by the erector, are mounted in such a way that the radial connection of the segments of the adjoining rings do not coincide. This is to avoid the formation of a weakened plane of the lining. To this end, the machine designers have provided 26 different combinations of the ring arrangement.

The erector is a feeder of rotary-ring type equipped with a vacuum lifting device for moving the lining segments in three directions, together with rotation thereof.

Each segment is set in the correct position by the feeder, and after the predetermined position is achieved, it is fixed to the previous segment. Then, actuators proper for the section being attached are based on it. After this step, the feeder releases the segment and the same cycle is repeated with the second ring. The same procedure is used for all elements of the ring. When the ring is complete and closed, a new boring cycle begins.

Connections between the individual segments of the lining are made by mounting sockets. The ring sealing is achieved by means of watertight seals located between adjacent segments of the ring and between the successive lining rings.

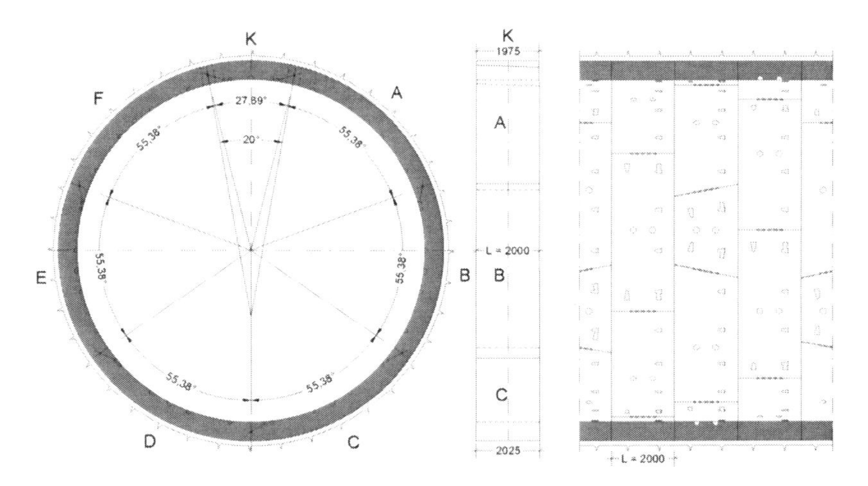

Figure 3. Diagram of the ring/design of the tunnel lining—OHL.

Figure 4. View of the tunnel interior during construction/photo: GIK.

4.5 *Boring advance and extension of cables and lines*

Together with the boring progress, periodically after every 6 meter drilled, the following elements necessary to carry out works associated with the construction of the tunnel are extended:

- piping that feeds and discharges muck and water and bentonite suspension mix
- water and air feeding connections
- track
- ventilation
- power supply (every 350 m).

4.6 *Stoppage in jet grouting block*

In order to *pass* under the river in a safe manner, a concrete block has been made before boring under the river. This is an area on the ground reinforced on a distance of 8 m, and formed to a depth between 15.8 m to 32.8 ppm. The block was formed in the jet-grouting technology, i.e. the existing ground has been mixed with stabilizing grout and injected in the high pressure jets. The planned stoppage of the machine in this block lasted several days and allowed for the actual assessment of the worn tools and their replacement. Specialized groups of three divers, after passing through the hyperbaric chamber, were entering the excavation chamber. All elements have been examined, followed by a process of dismantling and the necessary replacement of worn or damaged tools was performed. For the realization of these works, a team of employees was provided consisting, inter alia, of 36 professional divers, 2 hyperbaric chamber operators and 2 hyperbaric medicine specialists.

General wear of the cutter head tools during the boring of the first and second tunnel tubes was comparable and not significant, and evaluated as average and uniform.

4.7 *Separation plant*

The purpose of the separation plant is to clear slurry, or bentonite suspension, used during tunnelling to suspend the face, transport of muck and cooling and lubrication of boring tools. The separation plant is a device, without which the TBM cutter head cannot work. The HSP 2400 separator is the largest unit of this type that has worked in Poland so far. Slurry together with the excavated ground goes to a separation plant.

The separation plant (STP) consists of the following components: slurry mixer type HK60, HSP2400 separator, centrifugal separator, slush pit, slurry piping and transfer pumps.

The slurry mixer produces about 40 m^3 per hour of boring fluid with the desired consistency in automatic or manual mode. The mixer silos can hold up to 75 tons of bentonite. The mixer is controlled from the touch panel.

The HSP2400 separator is of container design with the following dimensions in plan: 23 m by 9 m and a height of 10 m. It has 6 double-deck vibrating screens built-in, including two pre-screens and four fine filter sieves. The separator is equipped with 8 slurry pumps with a capacity of 350 m^3/h each. The pumps are driven by 200 kW electric motors.

A pump with a capacity of up to 2400 m^3 per hour is located directly at the separator, and is driven by an electric motor with 750 kW of power output. The pump transfers the clean drilling fluid to the TBM. At the end of boring each of the two tunnel tubes, the total distance between the separation plant and the starting shaft was 1,440 m + 1,070 m of the tunnel + 370 m. Another pump located in the machine just behind the cutter head and the pump located in the start shaft pump the slurry enriched with the muck from the machine onto the highest floor of the separator. After passing through the distribution box and damping boxes, the slurry goes on the pre-screens, which receive broken boulders and stones with edge lengths from 4 to 14 mm. Finer fractions together with the pre-purified slurry get to the hydrocyclone battery and fine filter sieve. Cyclones cleanse the slurry off the particles as small as 30 microns. Dehydrated muck goes through two belt conveyors to the internal landfill, from where, it is exported to a target landfill in trucks.

Purified slurry goes into circulation again. Within an hour, the separator receives about 500 tons of muck (crushed boulders, stones, gravel and sand).

(I would write that in the whole process of boring 2 tunnel tubes, about 500 thousand m^3 of material has been excavated and separated.)

The separation plant also includes two slush pits with a capacity of 800 m^3 each. In one of them, clean boring fluid is stored, ready to be put into circulation, and the second slush pit is used to store sludge designed for regeneration.

Disc diameter of the cutter head (12.56 m) is greater than the thickness of most layers encountered on the route of the tunnel. Therefore, excavated material received at a given time by the separation plant was coming from several separate layers specified in the geotechnical documentation, which meant that the sands and gravels were mixed with peat, clay and warp. Such a diversity of soils required special attention and care for the drilling fluid parameters from the side of the separation plant crew.

Large number of stones, pebbles and boulders was a surprise. Fragments of broken boulders with their edge lengths up to 35 cm were collected on the pre-screen. The high content of stones was also the reason for a large number and different types of damage to the cutter head cutting tools. The techniques of soil sampling during geotechnical surveys do not allow for a full assessment of the number and size of the stones found in the ground.

4.8 *Summary of the tunnel boring*

The analysis of tunnelling required considering the following three core activities at its implementation:

- Development and implementation of a detailed boring method for the identified hydrogeological conditions and localized obstacles in the form of quays, the Martwa Wisła River and Orlen fuel depot.
- Development and implementation of methods of production and transportation of precast reinforced permanent lining elements, so-called tubing elements.
- Calculation of the expected ground settlement and localized obstacles and measurement of these settlements during the boring and construction of the tunnel tubing.

Settlement of ground and structures above the bored tube was an extremely important parameter for all involved in the implementation of the first tunnel tube. The results of measurements of ground settlement in the area of boring of both tunnel tubes under the Martwa Wisła indicate a relatively good estimation of settlement in the project analysis. The actual measurements carried out during the boring of the first and second tunnel tubes showed no settlements that could be considered critical.

5 FABRICATION OF TUBING ELEMENTS

To make precast elements in accordance with the design documentation, concrete class C45/55 with a degree of water resistance W8, extent of frost resistance F150 and absorbability less than 5% was selected. The tubing production process started in the aggregate warehouse—the correct particle size and grade of aggregates is a prerequisite for obtaining a good concrete. Reinforcement of the precast elements is the second important factor determining the flexural strength of the lining—reinforcement in the form of finished baskets was placed in the mould.

Moulds were made as steel welded structures, with a displacement tolerance less than 1 mm. This accuracy results from the documentation requirements. The precast elements are laid *dry*, and the stability of the structure is achieved by interaction between the tubing and the ground. Pressure and resistance provide a uniform distribution of compressive stress in the tubing elements and seizure on dry contact surfaces. Injection of concrete between the subsoil and the tubing, and a system of rubber seals are additional filling and sealing elements. The contact surfaces of tubing must fit together perfectly.

Ready precast elements were subjected to a final inspection, and then they were divided into sets (1 set included a full ring) and stored awaiting transport to the construction site.

6 IMPLEMENTATION OF TRANSVERSE PASSAGES

Cross-connections between the tubes at intervals of about 175 m have been designed for the tunnel. Taking into account the environmental conditions, soil permeability, and security of the investment implementation, the cross-connections will be made in the ground freezing process. This production technology is used to temporarily strengthen the ground, using only the physical properties of soil and water, without compromising the soil initial properties in a permanent manner.

Process of soil freezing with brine comprises subsequent works, which will be repeated for each of the transverse access, such as:

- Boring (partial) through the tunnel lining for the installation of the casing tube and preventer, which prevent uncontrolled inflows of water and soil into the tunnel during boring and installation of tubes.
- Boring through the remaining part of the tunnel lining using the preventer.
- Boring in ground and installation of freezing casings, temperature measurement and drainage pipes.
- Connection of freezing casings to the freezing units located in the main tunnel (closed circuit of brine distribution.)
- Installation of temperature sensors for temperature measurement inside the freezing casings (also sensors in the northern tunnel lining), connecting them to the data loggers and the main reader and database in the engineering office.
- Ground freezing in accordance with the project parameters and requirements.
- Maintaining freezing during the implementation of the cross-connection that is stopped after the completion of the cross-section permanent lining.

During the construction of the transverse passages, freezing units (chillers) will be applied, cooling the brine to a predetermined temperature. The brine is distributed in freezing casings in a closed circuit, and the main refrigerant, usually ammonia, is circulated inside the unit. Saline, or a salt solution (salt chloride) (so-called secondary refrigerant) is used as the frozen medium circulating through the freezing casings. By varying the concentration of salt in the solution, various melting points can be obtained.

Completion of the freezing process at one transverse passage can be determined by stating achievement of the design temperature for the project on all thermometers and a simultaneous achievement of a rise in the filtration water pressure in the unfrozen ground within the transverse passage.

7 SUMMARY

Detailed recognition of very difficult ground conditions allowed us to control the operation of the TBM at optimal parameters taking into account the condition and the type of ground. This allowed both safe and efficient implementation of the procedures provided in the documentation. Boring of the first meters of the first tunnel required testing the machine under real conditions.

Based on the experience from tunnelling, we can generally state that both types of soil found in the opening of the bored tunnel as well as the layout of geological layers shown in the documentation prepared on the stage of the project development were in line with the real conditions identified during the boring of the first and second tunnel tubes.

In addition, tunnelling of both tunnel tubes under the areas defined in the project as particularly sensitive, i.e. the port quays, the river bottom and the PKN Orlen depot, took place without problems and without significant ground settlement. Boring was carried out with maintaining the design face suspension pressures, but greater rhythm of boring was kept than originally anticipated in the project.

The effects achieved during tunnelling under the Martwa Wisła River result also from high-quality precast elements that make up the tunnel lining, i.e. tubing elements and precision of their installation.

Cooperation between the supervising team and the contractor, designer and scientific consultants ensured optimum and safe performance of the investment and enabled the implementation of the state-of-the-art and spectacular technical achievements into the engineering environment. We hope that the practical expertise and methods on how to implement such complicated and complex construction facility (i.e., construction of a tunnel) will allow us using the acquired experience in implementation of the next high priority projects.

REFERENCES

Construction project *The connection of the Airport and the Port of Gdańsk—Słowacki Route, Task IV The section from the »Polish Navy« junction to the »Ku Ujściu« junction; TUNNEL UNDER THE MARTWA WISŁA RIVER*, developed by Europrojekt Gdańsk Sp z o.o./SSF Ingenieure GmbH, Gdańsk, November 2010.

Instructions for the preparation and processing of the drilling fluid, OHL SA, Gdańsk 2013.

Materials and papers from the Symposium *Road tunnel under the Martwa Wisła River in Gdańsk* and the magazine *Inżynieria i Budownictwo* No. 1/2013.

Materials and papers from the Symposium *Road tunnel under the Martwa Wisła River in Gdańsk* and the magazine *Inżynieria i Budownictwo* No. 2/2014.

Methodology of the tunnelling direction implementation for the TBM system, OHL SA, Gdańsk 2013.

Monitoring of land and buildings within the TBM tunnelling method, OHL SA, Gdańsk 2013.

Monitoring project, OHL SA, Gdańsk 2013.

Report from the replacement of the cutting tools in the TBM cutter head, OHL SA, Gdańsk 2013.

Report of Geotechnical Parameters, OHL SA, Gdańsk 2013.

Work Quality and Organization Assurance. Ground Freezing. OHL SA, Gdańsk 2014.

Work Quality and Organization Assurance. Tunnelling, OHL SA, Gdańsk 2013.

Underground Infrastructure of Urban Areas 3 – Madryas et al. (Eds)
© 2015 Taylor & Francis Group, London, ISBN 978-1-138-02652-0

Author index